FUNDAMENTAL DC/AC CIRCUITS

PRENTICE HALL'S FUNDAMENTALS SERIES FOR ELECTRONIC TECHNOLOGY

Fundamental DC/AC Circuits: Concepts and Experimentation

Fundamental Electronic Devices: Concepts and Experimentation, Second Edition

Digital Electronics: Theory and Experimentation

Microprocessor Technology: Theory, Experimentation and Troubleshooting

FUNDAMENTAL DC/AC CIRCUITS

Concepts and Experimentation

Fredrick W. Hughes

Electronics Training Consultant

Prentice Hall, Englewood Cliffs, NJ 07632

Library of Congress Cataloging-in-Publication Data

Hughes, Fredrick W.,
Fundamental DC/AC circuits : concepts and experimentation /
Fredrick W. Hughes.
p. cm.
Includes index.
ISBN 0-13-058439-8
1. Electric circuits—Direct current—Programmed instruction.
2. Electric circuits—Direct current—Problems, exercises, etc.
3. Electric circuits—Alternating current—Programmed instruction.
4. Electric circuits—Alternating current—Problems, exercises, etc.
I. Title.
TK454.15.D57H84 1990
621.319'15—dc20 89-8651
CIP

Editorial/production supervision and
interior design: Ed Jones/Julie Boddorf
Cover design: Karen Stephens
Manufacturing buyer: Dave Dickey

Printed in the United States of America

10 9 8 7 6 5 4 3 2 1

ISBN 0-13-058439-8

Prentice-Hall International (UK) Limited, *London*
Prentice-Hall of Australia Pty. Limited, *Sydney*
Prentice-Hall Canada Inc., *Toronto*
Prentice-Hall Hispanoamericana, S.A., *Mexico*
Prentice-Hall India Private Limited, *New Delhi*
Prentice-Hall of Japan, Inc., *Tokyo*
Simon & Schuster Asia Pte. Ltd., *Singapore*
Editora Prentice-Hall do Brasil, Ltda., *Rio de Janeiro*

Dedication

This book is dedicated to those persons
whose talents and skills, through a concerted effort,
take an ordinary typewritten manuscript
and turn it into a work of art.
Without their assistance and promotion
no author could succeed. I extend my appreciation
to the following persons:

Alice Barr	*Mariann Hutlak*
Julie Boddorf	*Ed Jones*
Gregory M. Burnell	*Amy E. Ness*
Audrey Hellman	*Ruth E. Semper*

. . .and the many others behind the scenes.

Contents

Preface

In the last two decades the number of concepts, components, and products introduced by the electronics industry has been overwhelming. However, training institutions and courses of study are expected to graduate well-trained persons into this world of high-tech industry within the same reasonable training period. Although this may appear to be an impossible task, it can be accomplished. Today, importance must be placed on developing the skills required by industry at the job entrance level. At the same time, the entry-level worker must have the basic theoretical concepts of atomic structure and electrical principles to comprehend in detail the practical aspects of electronics. This book is the first in a series designed to provide the essential skills required to become an electronics technician.

To achieve this desired training, the student must become involved and be made responsible for self-learning from the start. With the very first unit in this book the reader is familiarized with the method of study. Subsequent units use the same format to develop in the reader a concise learning process. From Unit 2 on, exercises and experiments provide practical "hands-on" experience using test instruments. A learning process such as this is a timesaving method for acquiring basic skills while providing more time for methods of learning introduced by the instructor. In other words, the student learns the exact skills required for a beginning electronics technician. Other knowledge and skills offered by the instructor enhance the student's proficiency, which can lead to a higher entry level.

BOOK OBJECTIVES

When you have completed this book, you will be able to:

1. Apply basic atomic theory and electrical principles to practical electrical circuits.
2. Select proper wire sizes and explain the effects of electrical resistance.
3. Define basic electrical terms.

4. Identify and label discrete electrical components.
5. Use Ohm's law formulas for solving all the parameters of an electrical circuit.
6. Calculate the power dissipated by electrical components and circuits.
7. Solve the parameters for complex electrical circuits.
8. Construct electrical circuits and perform learning experiments.
9. Connect test equipment to electrical circuits and determine accurate readings.
10. Describe the principles of magnetism and electromagnetism and their application to generators, motors, and other electrical apparatus.
11. Define an ac sine wave and describe its uses.
12. Demonstrate the use of resistors, capacitors, inductors, and transformers in electrical circuits.
13. Read schematic diagrams and trace circuit paths.
14. Troubleshoot basic electrical circuits.

BOOK FORMAT

Each unit in this book follows a standard format consisting of seven sections. In the first section, the reader is introduced to the theory and concept of basic topics. Section 2 consists of definition exercises that enable the reader to begin to understand and use the language of electricity. Exercises and problems are given in section 3 to develop in the reader a working knowledge of electrical symbols, concepts, and circuit operations. These exercises include drawings, matching symbols, and various other methods to motivate and keep the interest of the reader. The problems develop the basic mathematical skills required of an electronics technician. Section 4 has experiments so that the reader can put the knowledge to use in a hands-on experience to increase manipulative skills. Troubleshooting applications are introduced in section 5 to prepare the reader immediately for the experiences found in industry. Section 6 is an instant review composed of concise statements about the topics to reinforce the reader's knowledge. The reader has instant feedback about how much was learned from the unit by performing the self-checking quiz in section 7. Answers to the experiments and quiz follow each unit. The first six units cover dc circuits, and the remaining six units deal mainly with ac circuits.

FOR THE STUDENT

Any person may use this book as a self-study course or in a classroom setting to acquire knowledge and skills of basic electrical circuits. The material is presented in an easy-to-read, straightforward manner that will provide the reader with a solid foundation for further study in electronics. The important thing to remember is to be serious about learning electronics and to spend time studying those topics that are more difficult for you.

FOR THE INSTRUCTOR

The book uses the individualized learning approach, which involves the reader in the activities of learning. This provides the instructor with student-centered instruction which does not require preparation.

In a lock-step type of course, students still have some capability of learning at their own rates with the use of this book. A student is able to perform other sections of a unit while waiting for personal instruction from the instructor. With this in mind, the book is well suited for multilevel classes (one instructor handling an entire electronics program) and open-ended courses where students can enter any time during the year. By placing more responsibility on students for their learning, schools are able to graduate better-prepared persons to enter the work force.

I wish you good luck and success in the electronics industry.

Fredrick W. Hughes

Unit A

Preliminary Review: Safety

SECTION A-1 BASIC POINTS ON ELECTRICAL SAFETY

1. *Safety* is the condition of being free of danger, personal risk, or injury and the protection of equipment and tools.
2. An *accident* is any unplanned event that occurs without warning which causes personal injury and/or damage to equipment and tools.
3. An *electrical shock* is the tingling sensation or muscular jolt that a person experiences when electrical current passes through a part of the body. An electrical shock can severely burn you or kill you (Figure A–1).
4. Electrical *insulating* materials do not allow electrical current to flow easily and can protect the human body from being shocked. Insulating materials include dry wood, rubber, leather, plastic, glass, dry cardboard, dry paper, and dry cloth.
5. Electrical *conducting* materials are nearly all metals, such as gold, silver, copper, iron, steel, aluminum, and tin. Wet or damp surfaces also conduct electrical current.
6. **Never** work on or operate electrical equipment when your body is wet or you have perspired sufficiently to cause the skin to be moist.
7. An *overloaded circuit* is an electrical circuit carrying too much current to be in a safe operating condition. This circuit produces heat, which can cause fires. An overloaded circuit occurs when too many electrical devices are connected to a single electrical outlet (Figure A–2). A safety device may operate on an overloaded circuit condition which "opens" and does not allow electrical current to flow in the circuit.
8. A *fuse* is an electrical safety device that "melts open" when a circuit is overloaded.
9. A *circuit breaker* is an electrical safety device that "pops open" when a circuit is overloaded. Once the overload condition has been removed and corrected, the circuit breaker can be reset for normal circuit operation.

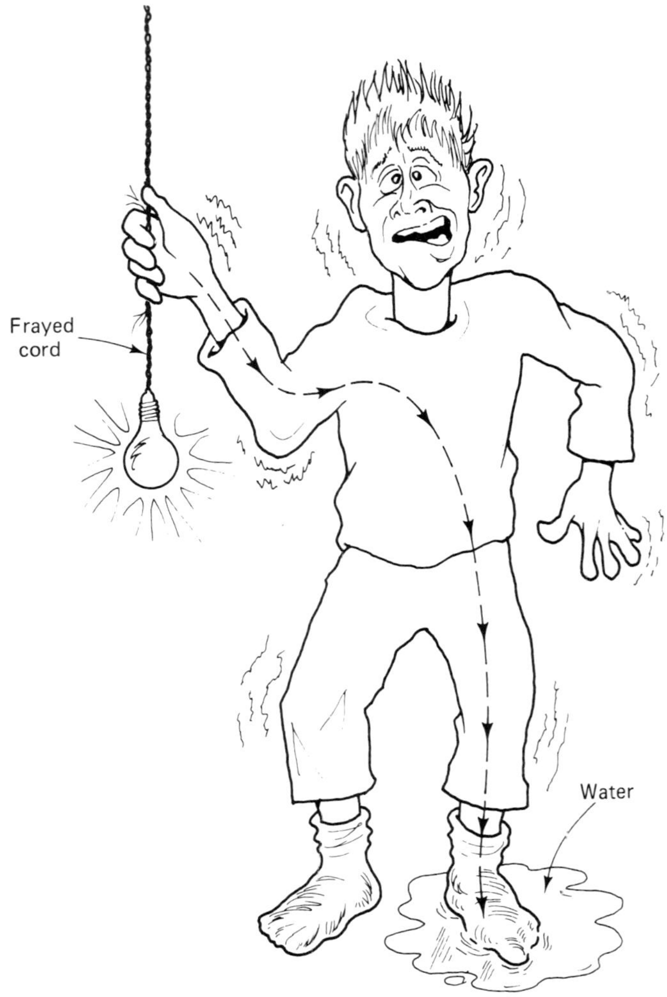

Figure A-1 Don't let this be you!

Figure A-2 Overload condition!

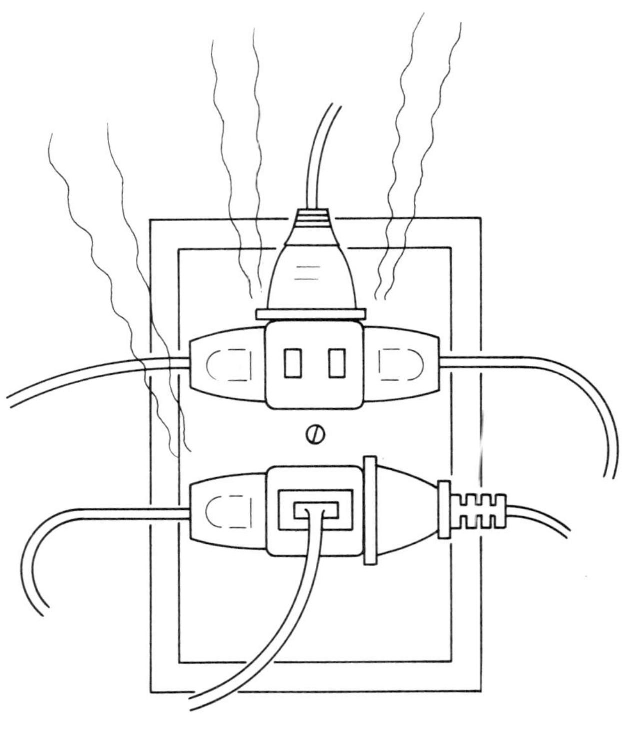

10. An *electrical outlet adapter* is a device that plugs into an electrical outlet and allows other electrical appliances to be used. It is better that this device not be used at all, or at least, only one for each pair of electrical outlets. Use of these adapters tends to cause an overloaded circuit condition.
11. Always remove an electrical cord from an electrical outlet by grasping the *plug end.* **Never** pull on the wire, because this action can damage the cord and possibly cause a safety hazard (Figure A–3).
12. Before use, examine all electrical cords on equipment and tools for cracks, frayed insulating material, and broken wires (Figure A–4).
13. In the event of an *electrical fire,* never attempt to put the fire out with water or water-type fire extinguishers, including the foam type. The water can cause more electrical current to flow and can damage equipment and parts still unaffected by the fire. Use **only** carbon dioxide (Co_2) or dry chemical fire extinguishers on electrical fires.
14. Be alert when working with electrical equipment. Have proper lighting and follow all written and verbal instructions given to you.
15. Use common sense and do **not** indulge in practical jokes or "horse-play" in an area that contains electrical apparatus.

The experiments in this manual do not use a voltage greater than 30 V (±15 V); therefore, the chance of getting an electrical shock is greatly reduced. However, all voltages have the potential to burn mate-

Figure A-3 A good way to break it!

Figure A-4 Repair before use!

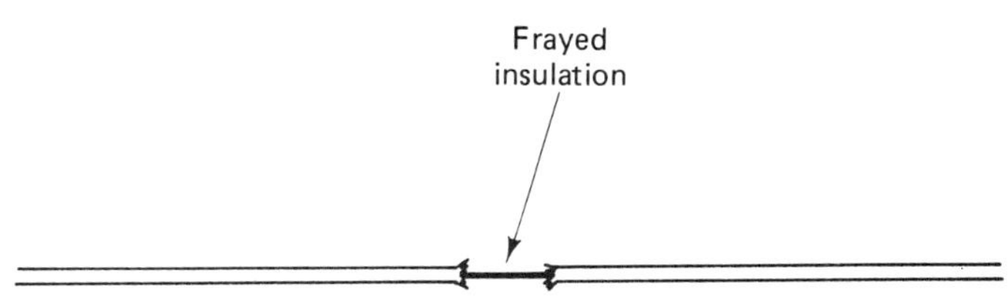

rials and start fires and to destroy electronic components, and they present hazards to the person performing the operations. Common sense and an awareness of electrical circuits are important whenever you are working on these experiments. An electronic technician or student may have to work with high voltages, power tools, and machinery. Before actual work is performed, sufficient instruction should be acquired in the proper use and safety requirements of all electronic devices.

SECTION A-2 CURRENT HAZARDS AND VOLTAGE SAFETY PRECAUTIONS

It takes a very small amount of current passing through the human body from an electrical shock to injure a person severely or fatally. The 60-Hz current values affecting the human body are as follows:

Current Value	*Effect*
1 mA (0.001 A)	Tingling or mild sensation
10 mA (0.01 A)	A shock of sufficient intensity to cause involuntary control of muscles, so that a person cannot let go of an electrical conductor
100 mA (0.1 A)	A shock of this type lasting for 1 second is sufficient to cause a crippling effect or even death
Over 100 mA	An extremely severe shock that may cause ventricular fibrillation, where a change in the rhythm of the heartbeat causes death almost instantaneously

The resistance of the human body varies from about 500,000 Ω when dry to about 300 Ω when wet (including the effects of perspiration). In this case, voltages as low as 30 V can cause sufficient current to be fatal (I = voltage/wet resistance = 30 V/300 Ω = 100 mA).

Even though the actual voltage of a circuit being worked on is low enough not to present a very hazardous situation, the equipment being used to power and test the circuit (i.e., power supply, signal generator, meters, oscilloscopes) is usually operated on 120 V ac. This equipment should have (three-wire) polarized line cords that are not cracked or brittle. An even better safety precaution is to have the equipment operate from an isolation transformer, which is usually connected to a workbench. To minimize the chance of getting shocked, a person should use only one hand while making voltage measurements, keeping the other hand at the side of the body, in the lap, or behind the body. Do not defeat the safety feature (fuse, circuit breaker, interlock switch) of any electrical device by shorting across it or by using a higher amperage rating than that specified by the manufacturer. These safety devices are intended to protect both the user and the equipment.

SECTION A-3
NEAT WORKING AREA

A neat working area requires a careful and deliberate approach when setting it up. Test equipment and tools should be set out on the workbench in a neat and orderly manner. Connecting wires from the test equipment to the circuit under test should be placed so as not to interfere with testing procedures. Before power is applied to a circuit, the area around the circuit should be cleared of extra wires, components, hand tools, and debris (cut wire and insulation).

SECTION A-4
HAND TOOL SAFETY PRECAUTIONS

Hand tools can be dangerous and cause severe injuries. Diagonal cutters, wire strippers, long-nose pliers, and crimping tools can pinch and cut. Use care in cutting wire since small pieces can become projectiles and hit another person in the face or eye.

Screwdrivers should be held properly so that they do not slip and puncture some part of the body. Do not use them as chisels or cutters.

A soldering iron should always be placed on a holder. Care must be used not to burn the body or other materials. Be careful of hot solder, which can splash and cause severe burns, especially to the eyes and face.

SECTION A-5
IN CASE OF ELECTRICAL SHOCK

When a person comes in contact with an electrical circuit of sufficient voltage to cause shock, certain steps should be taken as outlined in the following procedure:

1. Quickly remove the victim from the source of electricity by means of a switch or circuit breaker, by pulling the cord, or by cutting the wires with a well-insulated tool.
2. It may be faster to separate the victim from the electrical circuit by using a dry stick, rope, leather belt, coat, blanket, or any other nonconducting material.

   Caution! Do not touch the victim or the electrical circuit unless the power is off.
3. Call for assistance, since other persons may be more knowledgeable in treating the victim or can call for professional medical help while first aid is being given.
4. Check the victim's breathing and heartbeat.
5. If breathing has stopped but the victim's pulse is detectable, give mouth-to-mouth resuscitation until medical help arrives.
6. If the heartbeat has stopped, use cardiopulmonary resuscitation, *but only if you are trained in the proper technique.*
7. If both breathing and heartbeat have stopped, alternate between

mouth-to-mouth resuscitation and cardiopulmonary resuscitation (*but only if you are trained*).

8. Use blankets or coats to keep the victim warm and raise the legs slightly above head level to help prevent shock.
9. If the victim has burns, cover your mouth and nostrils with gauze or a clean handkerchief to avoid breathing germs on the victim and then wrap the burned areas of the victim firmly with sterile gauze or a clean cloth.
10. *In any case, do not just stand there*—do something within your ability to give the victim some first aid.

Unit 1

Basic Concepts of Electricity

INTRODUCTION

Learning any skill is dependent upon the knowledge of its terms and concepts. To facilitate a clear understanding of electrical devices and circuits, you must first understand the concepts of atomic theory, the principles of electrical charges, and the use of large numbers.

UNIT OBJECTIVES

Upon completion of this unit, you will be able to:

1. Define the terms *matter, energy, element, atom, nucleus, proton, neutron, electron,* and *compound.*
2. Explain the atomic theory.
3. Show the differences between various atoms.
4. Describe the relationship between valence shell, free electrons, and holes.
5. Explain the laws of electrical charges.
6. Describe the flow of electrons in a conductor.
7. List and explain various voltage sources.
8. Explain the construction and operation of a battery.
9. Compare the differences between dry cell and wet cell batteries.
10. Show how to connect batteries to increase voltage and current capabilities.
11. Calculate with numbers using scientific notation and the powers of 10.
12. Define the terms *giga, mega, kilo, milli, micro, nano,* and *pico.*
13. List the proper procedures for troubleshooting an electrical/electronic circuit.

SECTION 1-1 FUNDAMENTAL CONCEPTS

1-1A COMPOSITION OF MATTER

Examples of electricity can be pointed out by nearly everyone. Electricity produces light, operates radios, television sets, machinery, trains, automotive vehicles, computers, musical instruments, and many other devices used by people. It is generally known that *electricity consists of invisible positive and negative charges of matter that produce an invisible electrical current* that operates the numerous devices of the modern world. It is also known that small units of electricity are present in all *matter—which consists of gases, liquids, and solids* and therefore includes all life forms. The nature and amount of these units determine the ability of an electrical current to flow or pass through specific matter.

Energy is that invisible force which provides the capacity to do work. Energy is stored in matter, and scientists state that energy cannot be destroyed, only transformed. For example, water (a liquid) can be boiled to produce steam (a gas). The steam can be used to produce mechanical motion. Similarly, ice (a solid) can be melted to produce water. A basic understanding of the composition of matter enables a person to better understand the nature and operation of electrical and electronic circuits.

1-1a.1 The Atom

The atom is the smallest unit that forms the building blocks of all matter. It is a very small system within itself, and its concept is shown in Figure 1–1. There is a nucleus that contains *positively charged particles called protons and neutral or uncharged particles called neutrons. Electrons are negatively charged particles that travel around the nucleus in orbits,* similar to the way that planets travel about the sun. Electrons, which are much smaller than the nucleus, are maintained in orbit by the initial energy given them when they were created. There can be several orbits of varying distance from the nucleus of an atom. *These*

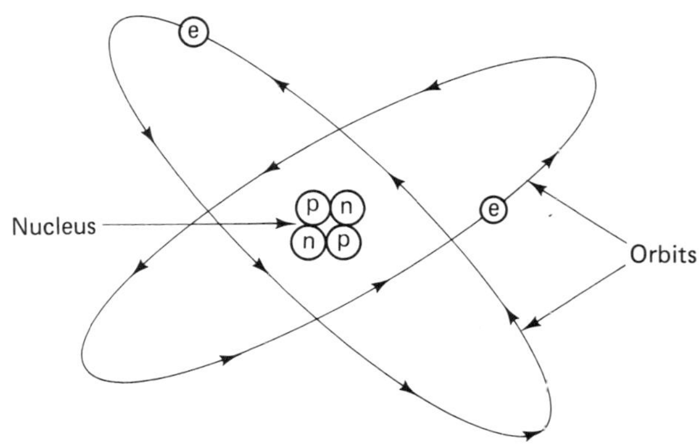

Figure 1-1 Atom and its parts: p, proton; n, neutron; e, electron.

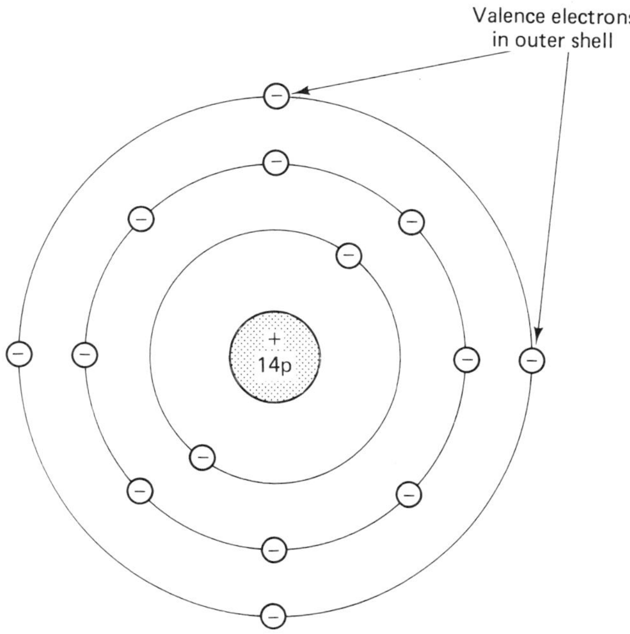

Figure 1-2 Atomic structure showing orbital shells (silicon atom).

orbits were created by various energy levels and are referred to as shells. There are also subshells. The shells can contain varying numbers of electrons, as shown in Figure 1–2. Each shell has the capability of containing a specific number of electrons, although they may not be filled. Those shells closer to the nucleus have a greater attraction between the electrons and nucleus, whereas the electrons in the outer shell have less attraction and can be pulled out of orbit to produce an electron flow.

This outer shell of an atom is called the valence shell and the electrons in this shell are called valence electrons. The ability to gain or lose an electron in the outer shell, called the *valence* of an atom, determines the chemical and electrical combining properties of the atom.

Regardless of the number of shells that an atom has with varying amounts of electrons in each shell, the valence or outer shell of each atom cannot contain more than eight electrons. The number of electrons, or the valence of an atom, determines whether the element is a good conductor or a good insulator. Materials such as gold, silver, and copper have a valence of 1, which means they can accept seven more electrons in the valence shell and are therefore excellent conductors. Elements such as boron, neon, and argon have a valence of 8 and are better insulators. Combinations of elements actually form practical insulators such as glass, plastic, and rubber.

A *negative valence* indicates an atom's tendency to gain electrons to fill or complete its outer shell. An atom that requires one electron to fill its outer shell is said to have a valence of -1. A *positive valence* indicates an atom's tendency to give up or donate electrons in its valence shell to other atoms. An atom that has an electron in its outer shell, which it will readily give to an other atom, is said to have a valence of $+1$.

The silicon atom shown in Figure 1–2 has a valence of ± 4. This means that it will donate its four existing valence electrons fairly easy, or it will accept four more electrons to its outer shell. For this reason, silicon is the major substance used in producing semiconductor materials.

1-1a.2 Atomic Elements

Elements are small particles of matter where all the atoms are the same and cannot be broken down or decomposed into simpler substances. Such examples are hydrogen, oxygen, silver, gold, copper, silicon, and uranium.

1-1a.3 Compounds

Compounds are the combination of two or more elemental atoms to form another material substance. A common example of this is water, which contains two parts of hydrogen to one part of oxygen. *The smallest unit of a compound is known as a molecule.* The familiar formula for water is written as H_2O.

1-1a.4 Atomic Number

In their neutral state, all atoms have equal numbers of protons and electrons. Look at the silicon atom shown in Figure 1–2. the +14 indicates that it has 14 protons in its nucleus. There are 14 planetary electrons in the three orbital shells. This atom has what is called an atomic number of 14. Therefore, *the atomic number indicates the same number of protons and electrons of an atom.*

1-1a.5 Atomic Weight

In the nucleus of most atoms, there are neutrons. For a time, neutrons were believed to have no practical purpose. It was found that bombarding the nucleus of the atom with electrons caused the neutrons to divide or split. This "splitting of the atom," referred to as *nuclear fission,* produces nuclear power. These neutrons have no relation to the atomic number, but add a small amount of mass to the atoms. *The atomic weight is then another classification of the approximate weight of the number of protons and neutrons in the nucleus of an atom.* Table 1–1 shows the relationships of a few atoms.

TABLE 1-1
Selected Elements and Their Values

Elements	*Symbol*	*Atomic Number*	*Atomic Weight*	*Atomic Valence*
Hydrogen	H	1	1.008	±1
Helium	He	2	4.003	0
Carbon	C	6	12.011	±4
Oxygen	O	8	16.004	−2
Silicon	Si	14	28.086	±4
Copper	Cu	29	63.54	+1
Silver	Ag	47	107.868	+1
Tin	Sn	50	118.69	+2
Gold	Au	79	196.967	+1
Uranium	U	92	238.03	+2 to +6

1-1b ELECTRICAL CHARGES

1-1b.1 Balanced Atom and Ions

As stated previously, a neutral or balanced atom has an equal number of protons and electrons. The magnitude of the negative charges on the electrons is exactly equal to the magnitude of the positive charges on the protons; therefore, the atom is electrically neutral. However, this neutral condition can easily be changed by external forces. The balanced condition of the atom can be upset by forces such as heat, light, electrostatic fields, chemical reactions, and magnetic fields. Figure 1–3 shows the effects of an atom losing and gaining an electron.

The electrons in the valence shell are "loosely bound" and sometimes referred to as *free electrons.* If an electron is removed from the

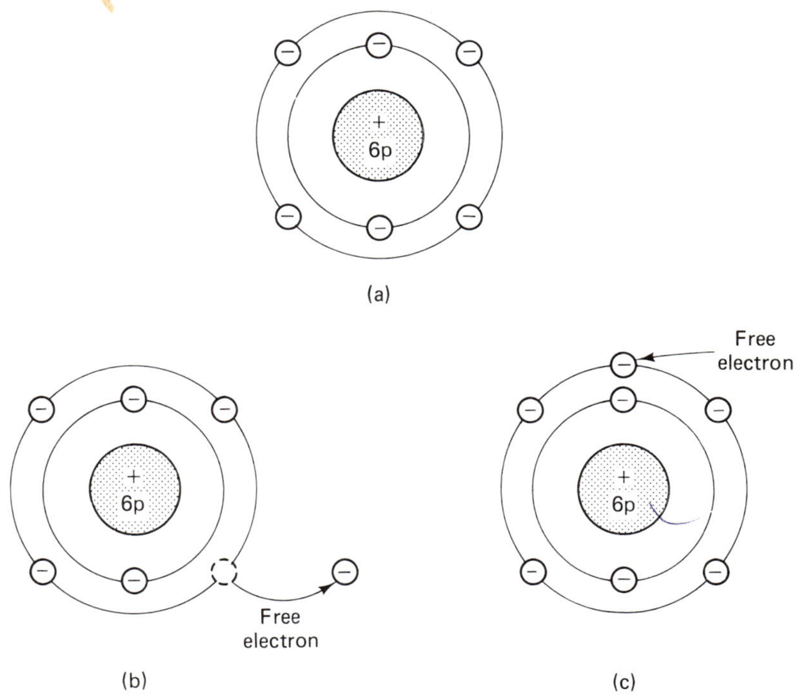

Figure 1-3 Balanced atom and electrical charges: (a) neutral atom; (b) positively charged atom; (c) negatively charged atom (carbon atom).

valence shell of an atom as shown in Figure 1–3b, the balanced condition of equal protons and electrons is upset and the net charge on the atom is positive. This is referred to as a *positive ion.*

If an electron is added to the valence shell as shown in Figure 1–3c, the balanced condition is also upset, but now there is an excess of electrons and the net charge on the atom is negative. This is referred to as a *negative ion.*

An atom that is no longer in a balanced condition is called an ion and the process of changing an atom to an ion is called ionization.

A positive ion has an invisible force about it that goes in an outward direction. A negative ion also has an invisible force about it, but it goes in the opposite, or inward, direction. These forces are referred to as *electrostatic fields.* A positive and a negative electrostatic field will attract each other, but electrostatic fields of the same charge will repel each other.

1-1b.2 Law of Electrical Charges

Figure 1–4 shows the effects of a concentration of charges on two bodies. When the bodies are suspended by a fine thread, they are able to move rather easily. If one body is positively charged and the other is negatively charged, they will be attracted to each other, as shown in Figure 1–4a. When they touch, electrons will flow from the negatively charged body onto the positively charged body until the two bodies are neutralized, or have the same equal charge. Now they will no longer attract or repel each other. If both bodies have the same charge, either positive or negative, they will repel each other, as shown in Figure 1–4b and c.

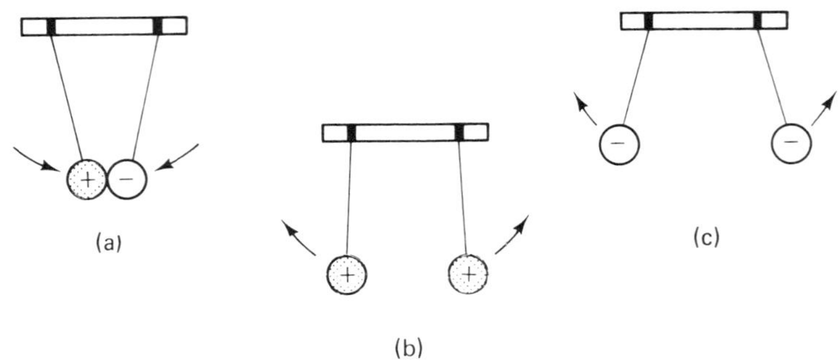

Figure 1-4 Law of electrical charges: (a) unlike charges attract; (b and c) like charges repel.

The law of electrical charges simply states that:

1. Like charges repel.
2. Unlike charges attract.

In other words, a negative charge repels another negative charge, a positive charge repels another positive charge, a negative charge attracts a positive charge, and a positive charge attracts a negative charge.

Charles A. Coulomb, a French engineer and scientist, worked with electrical charges and formulated what is called Coulomb's law of electrical charges. The unit of electrical charge is called a *coulomb* and consists of 6,250,000,000,000,000,000 electrons, often written as 6.25×10^{18}. (The use of 10^{18} is an example of powers of 10, a subject covered in a later section.) If a charge consists of 6.25×10^{18} electrons, it has a negative charge of 1 coulomb. However, if a charge has an absence or deficiency of 6.25×10^{18} electrons, it has a positive charge of 1 coulomb.

The relationship between two charged bodies has little practical use for the technician, but aids in our understanding of other phenomena, such as magnetism and electromagnetism. The force between the bodies is directly proportional to the charges on the bodies; that is, if a charge increases, the force will increase, and similarly, if a charge decreases, the force will decrease. However, the distance between the charged bodies affects the force inversely proportional to the square of the distance (the distance times itself). If the distance between the bodies increases by a factor of 2, the force will decrease by a factor of 4, or if the distance is halved, the force will increase by four times.

The force that exists between two charged bodies can be determined by the formula

$$F = \frac{kq_1 \times q_2}{d^2}$$

where F = force of attraction or repulsion between two charged bodies in units called newtons (a newton is defined as the force required to accelerate a mass of 1 kilogram at the rate of 1 meter per second)

k = constant number with a value of 9.0×10^9

q_1 = charge on one body (coulombs)

q_2 = charge on the second body (coulombs)

d^2 = square of the distance between the two bodies

1-1c CURRENT FLOW IN A CONDUCTOR

A good electrical conductor, such as silver or copper, has a large number of free electrons. These electrons drift about in a random motion between the valence bands of the atoms. This random drift is always present but has little practical use. The free electrons can be controlled to flow in a particular direction, as shown in Figure 1–5.

When a conductor is connected between a negative charge and a positive charge, the free electrons in the conductor are repelled by the negative charge and attracted by the positive charge. The resulting electron flow is then from negative to positive. Figure 1–5 shows only a probable single path of electrons, but remember that in actuality, there are millions, even billions, of electrons flowing.

As a step-by-step analysis, consider that a valence electron in the first atom at the right moves to the positive source. This leaves a *"hole"* (or *positive charge*). A valence electron in the second atom then moves over to fill this hole, which in turn leaves a hole in the second atom.

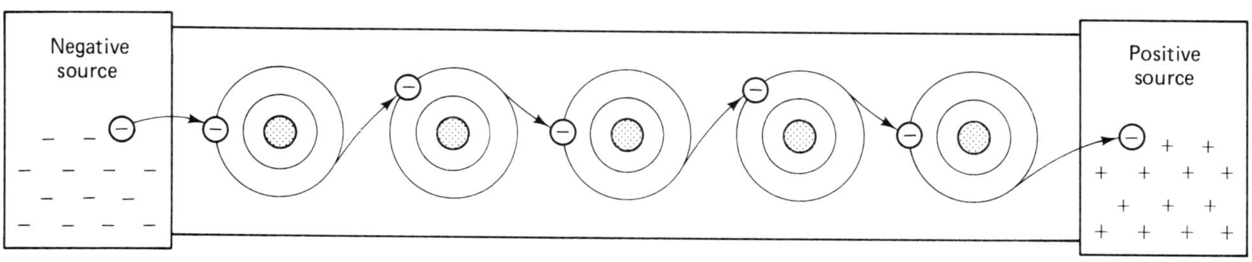

Figure 1-5 Electron flow among atoms in a conductor.

This action is repeated along the conductor until the last atom on the far left has a hole which is filled by an electron from the negative source. The electron flow will continue until the positive and negative sources are neutralized (no further charges are present).

If the positive and negative sources are a battery, the electron flow will last longer because of the chemical action producing electrons and positive ions within the battery. However, the large electron flow would deplete the battery very quickly and render it useless. This is an example of a *short circuit* and should never be attempted willingly. *Remember, never place a conductor or wire directly across the terminals of a battery.*

1-1d VOLTAGE SOURCES

Voltage is the accumulation of negative and positive ions into charges, which produces a force capable of creating an electron current flow for performing some task. This force, sometimes referred to as electrical pressure, is called *electromotive force,* usually abbreviated as EMF. For a voltage to exist, there must be a difference between two charges. In other words, one charge must be more positive than the other, or similarly, one charge has to be more negative than the other. This *difference* in charges creates the *potential* for allowing electrons to flow from negative to positive and thus the possibility of doing work. For this reason voltage is often referred to as *difference of potential* or *potential diffference.* Remember, even when no device is connected to a voltage source, there still exists the potential for doing work. A charged battery is an example of this.

There are six fundamental ways of producing voltage, as shown in Figure 1–6. In this section we describe these methods, although their applications in the electronics industry vary.

1-1d.1 The Triboelectric Effect (Static Electricity)

The triboelectric effect produces voltages from friction. One of the oldest methods, it was found that by rubbing different substances together, a static charge of a difference of potential would be developed. Figure 1–6a illustrates that a silk cloth, when rubbed on a glass rod, will accumulate electrons from the rod. The silk will be negatively charged and the glass rod, having fewer electrons, will be positively charged. When fur and a hard rubber rod are rubbed together, the fur will be positively charged and the rubber rod will be negatively charged.

In low-humidity conditions, a person walking with shoes across a nylon or wool rug will often build up on the body a charge of several hundred volts. Touching a metal object such as a door knob usually discharges the body, resulting in a tiny arc and a mild shock. This method of producing voltage is not efficient for normal applications.

1-1d.2 The Piezoelectric Effect

The piezoelectric effect produces voltage from pressure. Certain types of crystal substances, such as quartz and Rochelle salts, have positive and negative charges distributed through them, so that the net or overall charge is neutral. However, when pressure is applied to the crystal, electrons will accumulate on one side, which produces a potential difference across the crystal. When the pressure is relieved, the charge becomes neutral again. Figure 1–6b shows how metal plates are placed on the crystal to conduct the electrons. The voltage generated from crystals is very small and is generally used only in crystal microphones and phonograph pickups.

1-1d.3 The Thermoelectric Effect

The thermoelectric effect produces voltage from heat. Two dissimilar metals, such as copper and zinc, are joined together into a device called a *thermocouple,* as shown in Figure 1–6c. One end of the thermocouple

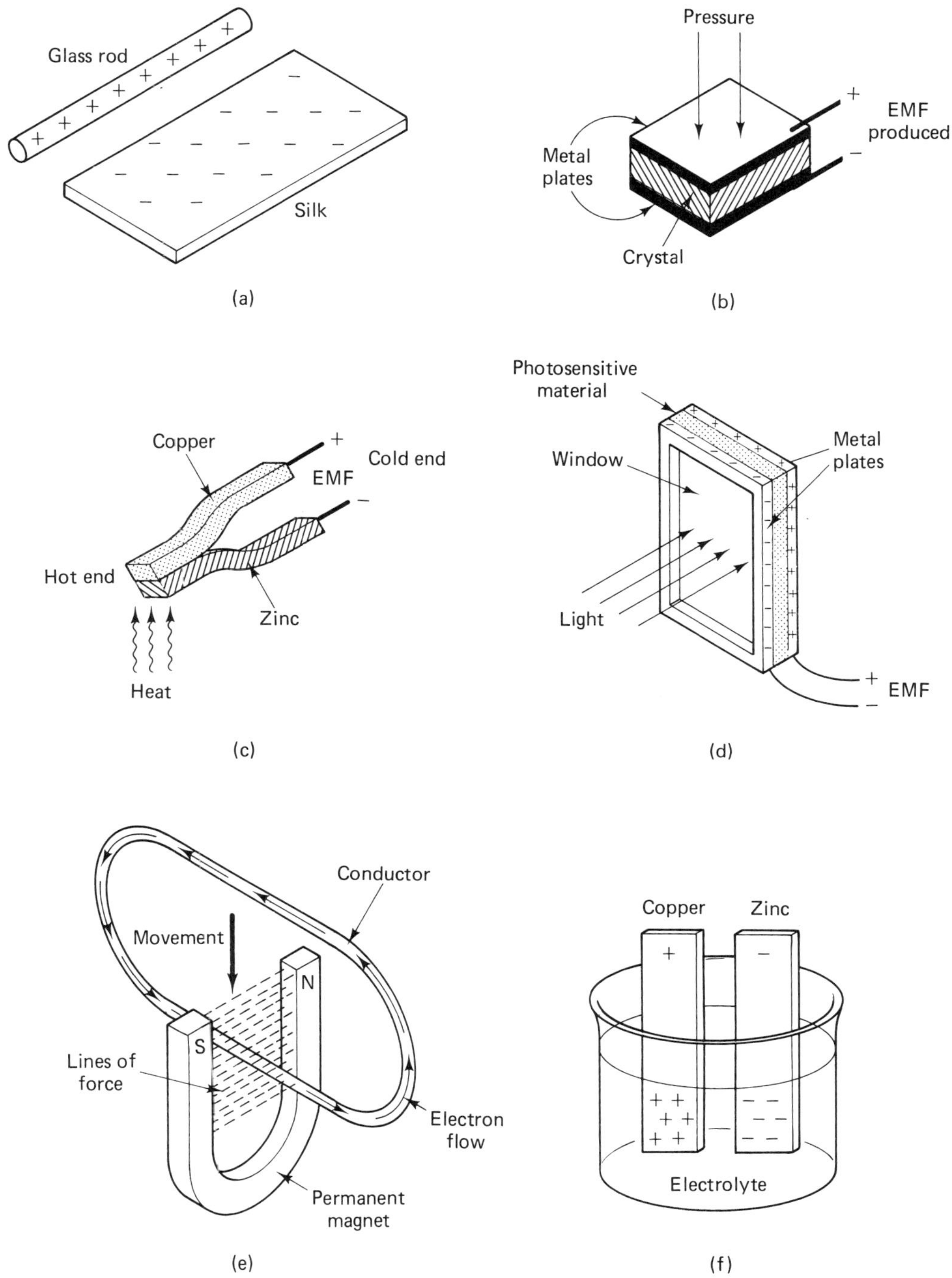

Figure 1-6 Basic methods of producing voltage: (a) friction; (b) pressure; (c) heat; (d) light; (e) magnetism; (f) chemical action.

is referred to as the hot end, where heat is applied, and the other end is the cold end, from which the voltage is taken. When the hot-end junction is heated, electrons flow from the copper material into the zinc material. The copper then becomes the positive potential and the zinc becomes the negative potential. Wires attached to the cold end are used to connect the voltage to an external device. Thermocouples are usually used in test instruments for temperature-to-voltage conversion and to maintain gas and electric burner systems using pilot heating.

1-1d.4 The Photoelectric Effect

The photoelectric effect produces voltage from light. Figure 1–6d shows the construction of a photo or solar cell device, which converts light energy to electrical energy. A photosensitive material such as selenium, germanium, cadmium, or cesium is placed betweeen two metal conductors. A window in one plate allows light to strike the photosensitive material, which in turn releases electrons from some of its atoms. The resulting action is a developed voltage of a few tenths of a volt capable of producing a few thousandths of an ampere or electrical current. This is the primary electrical source used in space satellites and space vehicles; however, it requires hundreds, up to thousands, of these cells connected together to produce a usable voltage source with acceptable current levels. At present, solar cells are expensive and are found primarily in low-voltage applications.

1-1d.5 The Electromagnetic Effect

The electromagnetic effect produces electricity when an electrical conductor moves through a magnetic field. Figure 1–6e shows this effect. A magnetic field consists of invisible lines of force that exist between the north and south poles of a permanent magnet. When the conductor is moved perpendicular through the lines of force to cause a cutting action, electrons are forced to move from one atom to another. If the conductor is a continuous loop or closed circuit, an electric current will flow. When the conductor moves downward, as shown in the figure, the current will flow in a counterclockwise direction. When the conductor moves upward, the current will flow in a clockwise direction. This action produces a fundamental alternating-current (ac) alternator or generator.

If the conductor is formed into many coils that will cut the lines of force simultaneously, a larger amount of current can be produced. This is the most practical method of producing large amounts of electricity in the world today. In other chapters we discuss in greater depth this principle of voltage generation and magnetism.

1-1d.6 The Electrochemical Effect

The electrochemical effect produces voltage from chemical action. The process involves changing chemical energy to electrical energy. Batteries are examples of this type of voltage source. A basic battery or voltage cell is shown in Figure 1-6f. A chemical solution referred to as an electrolyte, consisting of sulfuric acid and water, is placed in a container. A chemical action is produced, wherein the hydrogen atoms become positive ions and the sulfate molecules become negative ions. The number of ions is the same; therefore, the solution has no net charge. Two metallic bars called *electrodes* are placed into the solution. One electrode is made of copper, which has an excess of free electrons. These free electrons are attracted by the positive ions, resulting in a net positive charge on the copper electrode. The other electrode is made of zinc. With a similar chemical action, the zinc electrode gives up positive ions to the solution, resulting in a net negative charge. A difference of potential is now created between the two electrodes. If a conductor were placed across these electrodes, an electron current would flow from the negatively charged zinc electrode to the positively charged copper electrode. The current would always flow in the same direction and is referred to as *direct current,* or simply dc.

1-1d.7 The Dry Cell Battery

The common flashlight battery is a dry cell battery. The schematic symbol for a battery is shown in Figure 1-7a. The horizontal lines represent the plates or terminals of the battery. The longer line is the positive terminal and the shorter line is the negative terminal. This type of battery is referred to as a *dry cell,* because the electrolyte is not a pure liquid but rather, a moist paste. Figure 1-7b shows a cutaway view of a dry cell battery. The carbon rod is the positive electrode, which contacts the steel top cover used for external positive external connection. The zinc can or container is the negative terminal. Plastic or cardboard covers the zinc can and insulates the positive terminal from the negative terminal. The wax seal prevents the electrolyte paste from leaking

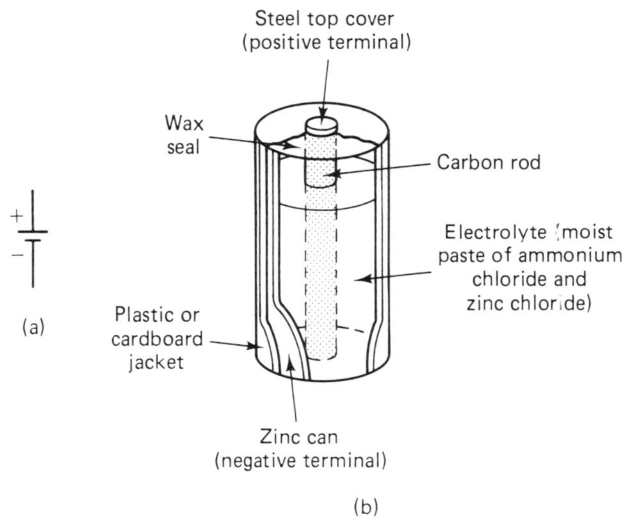

Figure 1-7 Dry cell battery: (a) electrical symbol; (b) construction view.

out, which allows the battery to be mounted in any position. The chemical action of this battery produces about 1.5 V for a single cell. There are several battery sizes that produce the same voltage. The size of the battery determines how much current can be supplied, and of course, a larger battery should last longer. A D-cell battery can supply more current than a C-cell or AA-cell battery, yet they all produce 1.5 V. Therefore, the voltage produced by a dry cell battery is determined by the materials used in the electrolyte and electrodes.

The chemical action of this battery will continue to produce a sufficient voltage for about two years, even if the battery is not used. The electrolyte paste eventually dries out and the battery loses its ability to produce EMF. It is said that this type of battery has a shelf life of about two years. *This type of dry cell, which cannot be recharged, is referred to as a primary cell.*

1-1d.8 The Wet Cell Battery

A wet cell battery can be recharged and is referred to as a secondary cell. This type of battery is found in virtually all automotive vehicles in the form of a *lead-acid battery.* Figure 1-8 shows the construction of a wet

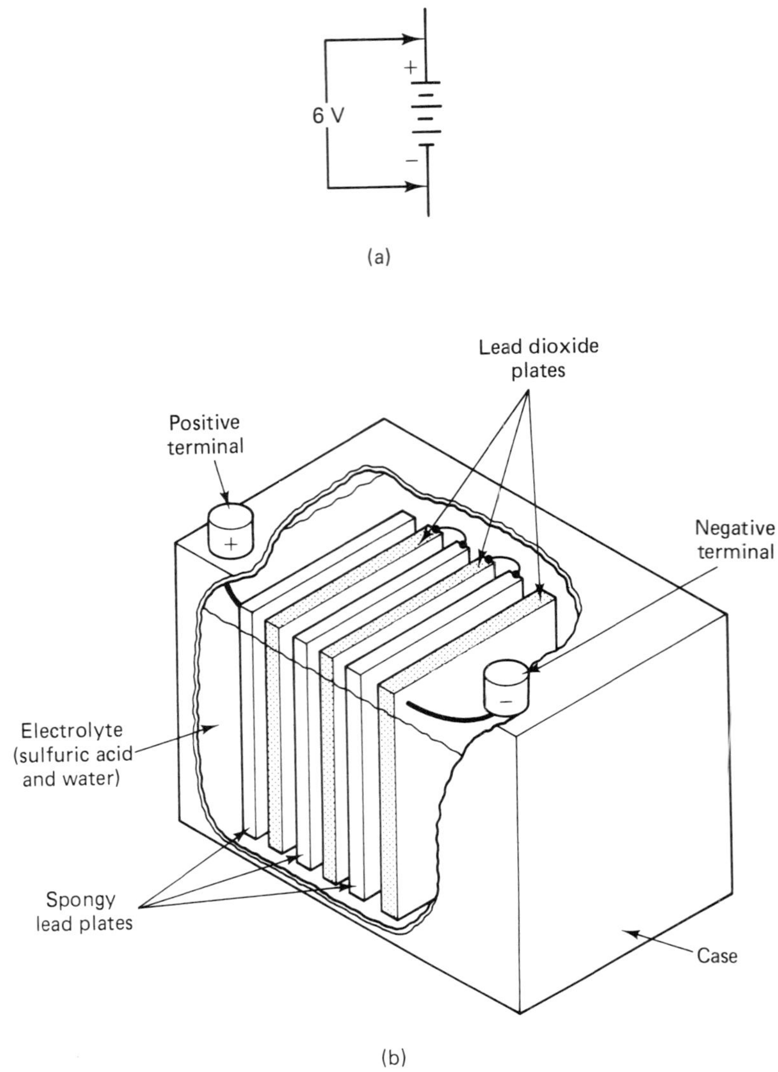

Figure 1-8 Wet cell battery: (a) electrical symbol; (b) construction view.

cell battery. A single two-electrode wet cell produces about 2.1 V. These cells are connected together in three-cell and six-cell combinations to provide the necessary 6.3- and 12.6-V batteries, respectively. The cells are made of plates that are connected internally within the battery. The negative electrodes or plates are made of soft spongy lead and the positive plates are made of lead dioxide (or lead peroxide). The electrolyte solution consists of three parts of concentrated sulfuric acid to eight parts of distilled water.

Actually, this battery converts chemical energy to electrical energy while it is being used. Upon discharge, when the battery is supplying current to an external device, the chemical action tends to produce a more positive charge on the lead dioxide and a negative charge on the spongy lead. However, lead sulfate, a whitish material, is produced and tends to collect on the electrodes, which reduces the chemical action of the battery. Also, more hydrogen is released, diluting the sulfuric acid, thereby making the electrolyte weaker. This discharging action will continue until a point is reached where less current is available from the battery, and voltage at the terminals decreases.

To recharge the battery, the current must be reversed. A battery charger must have a slightly higher voltage than the voltage of the battery to be charged and be connected correctly. Proper connection of the charger is with the positive lead to the positive terminal of the battery and the negative lead to the negative terminal. During recharge, the lead sulfate on the electrodes is decomposed and returns to the electrolyte, which increases the sulfuric acid content. Thus the charging action restores the electrodes and the electrolyte to the original condition. With proper charging procedures and maintenance, a lead-acid battery is capable of being charged and discharged many times.

1-1d.9 Increasing Voltage and Current

Many battery-operated devices require more voltage and/or current than can be supplied by a single battery. Techniques for connecting batteries to increase voltage and current are shown in Figure 1-9. To increase voltage, batteries are connected in a series-aiding manner, as shown in Figure 1-9a. A *series connection* occurs when one end of a component is connected to one end of another component. The negative terminal of one battery is connected to the positive terminal of the next battery, and so on. If four 1.5-V batteries are connected in this manner, each voltage is additive, for a total voltage of 6 V. The same current flows through all components of a series circuit. If one of the batteries was connected in reverse, it would oppose the total voltage and the resulting voltage would be 4.5 V; however, this is usually not the practice in real applications.

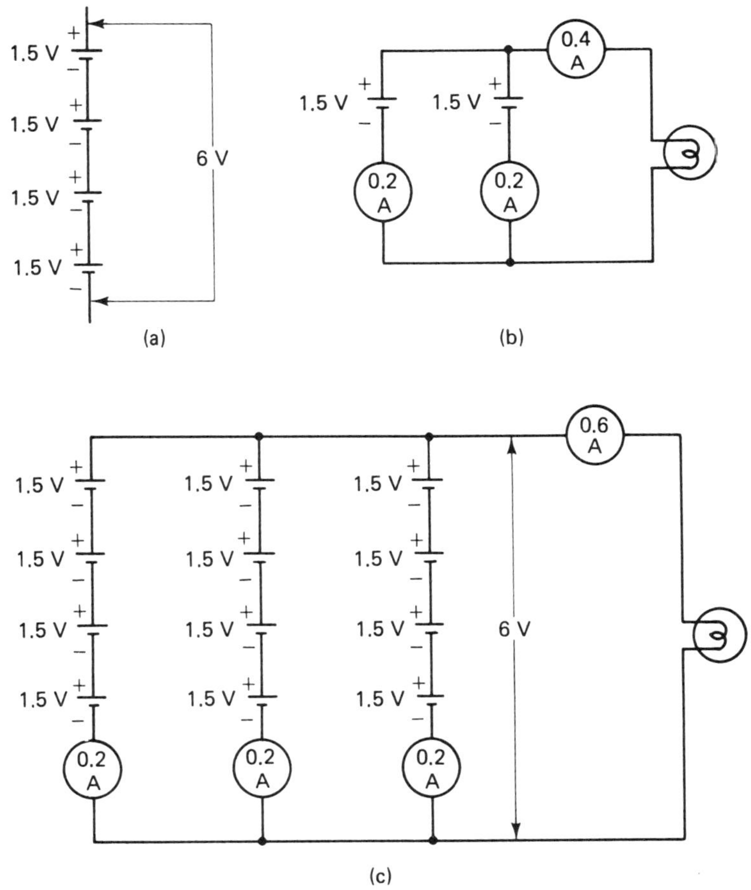

Figure 1-9 Voltage cell connections to increase capabilities: (a) in series to increase voltage; (b) in parallel to increase current; (c) in series–parallel to increase voltage and current.

To increase the current capabilities of a voltage source, batteries are connected in parallel, as shown in Figure 1-9b. A parallel connection occurs when the ends of each component are connected together. The positive terminal of one battery is connected to the positive terminal of a second battery and their negative terminals are connected together. The current flow of each battery is additive. If one battery can supply 0.2 A (the ampere is a unit of current flow), two batteries connected in parallel can supply a total current of 0.4 A. The circles in series with the components in the circuit indicate the currents flowing in the circuit and represent current or ampere meters. Connecting batteries in parallel is similar to increasing the number of plates or plate area, which produces more current. The voltage across the batteries is the same 1.5 V and it *must be remembered never to connect batteries with different voltages in parallel.*

To increase voltage and current capabilities, batteries can be connected in series and parallel, as shown in Figure 1-9c. Assume that a particular lamp requires 6 V and will use 0.6 A of current, but we have only 1.5-V batteries capable of 0.2 A. Four batteries connected in series will produce the required 6 V, but supplies only 0.2 A of current. Therefore, two more branches of the series connection are needed to supply the required 0.6 A of current. These three branches are connected in parallel. The total number of batteries required for this circuit can be found from the following formulas:

$$\text{required number of batteries for voltage (one branch)} = \frac{\text{total voltage } (V_T)}{\text{voltage of one battery } (V_B)} = \frac{6\text{ V}}{1.5\text{ V}} = 4_{\text{batt}}$$

$$\text{required number of branches for total current} = \frac{\text{total current } (I_T)}{\text{current of one branch } (I_B)} = \frac{0.6\text{ A}}{0.2\text{ A}} = 3_{\text{branches}}$$

$$\text{total number of batteries required} = 4_{\text{batt}} \times 3_{\text{branches}} = 12_{\text{batt}}$$

This circuit arrangement requires a total of 12 batteries.

1-1e SCIENTIFIC NOTATION AND UNITS OF MEASUREMENT

Numbers used in electricity and electronics usually indicate extremely large and very small quantities, such as 6,000,000 and 0.000001. Writing out and calculating with numbers such as these is slow and cumbersome and subject to making errors. A shorthand number system called scientific notation, and specified units of measurement, simplify calculations with numbers.

1-1e.1 Scientific Notation

Scientific notation expresses numbers in an abbreviated form using the powers of 10 as its base system. For example, the number 1000 means that 1 is multiplied by 10 three times ($1 \times 10 \times 10 \times 10$). The scientific notation of 1000 is 1×10^3, where $10^3 = 1000$. Remember, the power of a number indicates how many times the number is multiplied by itself to express the real number. For example, $10^3 = 10 \times 10 \times 10 = 1000$.

1-1e.2 Converting to Powers of 10

Converting large whole numbers to powers of 10 is relatively easy. For example, let's convert the number 3400 to powers of 10.

1. Express the number as a number between 1 and 9 by moving the decimal point *left* between the 3 and the 4:

   3.400 implied decimal point

2. Count the number of decimal places to the left that the point was moved and use this number as the exponent or power of 10:

   3.4×10^3 ← exponent or power

   Examples of other whole numbers using powers of 10 are

$$260 = 2.6 \times 10^2$$
$$78{,}000 = 7.8 \times 10^4$$
$$5{,}900{,}000 = 5.9 \times 10^6$$

Converting small fractional numbers to powers of 10 is also easy. For example, let's convert the number 0.0025 to powers of 10.

1. Express the number as a number between 1 and 9 by moving the decimal point *right* between 2 and 5:

   0 002.5
   decimal point

2. Count the number of decimal places to the right that the point was moved, and use this number as the negative exponent or power of 10:

   2.5×10^{-3} ← negative exponent or power indicates a fractional number

   Examples of other fractional numbers using powers of 10 are

$$0.064 = 6.4 \times 10^{-2}$$
$$0.00083 = 8.3 \times 10^{-4}$$
$$0.00000750 = 7.95 \times 10^{-6}$$

1-1e.3 Addition and Subtraction Using Powers of 10

Addition and subtraction is possible using powers of 10 provided that the exponents are the same. For example, if 2.5×10^3 is to be added to 3.4×10^2, one of the numbers must have its decimal point shifted so that the exponents are the same. In the first case:

$$\begin{array}{rr} 2.5 \times 10^3 = & 25 \times 10^2 \\ + & 3.4 \times 10^2 \\ \hline & 28.4 \times 10^2 \end{array}$$

In the second case:

$$3.4 \times 10^2 = \begin{array}{r} 2.5 \times 10^3 \\ +\ \underline{0.34 \times 10^3} \\ 2.84 \times 10^3 \end{array}$$

Then to check the answer: $28.4 \times 10^2 = 2840.$ and $2.84 \times 10^3 = 2840.$ Notice that in the first case the decimal point is shifted right to make the exponent 2, and in the second case the decimal point is shifted left to make the exponent 3.

The negative exponents for fractional numbers must have the same numerical value for adding and subtracting. In the following example, notice the direction the decimal point is shifted to make the exponents the same.

$$\begin{array}{rcrcrl} 6.3 \times 10^{-2} & = & 63.0 \times 10^{-3} & = & 6.30 \times 10^{-2} & \\ -\ \underline{1.2 \times 10^{-3}} & = & -\ \underline{1.2 \times 10^{-3}} & = & -\ \underline{0.12 \times 10^{-2}} & \\ & & 61.8 \times 10^{-3} & & 6.18 \times 10^{-2} & = 0.0618 \end{array}$$

1-1e.4 Multiplication Using Powers of 10

Powers of 10 are particularly useful with multiplication and division with simplifying calculations. To multiply two numbers expressed in powers of 10, you simply multiply the coefficients and then algebraically add the exponents. For example:

$$\begin{aligned} & (4.5 \times 10^3) \times (2 \times 10^2) && = 4500 \times 200 \\ = & (4.5 \times 2) \times (10^3 \times 10^2) && \\ = & 9.0 \times 10^{3+2} && \\ = & 9.0 \times 10^5 && = 900{,}000 \end{aligned}$$

In the case of a negative power:

$$\begin{aligned} & (3.6 \times 10^3) \times (4 \times 10^{-2}) && = 3600 \times 0.04 \\ = & (3.6 \times 4) \times (10^3 \times 10^{-2}) && \\ = & 14.4 \times 10^{3+(-2)} && \\ = & 14.4 \times 10^1 && = 144 \end{aligned}$$

1-1e.5 Division Using Powers of 10

Division, using powers of 10, is similar to multiplication. To divide numbers expressed in powers of 10, you first divide the coefficients and then algebraically subtract the exponent in the divisor from the exponent in the dividend. For example:

$$\begin{aligned} & (6.9 \times 10^4) \div (3 \times 10^2) && = 69{,}000 \div 300 \\ = & (6.9 \div 3) \times (10^4 \div 10^2) && \\ = & 2.3 \times 10^{4-2} && \\ = & 2.3 \times 10^2 && = 230 \end{aligned}$$

In the case of a negative exponent:

$$
\begin{aligned}
& (8 \times 10^{3}) \div (2 \times 10^{-2}) && = 8000 \div 0.02 \\
= & (8 \div 2) \quad \times (10^{3} \div 10^{-2}) && \\
= & \quad 4 \quad \times 10^{3-(-2)} && \\
= & \quad 4 \quad \times 10^{5} && = 400{,}000
\end{aligned}
$$

1-1e.6 Units of Measurement

In the electrical/electronics field, certain powers of 10 are used frequently. To facilitate the use, calculation, and understanding of these powers, they have been assigned names and symbols. These are called units of measurement; some of the most common follow:

Power of 10	*Prefix*	*Symbol*	*Sample Use*
10^{9}	giga	G	gigahertz
10^{6}	mega	M	megaohm
10^{3}	kilo	k	kilowatt
10^{-3}	milli	m	milliampere
10^{-6}	micro	μ	microampere
10^{-9}	nano	n	nanosecond
10^{-12}	pico	p	picofarad

Table 1-2 shows the numerical name, numerical value, powers of 10, and some prefixes and symbols for whole and fractional numbers.

1-1f BASIC ELECTRICAL/ELECTRONIC HAND TOOLS

Basic electrical/electronic hand tools are necessary for proper construction and efficient maintenance of electronic circuits. Figure 1-10 shows these basic tools, which should be included in a tool kit. Several *standard blade-type screwdrivers* (a) and *Phillips-type screwdrivers* (b) must be in a tool kit. *Nut drivers* (c) are similar to screwdrivers but are used to "speed up" assembly and disassembly of nuts on screws and bolts. Some connections do not use solder but are wire wrapped and require a special tool called a *wire wrapping and unwrapping tool* (d). *Long-nose pliers* (e) are used to form proper bends in component leads, to hold components in place while soldering, and to reach hard-to-get-to places in a circuit. *Diagonal cutters* (f) are used for cutting component leads and wires. Once expertise is developed, they can be used for stripping insulation from wire so that wires are not nicked or cut. They should never be used to cut metal or screws and bolts. This will nick and dull the cutting edges. *Wire strippers* (g) are best to use in removing insulation from wire. They are adjustable and can be set very accurately so as not to damage the wire. A *crimping tool* (h) is used to attach solderless terminals and connectors to the ends of wire.

Figure 1-11 shows the tools and accessories needed for soldering electrical/electronic circuits. When many connections must be soldered at one time, a *soldering iron* (a) of usually less than 50 watts can be

TABLE 1-2
Numerical Values and Units

Numerical Name	*Numerical Value*	*Power of 10*	*Prefix*	*Symbol*
Trillion	1,000,000,000,000	10^{12}	tera	T
100 billion	100,000,000,000	10^{11}		
10 billion	10,000,000,000	10^{10}		
Billion	1,000,000,000	10^{9}	giga	G
100 million	100,000,000	10^{8}		
10 million	10,000,000	10^{7}		
Million	1,000,000	10^{6}	mega	M
100 thousand	100,000	10^{5}		
10 thousand	10,000	10^{4}	myria	my
Thousand	1,000	10^{3}	kilo	k
Hundred	100	10^{2}	hecto	h
Ten	10	10^{1}	deca	da
Unit	1	10^{0}		
Tenth	0.1	10^{-1}	deci	d
Hundredth	0.01	10^{-2}	centi	c
Thousandth	0.001	10^{-3}	milli	m
10-thousandth	0.0001	10^{-4}		
100-thousandth	0.00001	10^{-5}		
Millionth	0.000001	10^{-6}	micro	μ
10-millionth	0.0000001	10^{-7}		
100-millionth	0.00000001	10^{-8}		
Billionth	0.000000001	10^{-9}	nano	n
10-billionth	0.0000000001	10^{-10}		
100-billionth	0.00000000001	10^{-11}		
Trillionth	0.000000000001	10^{-12}	pico	p
Quadrillionth	0.000000000000001	10^{-15}	femto	f
Quintillionth	0.000000000000000001	10^{-18}	alto	a

plugged in and left on. The *soldering iron holder* (b) provides a safe place for a hot iron for the protection of materials and the safety of the user between soldering connections. A moist pad or sponge can be placed nearby to keep the tip of the iron clean. *Soldering aids* (c) are small chisel-like picks used to remove "solder shorts" between connections and to clean up solder joints. A *soldering gun* (d) is fast for making small repairs and requires very little warm-up time; however, they cannot be used on some solid-state devices. Solder *flux* (e) is a paste-like substance placed on solder connections before soldering. During the soldering process, contaminants are burned away, leaving a better solder joint connection. A *desoldering tool* (f) is used to remove solder from connections to facilitate the removal of components. The solder connection is first heated and the rubber vacuum bulb is squeezed as the nozzle of the tool is placed on the connection. The bulb is then released and the solder is drawn (sucked) up into the tool. With less solder at the connection, the component is easier to remove from the circuit.

1-1g BASIC ELECTRONIC HARDWARE

Specific types of hardware are associated with electronic circuits. The following figures and descriptions pertain to basic electronic hardware.

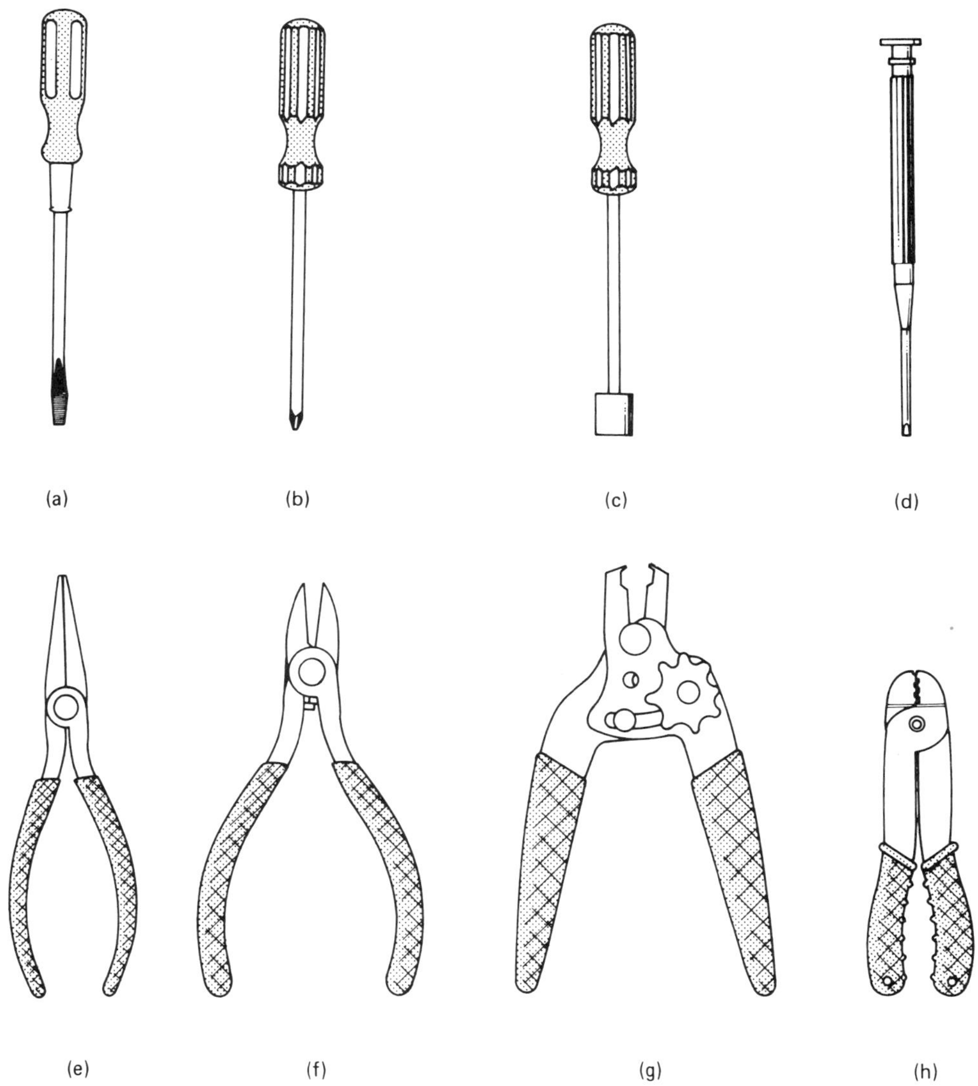

Figure 1-10 Basic hand tools: (a) blade screwdriver; (b) Phillips screwdriver; (c) nut driver; (d) wire wrapping/unwrapping tool; (e) long-nose pliers; (f) diagonal cutters; (g) wire strippers; (h) crimper (solderless terminals).

Figure 1-12 shows circuit boards. A *printed circuit* (PC) *board* connects discrete components together by means of thin ribbon conductors, which are permanently bonded to a board made of laminated plastic. The boards are mass-produced by means of a photographic and etching process. *Perforated board ("perfboard")* or *vector board* is made of phenolic or epoxy glass and contains numerous holes. The boards come with various hole sizes and "on center" dimensions. Circuits are built on these boards and they are often used for engineering prototype and initial testing.

Some circuits use *transistor and IC sockets,* shown in Figure 1-13a-c. These facilitate replacing components and make testing procedures easier. *Heat sinks,* shown in Figure 1-13d-f, are used with power transistors to draw the heat away from the device, which improves circuit operation and stability and increases the life span of the device.

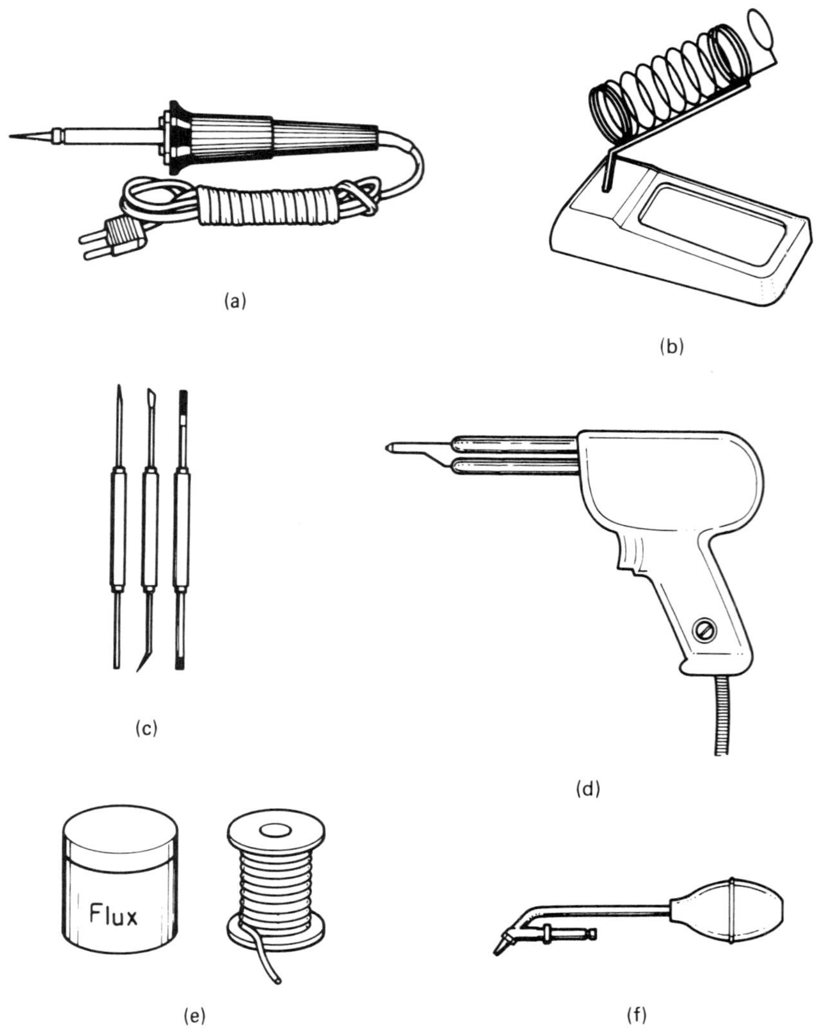

Figure 1-11 Soldering tools: (a) soldering iron; (b) iron holder and cleaning sponge; (c) soldering aids; (d) soldering gun; (e) solder flux and rosin core solder; (f) desoldering tool.

Printed-circuit-board edge connectors provide wire connections between PC boards in an electronic system and are shown in Figure 1-14. These connectors are mounted into the metal chassis and then the PC boards are pushed into them. They are of two types, wire-wrap and solder-lug, and provide quick-disconnect features for changing PC boards.

Terminal strips, shown in Figure 1-15, are used as tie points for wire and component connections. The terminals are mounted on insulating material, such as Bakelite, phenolic, or porcelain. A solder terminal strip (a) may use one terminal for chassis or ground connection. The barrier terminal strip (b) uses screw connections and the wires have solderless terminals.

Miscellaneous hardware is usually associated with fasteners and devices for constructing rigid and dependable circuits, shown in Figure 1-16. *Metal screws and nuts* (a, b, c, f) are used for bolting. *Sheet-metal screws* (d, e) are used to self-tap for securing chassis parts. *Serrated lock washers* (h, i) are used to "dig into" the chassis to lock a screw and sometimes to provide a good electrical connection. *Split-lock washers*

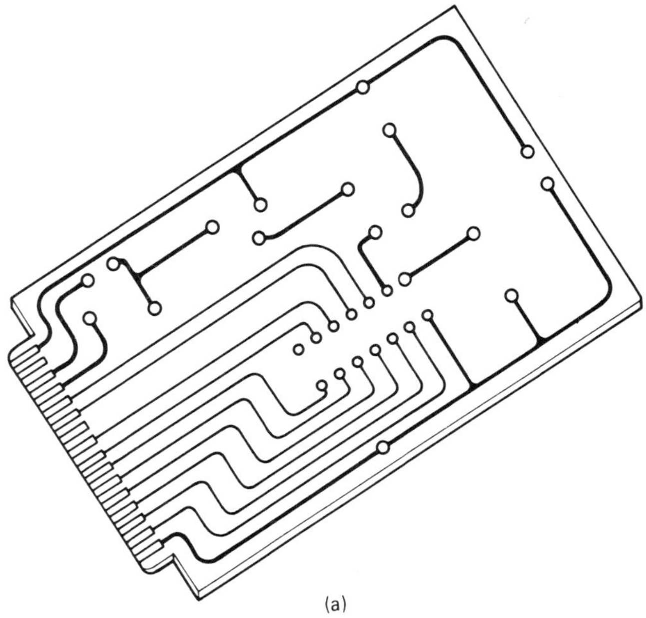

(a)

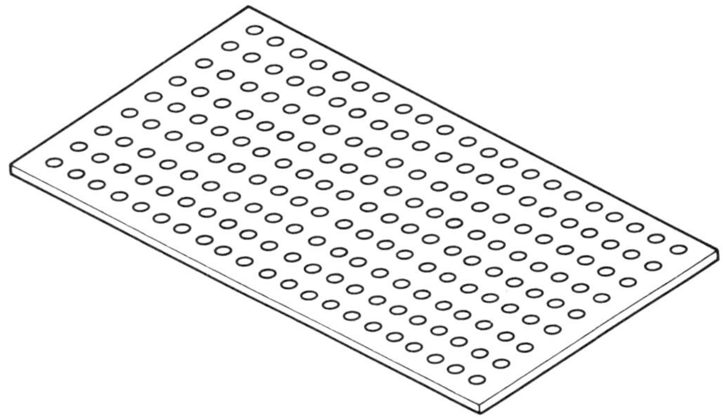

(b)

Figure 1-12 Circuit boards: (a) printed circuit board; (b) perforated board (''perfboard'') or vector board.

Figure 1-13 Sockets and heat sinks: (a) small-signal transistor socket; (b) power transistor socket; (c) IC socket; (d) heat sink pushes over transistor body; (e) transistor bolts to heat sink; (f) larger fins radiate more heat.

(a)

(b)

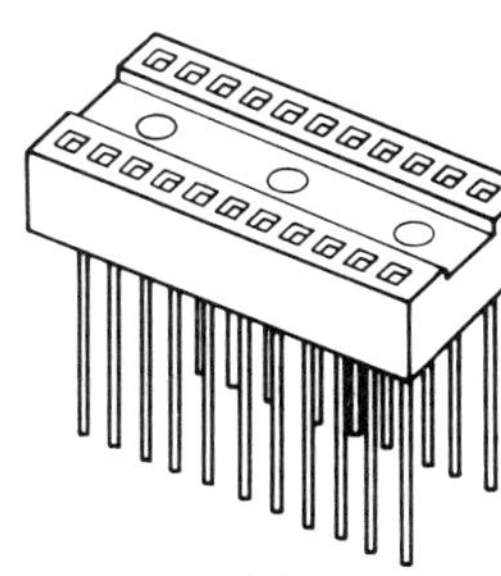

(c)

(d)

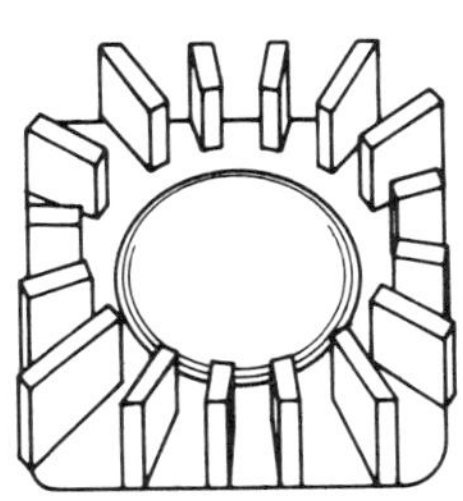

(e)

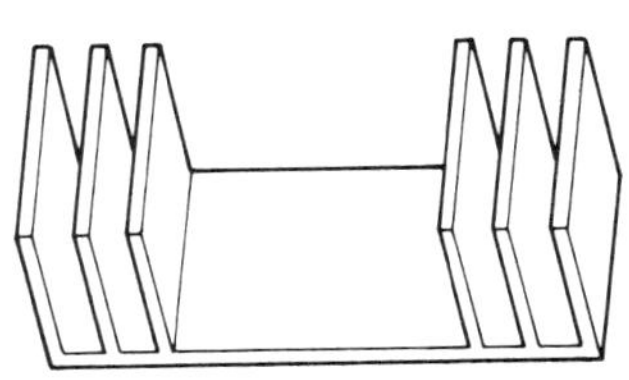

(f)

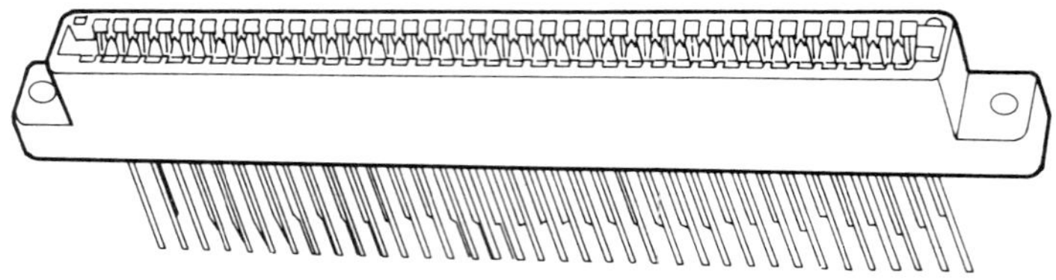

(a)

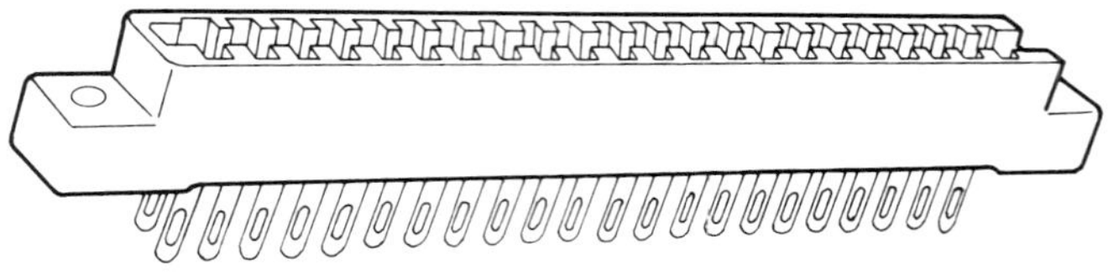

(b)

Figure 1-14 Printed-circuit-board edge connectors: (a) wire-wrap type; (b) solder-lug type.

(k) are used to lock a screw where a good electrical connection is not essential. *Flat washers* (l) are used for spacing and also not to cause marring of a chassis or panel. *Fiber washers* (m) are used for insulating purposes. *Solder lugs with a serrated washer feature* (n) are used to provide a chassis or "ground" connection for wires. A *cable clamp* (o) fits around a cable or harness to secure it to the chassis. *Rubber grommets* (p) are used in holes in the chassis where wires must pass through. This prevents the insulation of the wires from becoming cut and causing shorts or circuit problems. *Solderless terminals* (q, r) are used on the ends of wire that must use screw-type terminals. *Alligator clips* (s) are useful as clip leads for testing, troubleshooting, and experimenting with circuits. They may also be used as lead heat sinks when soldering solid-state components. *Component mounting terminals for PC and vector boards* (t) are used to provide circuit ruggedness and dependability.

Figure 1-15 Terminal strips: (a) solder terminal; (b) barrier (screw-type) terminal.

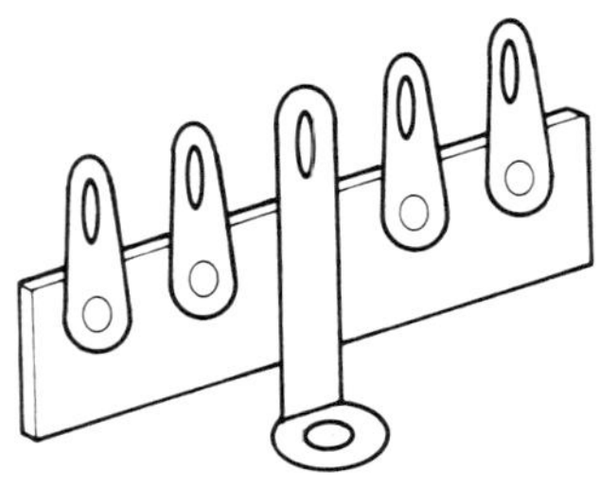

(a)

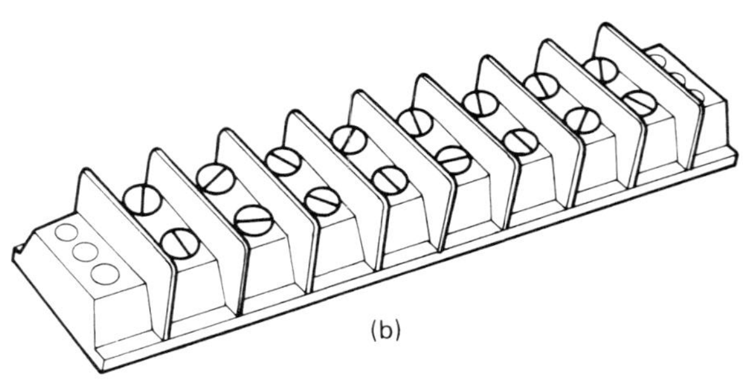

(b)

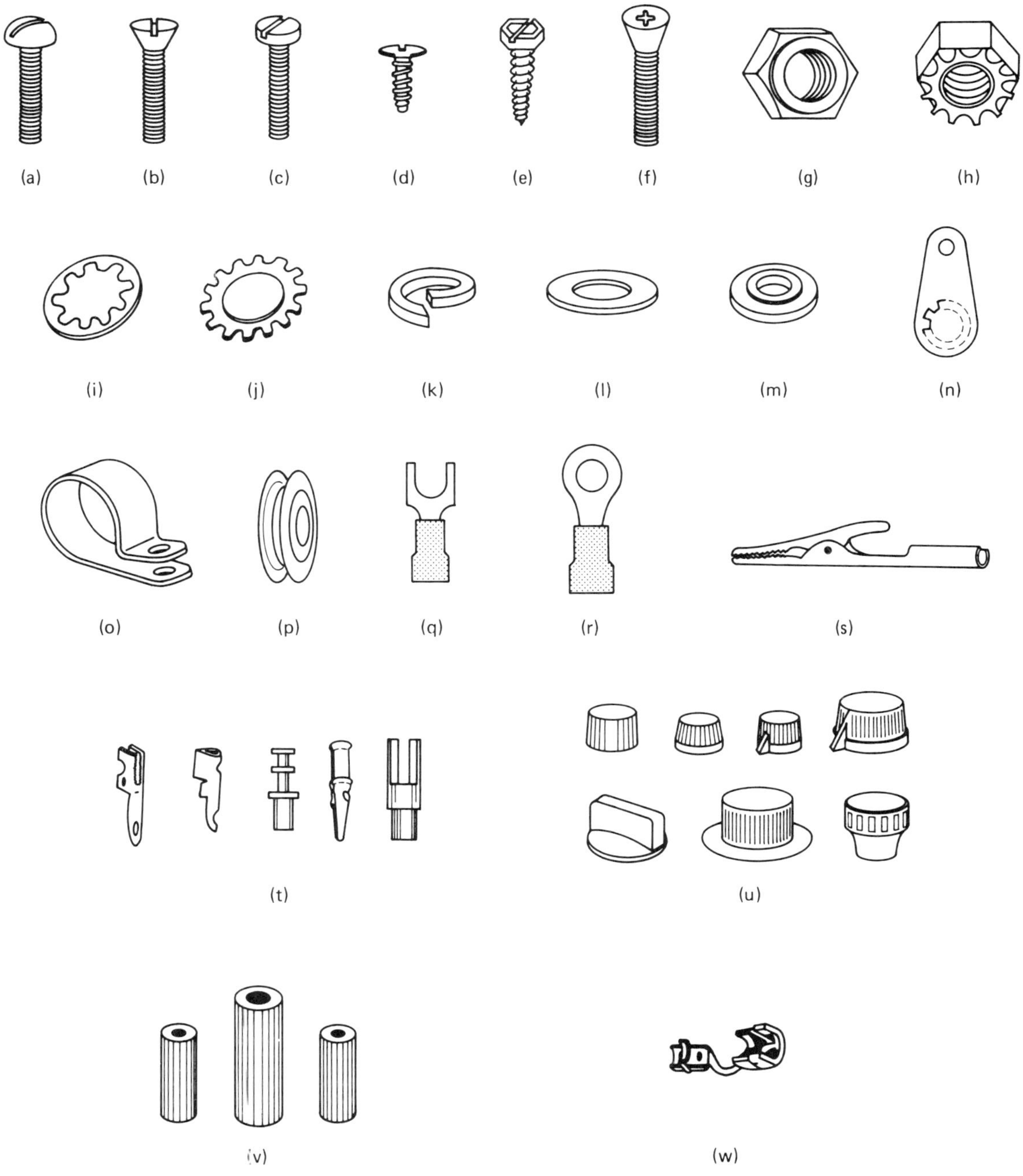

Figure 1-16 Miscellaneous hardware: (a) round-head screw; (b) oval-head screw; (c) binder-head screw; (d) sheet-metal screw; (e) sheet-metal hex-head screw; (f) Phillips-head screw; (g) hex nut; (h) hex nut with lock washer; (i) internal-tooth lock washer; (j) external-tooth lock washer; (k) split-ring lock washer; (l) flat washer; (m) fiber washer with shoulder; (n) solder lug; (o) cable clamp; (p) rubber grommet; (q) solderless terminal—spade tongue; (r) solderless terminal-ring tongue; (s) alligator clip; (t) PC and/or ''vector'' board component mounting terminals; (u) control knobs; (v) spacers or stand-offs; (w) strain relief.

Control knobs (u) of various sizes are needed for switches, potentiometers, and tuning capacitors associated with circuit operation. *Spacers* or stand-offs (v) are used for special component applications. They may be made of metal, phenolic, or ceramic. A *strain relief* (w) has a similar purpose as a rubber grommet and provides a dependable connection for the line cord to a piece of equipment. It fits into the hole of a chassis while applying pressure to the line cord that is passing through the hole.

SECTION 1-2 DEFINITION EXERCISES

Write a brief description for each of the following terms.

1. Matter ____________________

____

2. Energy ____________________

____

3. Atom ____________________

____

4. Nucleus ____________________

____

5. Proton ____________________

____

6. Neutron ____________________

____

7. Electron ____________________

____

8. Energy shells ____________________

____

9. Valence electrons ____________________

____

10. Negative valence ______________________________

____

11. Positive valence ______________________________

____

12. Element ______________________________

____

13. Compound ______________________________

____

14. Molecule ______________________________

____

15. Atomic number ______________________________

____

16. Atomic weight ______________________________

____

17. Ions ______________________________

____

18. Ionization ______________________________

____

19. Electrostatic field ______________________________

____

20. Law of electrical charges ______________________________

____

21. Electron flow ______________________________

____

22. Coulomb ______________________________

____

23. Voltage ____

____

24. EMF ____

____

25. Difference of potential ____

____

26. Triboelectric effect (static electricity) ____

____

27. Piezoelectric effect ____

____

28. Thermo electric effect ____

____

29. Photoelectric effect ____

____

30. Electromagnetic effect ____

____

31. Electrochemical effect ____

____

32. Dry cell battery ____

____

33. Primary cell ____

____

34. Wet cell battery ____

____

35. Secondary cell ______________________________

____

36. Scientific notation ______________________________

____

37. Powers of 10 ______________________________

____

38. Giga ______________________________

____

39. Mega ______________________________

____

40. Kilo ______________________________

____

41. Milli ______________________________

____

42. Micro ______________________________

____

43. Nano ______________________________

____

44. Pico ______________________________

____

45. Printed circuit board ______________________________

____

46. Perfboard ______________________________

____

SECTION 1-3
EXERCISES AND PROBLEMS

Complete this section before beginning the next section.

1. Referring to Figure 1-1, draw the atom and label its parts.

2. Referring to Figure 1-2, draw the atom and its energy shells.

3. Fill in the blanks in the following statements.

a. A positive ion has an invisible force about it that goes in an ____________ direction.

b. A negative ion has an invisible force about it that goes in an ____________ direction.

c. The unit of electrical charge is called the ____________ and consists of 6.25×10^{18} ____________ .

d. The law of electrical charges states that like charges ____________ and unlike charges ____________ .

e. Electron flow is from ____________ to ____________ .

f. Valence electrons can become ____________ electrons and move from atom to atom.

g. A "hole" is defined as the absence of an ____________ and results in a ____________ charge.

4. Write the formula for the force that exists between two charged bodies, calculate the following problems, and answer the questions.

a. $F =$

b. $q_1 = 2$ coulombs
$q_2 = 3$ coulombs
$d = 4$ cm
$F =$ __________

c. $q_1 = 2$ coulombs
$q_2 = 3$ coulombs
$d = 2$ cm
$F =$ __________

d. $q_1 = 2$ coulombs
$q_2 = 3$ coulombs
$d = 8$ cm
$F =$ __________

e. When the distance between two charged bodies decreases by 2 (halved), the force ____________ four times.

f. When the distance between two charged bodies increases by 2 (doubled), the force ____________ four times.

5. Match the term in column A with its proper voltage producing effect in column B.

Column A	*Column B*
____ **a.** Triboelectric effect	**1.** magnetism
____ **b.** Piezoelectric effect	**2.** light
____ **c.** Thermoelectric effect	**3.** pressure
____ **d.** Photoelectric effect	**4.** heat
____ **e.** Electromagnetic effect	**5.** chemical
____ **f.** Electrochemical effect	**6.** friction

6. Draw the schematic symbol for the following batteries.

a. single-cell battery

b. four-cell battery

7. Using 1.5-V batteries capable of producing 0.2 A of current, draw the battery schematic symbol connections needed to produce the following voltage sources (refer to Figure 1-9).

a. voltage = 4.5 V
current = 0.2 A

b. voltage = 1.5 V
current = 0.6 A

c. voltage = 6 V
current = 0.4 A

d. voltage = 9 V
current = 0.6 A

8. Express the following numbers as a number between 1 and 9 using powers of 10.

a. 670 = ____________ b. 5800 = ____________

c. 125 = ____________ d. 230,000 = ____________

e. 4,200,000 = ____________ f. 0.0034 = ____________

g. 0.0123 = ____________ h. 0.000062 = ____________

i. 0.000000046 = ____________ j. 0.792 = ____________

9. Express the following numbers in their full value.

a. 5.4×10^3 = ____________ b. 6.25×10^{-3} = ____________

c. 3.3×10^2 = ____________ d. 48×10^4 = ____________

e. 2.4×10^6 = ____________ f. 75×10^{-4} = ____________

g. 0.98×10^2 = ____________ h. 0.83×10^{-2} = ____________

i. 768×10^0 = ____________ j. 2.9×10^1 = ____________

10. Add the following numbers using powers of 10.

a. $3.5 \times 10^3 + 1.2 \times 10^2$

b. $68 \times 10^2 + 42 \times 10^4$

c. $0.98 \times 10^{-2} + 7.34 \times 10^{-3}$

d. $2.1 \times 10^6 + 6.3 \times 10^3$

11. Subtract the following numbers using powers of 10.

a. $5.6 \times 10^3 - 1.2 \times 10^2$

b. $7.8 \times 10^2 - 0.33 \times 10^3$

c. $62.5 \times 10^{-2} - 89.3 \times 10^{-3}$

d. $0.345 \times 10^5 - 9.21 \times 10^3$

12. Multiply the following numbers using powers of 10.

a. $2.6 \times 10^3 \times 5.4 \times 10^2$

b. $12.3 \times 10^4 \times 3.2 \times 10^2$

c. $8.9 \times 10^{-3} \times 2.2 \times 10^{-2}$

d. $6.7 \times 10^{-3} \times 4.2 \times 10^3$

13. Divide the following numbers using powers of 10.

a. $3.5 \times 10^4 \div 7 \times 10^2 =$ ____________

b. $68 \times 10^3 \div 2 \times 10^{-2} =$ ____________

c. $21.44 \times 10^{-2} \div 3.2 \times 10^2 =$ ____________

d. $17.48 \times 10^{-3} \div 2.3 \times 10^{-2} =$ ____________

14. List the powers of 10 assigned to the following units of measurement.

a. kilo = ____ **b.** mega = ____ **c.** giga = ____ **d.** milli = ____
e. micro = ____ **f.** nano = ____ **g.** pico = ____

15. Referring to Figure 1-10, list the basic hand tools that are necessary for an electronics tool kit.

a. **b.**

c. **d.**

e. **f.**

g. **h.**

16. Referring to Figure 1-11, list the desired soldering devices that may be found in an electronics tool kit.

a. **b.**

c. **d.**

e. **f.**

SECTION 1-4 EXPERIMENTS

EXPERIMENT 1. Static Electricity

Objective: To show the effects of static electricity developed from friction and prove the law of electrical charges.

Introduction: As a result of friction, some materials will develop an electrical charge. A glass rod when rubbed with silk will build up a positive charge. Other materials will also work well with the glass rod, such as 100% acrylic and some nylon and Dacron polyesters. A glass medical thermometer can be substituted for the glass rod.

Materials Needed:

1 Book
2 Pencils
Fine thread
Small amount of tin foil
Glass rod or thermometer
1 Piece of material of silk, acrylic, nylon, or Dacron polyester

Procedure:

1. Tie a piece of thread 3 to 4 in. long to the end of each pencil.
2. Form a small tinfoil ball about $\frac{1}{8}$ in. in diameter over the end of each thread. There is now a small tinfoil ball attached to the end of the thread on each pencil.
3. Arrange the pencils and book on the edge of a desk or table, as shown in Figure 1-17.

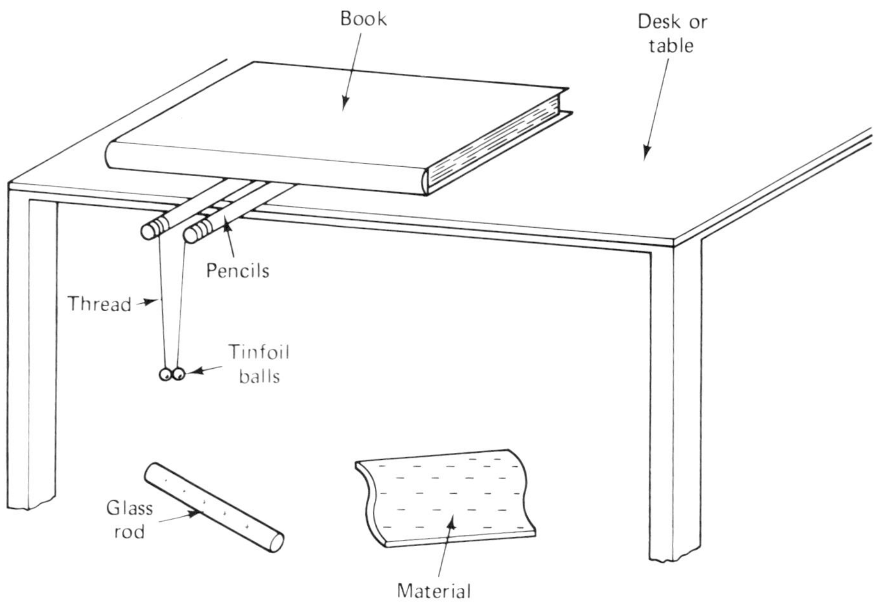

Figure 1-17 Static electricity and law of electrical charges.

4. Move the pencils as close as possible, so that the balls touch. Rotate the pencils accordingly so that the balls are adjacent to each other.
5. Rub the glass vigorously on the material.
6. Remove the rod from the material and place it on the two small balls. Some of the charge on the rod will transfer to the balls. This procedure in steps 5 and 6 may have to be done several times for a sufficient charge to build up on the balls. At some point in this procedure, the two balls will be repelled by the rod. This action demonstrates the law of electrical charges, which states that "like charges repel." When the rod is removed, the balls themselves should still repel each other somewhat. It is more difficult to prove the electrical law that "unlike charges attract" because the charges are neutralized when the balls touch.

Fill-In Questions:

1. The glass rod develops a ____________ charge when it is rubbed on the material.

2. Like charges ____________ .

3. Unlike charges ____________ .

4. The two balls repel each other because they have the ____________ charge.

SECTION 1-5 TROUBLESHOOTING APPLICATION: PROCEDURE AND PRACTICES FOR EFFICIENT TROUBLESHOOTING AND TESTING

The troubleshooting application sections in this book are intended to develop you with an analytical approach to electronic circuit operation. The electronic technician is hired not so much for knowledge alone but for the equipment that is repaired and maintained. When troubleshooting equipment problems, there are a few steps that should be remembered and practiced. The same steps should be used by students when they are constructing and performing laboratory projects or experiments.

Basic Troubleshooting Procedures

1. Understand the function of the device or circuit. What is it supposed to do?
2. Understand the device or circuit's basic operation. How does it work?
3. Have the proper schematic diagram for the device or circuit.
4. Have the proper tools and test equipment.
5. Keep work neat and free of debris.
6. Observe and practice safety rules.
7. Make sure that test equipment is connected correctly.
8. Check the obvious:
   a. Look for burned, broken, or damaged parts.
   b. Is the line cord plugged in?
   c. Is the device turned on?
   d. Is power getting to the device?
9. Keep a log on problems, symptoms, and remedies.
10. After the repair is completed, make a thorough test of operation and performance of the device.

SECTION 1-6 INSTANT REVIEW

- Matter is anything that occupies space, and consists of gases, liquids, and solids.
- Energy is that invisible force which provides the capability to do work.
- The atom is the smallest unit that forms the building blocks of all matter.
- The nucleus of an atom consists of protons and neutrons.
- A proton is a positively charged particle.
- A neutron is an uncharged particle.
- An electron is a negatively charged particle that orbits about the nucleus of an atom.
- The energy level at which electrons orbit the nucleus of an atom are called shells.
- The outer shell of an atom is called the valence shell.

- Valence electrons can easily become free electrons.
- The absence of an electron leaves a hole or positive charge.
- The atomic number is the equal number of electrons and protons of an atom.
- The atomic weight is the weight of the nucleus of an atom, including protons and neutrons.
- An element is the smallest particle of matter, wherein all the atoms are the same and cannot be broken down.
- A compound is a combination of two or more elements.
- The smallest unit of a compound is the molecule.
- An atom is normally electrically balanced.
- An atom with a positive net charge is called a positive ion.
- An atom with a negative net charge is called a negative ion.
- The invisible force about a positive ion goes in an outward direction.
- The invisible force about a negative ion goes in an inward direction.
- The invisible force about an ion is called an electrostatic field.
- The law of electrical charges states that (1) like charges repel and (2) unlike charges attract.
- Electrons flow from negative to positive.
- The accumulation of positive and negative ions produces an electrical pressure called voltage.
- Voltage is referred to as electromotive force, abbreviated EMF.
- Voltage is also referred to as a difference of potential.
- The triboelectric effect produces voltage from friction.
- The piezoelectric effect produces voltage from pressure.
- The thermoelectric effect produces voltage from heat.
- The photoelectric effect produces voltage from light.
- The electromagnetic effect produces voltage from magnetism.
- The electrochemical effect produces voltage from chemical action.
- A battery consists of two dissimilar metals called terminals placed into an electrolyte solution of acid.
- One terminal of a battery will become positively charged and the other will become negatively charged.
- The dry cell battery is called a primary cell and is not rechargeable.
- The wet cell battery is called a secondary cell and is rechargeable.
- Batteries connected in series-aiding produce a higher voltage.
- Batteries connected in parallel have a higher current capability.
- Scientific notation is a shorthand method of calculating large numbers based on the powers of 10.
- Units of measurement commonly used in electronics are

$$\text{giga} = 10^9$$
$$\text{mega} = 10^6$$
$$\text{kilo} = 10^3$$

milli = 10^{-3}
micro = 10^{-6}
nano = 10^{-9}
pico = 10^{-12}

SECTION 1-7 SELF-CHECKING QUIZ

Circle the most correct answer for each question.

1. Energy is the:

a. rate of doing work

b. force that causes electrons to move from negative to positive

c. invisible force that provides the capability to do work

d. none of the above

2. An atom consists of:

a. protons, electrons, and molecules

b. protons, neutrons, and electrons

c. neutrons, electrons, and ions

d. electrons, protons, and elements

3. The valence shell of an atom is the:

a. inner energy shell

b. outer energy shell

c. result of ionization

d. none of the above

4. Valence electrons:

a. can become "free" electrons

b. are found in the valence shell

c. answers a and b are correct

d. none of the above

5. The smallest particle of matter where all the atoms are the same is called:

a. a molecule **b.** an atom

c. an ion **d.** an element

6. A compound is a combination of two or more elements.

a. True **b.** False

7. The smallest unit of a compound is known as a molecule.

a. True **b.** False

8. The atomic number indicates the approximate weight of protons and neutrons in the nucleus of an atom.

a. True **b.** False

9. A positive ion is an unbalanced atom with a deficiency of electrons.

a. True **b.** False

10. Electron flow is from a positive charge toward a negative charge.

a. True **b.** False

11. If the distance between two charged bodies is decreased by one half, the force between them will:

a. increase by four times

b. decrease by four times

c. increase by two times

d. decrease by two times

12. If a positively charged body and a negatively charged body are brought close together, they will attract each other.

a. True

b. False

13. The invisible force about an ion is called:

a. an ion field
b. a magnetic field
c. an electrostatic field
d. a neutron field

14. Voltage is also referred to as:

a. electromotive force
b. EMF
c. difference of potential
d. all of the above

15. The triboelectric effect produces voltage from:

a. pressure
b. friction
c. chemical action
d. light

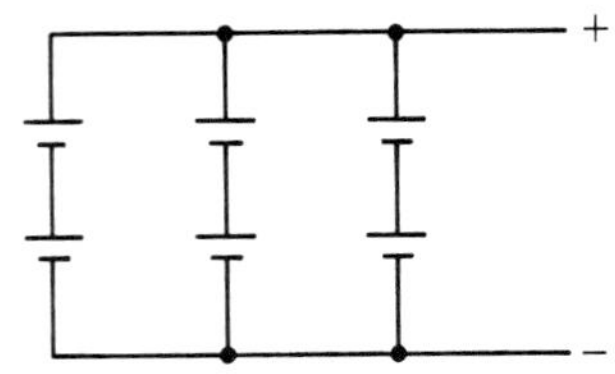

Figure 1-18

16. Referring to Figure 1-18, if each battery is 1.5 V at 0.25 A, the total voltage is ____________ and the total current capability is ____________ .

a. 9 V, 0.25 A **b.** 6 V, 0.75 A.
c. 3 V, 0.75 A **d.** 1.5 V, 0.75 A

17. A battery that is capable of being recharged is referred to as a primary cell.

a. True **b.** False

18. The sum of 3.4×10^2 added to 4.7×10^3 is:

a. 8.1×10^5 **b.** 5.04×10^1
c. 50.4×10^2 **d.** 3.87×10^3

19. The difference of 7.6×10^{-3} subtracted from 3.9×10^{-2} is:

a. 3.7×10^{-2} **b.** 37.0×10^{-1}
c. 31.4×10^{-2} **d.** 31.4×10^{-3}

20. The product of 4.5×10^3 times 3.6×10^4 is:

a. 16.2×10^{12} **b.** 16.2×10^1
c. 162×10^6 **d.** 1.62×10^7

21. The quotient of 1.2×10^3 divided by 2.5×10^6 is:

a. 0.48×10^{-3} **b.** 4.8×10^3
c. 48×10^9 **d.** 480×10^{-9}

22. The prefix assigned to 10^3 is:

a. milli **b.** mega
c. pico **d.** kilo

23. The prefix is assigned to 10^{-6} is:

a. pico **b.** kilo
c. micro **d.** nano

24. The prefix milli represents the power of:

a. 10^3 **b.** 10^{-3}
c. 10^{-6} **d.** 10^9

25. The prefix mega represents the power of:

a. 10^6 **b.** 10^3
c. 10^{-3} **d.** 10^{-6}

ANSWERS TO EXPERIMENT AND SELF-CHECKING QUIZ

Experiment 1: **(1)** positive **(2)** repel **(3)** attract **(4)** same

Self-Checking Quiz:

(1) c	**(2)** b	**(3)** b	**(4)** c	**(5)** d	**(6)** a	**(7)** a
(8) b	**(9)** a	**(10)** b	**(11)** a	**(12)** a	**(13)** c	**(14)** d
(15) b	**(16)** c	**(17)** b	**(18)** c	**(19)** d	**(20)** c	**(21)** a
(22) d	**(23)** c	**(24)** b	**(25)** a			

Unit 2

Basic Electrical Circuits

INTRODUCTION To test and repair electrical circuits you must understand all of the parts of the circuit and be able to use the test equipment that is required.

UNIT OBJECTIVES Upon completion of this unit, you will be able to:

1. Define *voltage, resistance,* and *current.*
2. Explain a basic series electrical circuit.
3. Read and interpret a multimeter.
4. Measure resistance, voltage, and current with a multimeter.
5. Test a series circuit with a multimeter.
6. Explain the conditions of open circuit and short circuit.
7. Explain a basic parallel electrical circuit.
8. Define basic electrical components such as switches and lamps.
9. Construct and test series and parallel circuits.

SECTION 2-1 FUNDAMENTAL CONCEPTS

2-1a DEFINING BASIC ELECTRICAL TERMS

An electrical circuit can be defined as an electrical path between two or more devices for the purpose of carrying electrical current to produce some useful function. There are three important terms associated with an electrical circuit:

1. *Voltage:* the force that causes electrical current flow in the circuit, measured in units called volts. The *volt* is defined as the force required to produce 1 ampere of current through a resistance of 1 ohm. The voltage of a battery produces current that flows in one direction and is referred to as *direct-current* (dc) *voltage.* It must be remembered that voltage is the force that produces current flow. Voltage *does not* flow but is the force that moves electrons in a conductor. Voltage is always in parallel or "across" a circuit component. Voltage is indicated by V or E.
2. *Resistance:* the opposition to current flow of the components in the circuit, measured in units called ohms. The *ohm*, symbolized by the Greek capital letter omega (Ω), is defined as the value of electrical resistance that will allow 1 ampere of current to flow through a material when 1 volt is applied. The conducting wires are assumed to have no, or zero, resistance, whereas the device in a circuit may have from several up to millions of ohms of resistance. Remember, resistance is stationary and *does not* move. It limits the amount of electrons flowing in a circuit. Resistance is indicated by R.
3. *Current:* the flow of electrons through the circuit caused by the force of the voltage, measured in units called amperes. The *ampere* is defined as the current flow produced when 1 volt is applied across 1 ohm of resistance. It is also defined that a current of 1 ampere is produced when 1 coulomb (6.25×10^{18} electrons) flows past a given point in 1 second. Remember, current *does* the flowing "in" or through the conductors and devices of a circuit. Current is indicated by I.

2-1b THE BASIC ELECTRICAL SERIES CIRCUIT

A basic electrical circuit consists of a power source, a load, and conductors for connecting the components, as shown in Figure 2-1. The power source is a 1.5-V battery, the load, which is an electrical device that performs a useful function; in this case, a lamp and the conductors are connecting wires. A switch is also used to control the flow of electrical current in the circuit.

In the pictorial diagram of Figure 2-1a, when the switch lever is placed to the left, the switch is open and no current flows in the circuit.

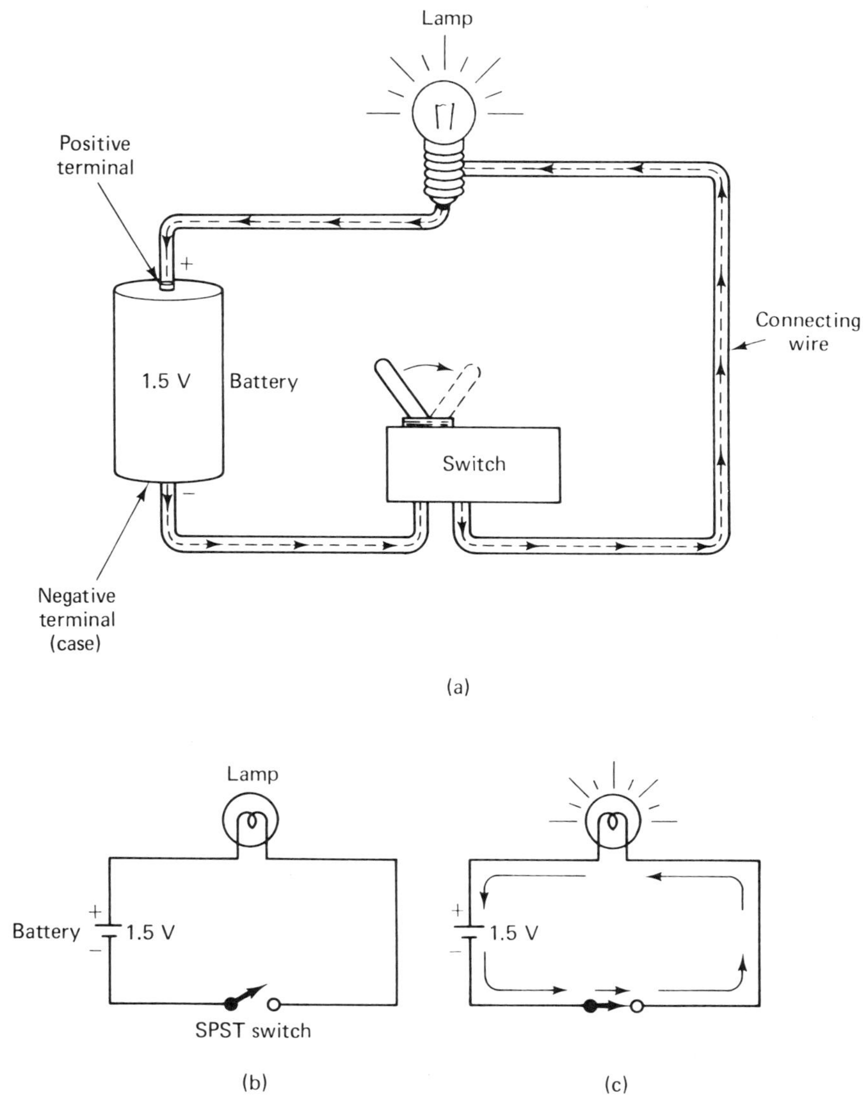

Figure 2-1 Basic electrical circuit: (a) pictorial diagram; (b) schematic diagram, open circuit; (c) schematic diagram, closed circuit.

When the switch lever is placed to the right, the switch is closed and electrons leave the negative terminal of the battery and flow through the switch, then through the lamp to the positive terminal of the battery. The electrons flowing through the resistance of the lamp create heat and produce light—hence, a typical flashlight circuit.

Figure 2-1b and c show what is called the schematic diagram (electrical drawing). Notice the symbols for the battery, switch, and lamp. As a technician, you should be able to draw these symbols from memory. In Figure 2-1b the switch is open and the resistance of the circuit from the viewpoint of the battery is infinite. Therefore, no current flows in the circuit and the lamp is off or not glowing. This condition is referred to as an *open circuit.* In Figure 2-1c the switch is closed and current flows through the circuit, as indicated by the arrows, which in turn causes the lamp to glow. The battery now "sees" a lower resistance, which depends on the resistance of the lamp. The closed switch is assumed to have no or zero resistance. This condition is referred to as a *closed circuit.*

This circuit is referred to as a *series circuit,* because electrons leaving the negative terminal of the battery flow through the switch and then through the lamp to reach the positive terminal of the battery. Since the same current flows through the switch and then the lamp, it is said that the switch and lamp are in series.

2-1c USE OF A MULTIMETER

Some means of testing inoperative circuits must be used in order to save time, if not money. For instance, with the basic electrical circuit of Figure 2-1b, when the switch is closed, the lamp does not glow. The question arises: Is the lamp burned out, the switch faulty, or the battery run down? New components can be purchased for each of the three and then each one substituted (replaced) in the circuit until the faulty component is located. If only one component is bad, extra money will not have to be spent replacing the other two new components.

Meters are used to check the operation of circuit components and actual circuit performance. Three meters are used to check the total performance of a circuit. An *ohmmeter* is used to measure the resistance of the lamp and operation of the switch. A *voltmeter* is used to measure the voltage of the battery. An *ammeter* is used to measure the current flowing in the circuit. These three meters are usually built into a single unit called a *multimeter*, shown in the generalized drawings of Figure 2-2.

Figure 2-2 Multimeters: (a) analog VOM; (b) digital VOM (DVM).

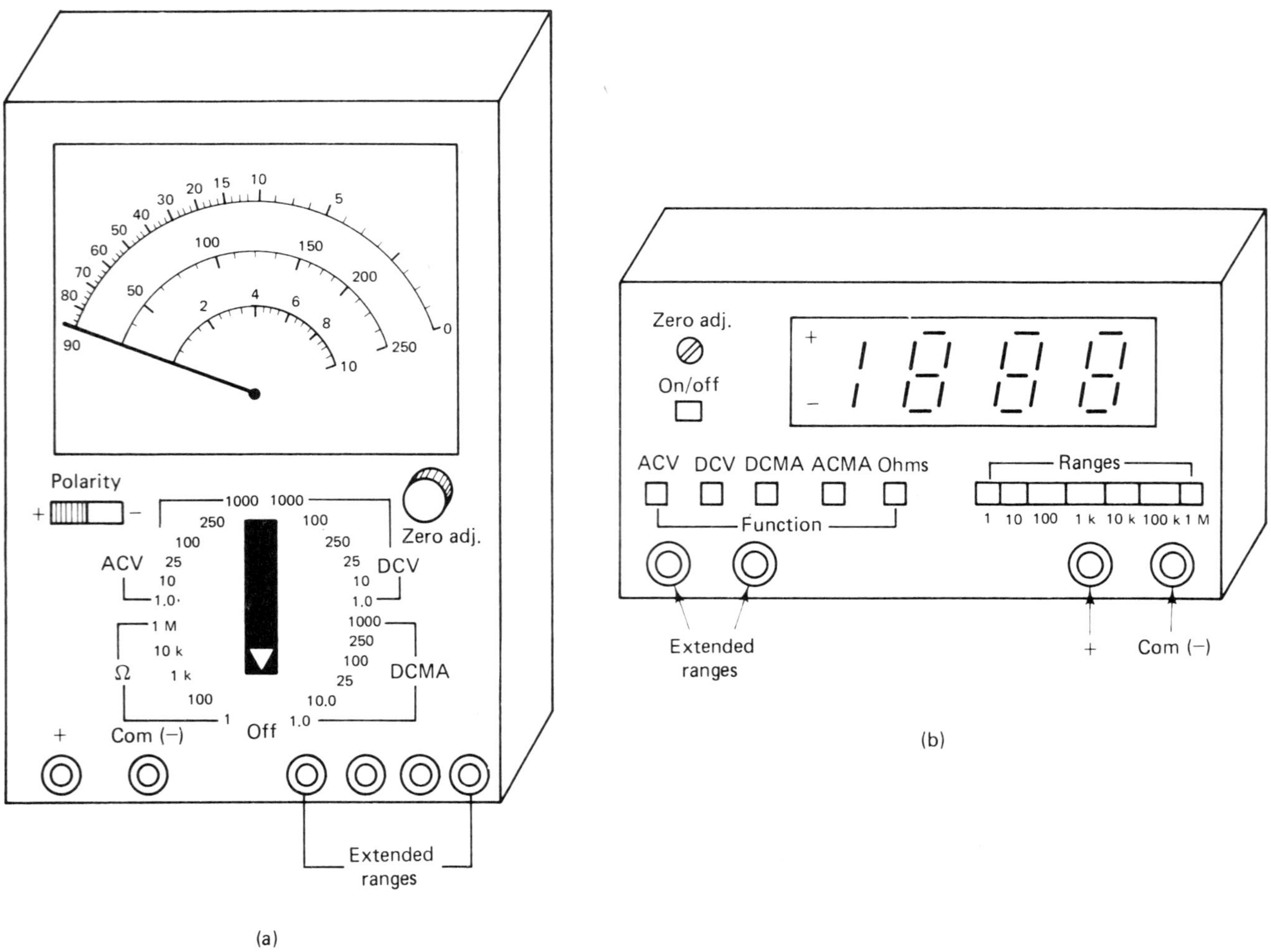

An analog multimeter (often referred to as a *volt-ohm-ammeter* or VOM) has a set of scales on which a needle or pointer indicates a quantity being measured (Figure 2-2a). The digital VOM or digital voltmeter (DVM) uses seven-segment numerical readout indicators (Figure 2-2b). Each of these meters has a function switch or switches for selecting the measurement of resistance, voltage, or current. Generally, resistance is indicated by ohms or its symbol, Ω. Direct-current voltage, such as for measuring a battery, is labeled DCV. Direct current is given as DCMA, with the "MA" indicating milliamperes. Also included is a section for measuring effective alternating current voltage, given as ACV. Usually, an analog VOM does not measure alternating current, but most DVMs can and may be indicated as ACMA. The function switch may include ranges (specific limits to be measured) or have separate range switches.

There will be jacks for connecting test leads. The two most used jacks are the + jack and the common or − jack. Other jacks are used for extended ranges. A zero adjust control is used to adjust the meter for a reading of zero when resistance is being measured. This control is used to compensate for weakened battery voltage or internal circuitry required by the meter in the resistance measuring function.

A polarity switch may be included with the analog VOM, which simply reverses, electrically, the test lead jacks. If the leads are connected incorrectly in a circuit, the pointer will swing off the meter scale the wrong way. By moving this switch to the other position, the pointer will move correctly up the scale. The DVM usually has a + or − indicator that lights up to show proper polarity of the test leads. The following descriptions will show you how to read an analog VOM correctly.

2-1c.1 Measuring Resistance

When measuring resistance, you must be able to read the ohms scale correctly, as shown in Figure 2-3. The ohms scale is the top one shown on the analog meter face. Pay attention to this scale and ignore the others at this time. Notice that this scale reads from right to left, zero on the right and ascending numbers as it progresses left.

The function/range switch is set on 1 k *(k stands for "kilo" and means 1000).* This means that the numbers on the ohms scale represent thousands: for example, 2 = 2000, 5 = 5000, 10 ≐ 10,000, and so on. In other words, the range switch acts as a multiplier of the numbers read on the scale. Other ranges include

$$\begin{aligned} 1 &= \times\ 1 \\ 100 &= \times\ 100 \\ 10\text{ k} &= \times\ 10{,}000 \\ 1\text{ M} &= \times\ 1{,}000{,}000 \text{ (M, mega, stands for 1 million)} \end{aligned}$$

The tip of the pointer is resting between 50 and 70 of the ohms scale. Midway between 50 and 70 is a large line indicating 60; however, the pointer is resting on a smaller line halfway between 50 and the 60 line, which would indicate 55. Therefore, 55 times the range switch setting of 1 k equals 55 kΩ or 55,000 Ω. If the range switch were set on 10 kΩ, 55 times 10 k would indicate a meter reading of 550 kΩ or 550,000 Ω. Similarly, if the range switch were set on 1, the quantity 55 times 1 would indicate a meter reading of 55 Ω.

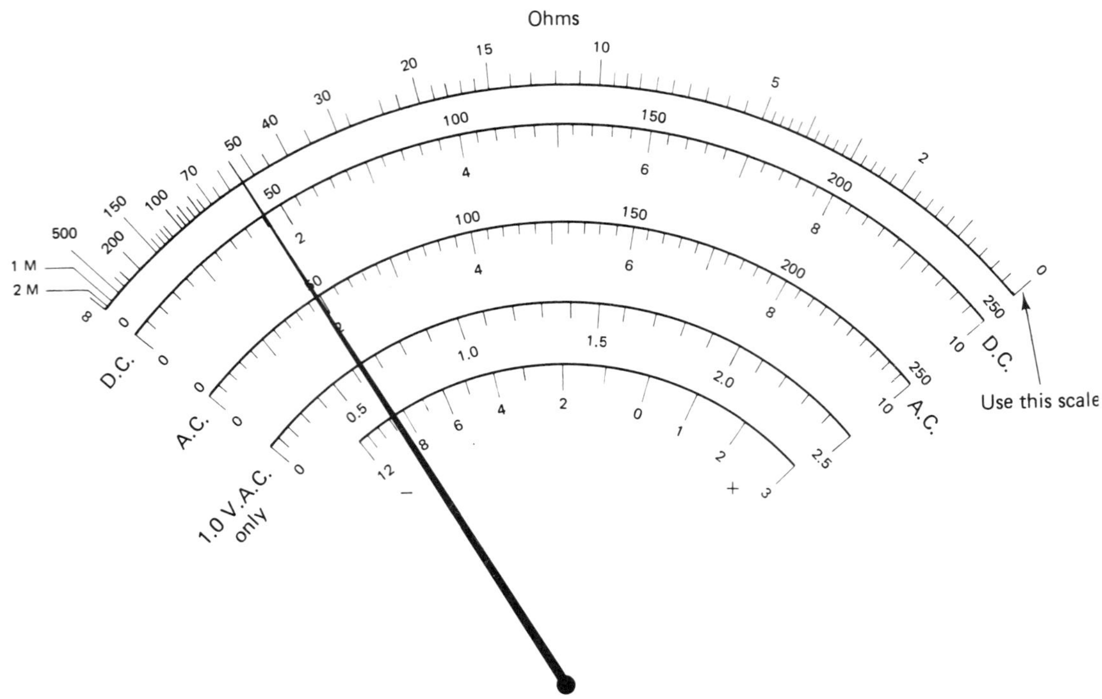

Figure 2-3 Reading ohms (range switch set at 1 k pointer indicates 55 kΩ).

Notice the right side of the ohms scale. Halfway between 0 and 2 there is a large line indicating 1. There are four smaller division lines indicating values less than 1. Starting from 0, it takes five total lines to reach line 1. Therefore, 1 divided by 5 will indicate that each small line is equal to 0.2. If the pointer was resting on the third line from 0, the reading would be 3 times 0.2 equals 0.6. With the range switch set on 100, the meter would be indicating that 0.6 times 100 equals 60 Ω.

A simple formula can be used to remember how to find the value of a subdivision on any meter scale:

$$\text{value of each subdivision} = \frac{\text{larger number} - \text{smaller number}}{\text{total subdivisions between numbers}}$$

As an example, from number 5 to number 10, it takes 10 subdivisions. The difference between 5 and 10 is 5; therefore, 5 divided by 10 indicates that each subdivision is equal to 0.5.

$$\text{value of each subdivision} = \frac{10 - 5}{10} = \frac{5}{10} = 0.5$$

If the pointer was resting 3 subdivisions from the number 5 toward the number 10, the meter reading would be 3 times 0.5 equals 1.5 plus the number 5, for a total of 6.5. With the range switch set on 100, the meter would be indicating that 6.5 times 100 equals 650 Ω. Follow this procedure any time that the pointer falls between any given numbers on the ohms scale.

The most accurate ohm readings occur near the middle of the scale. When making a resistance measurement, always try to get the pointer toward this area by changing the range switch.

2-1c.1a Measuring the Resistance of a Lamp

Figure 2-4a shows a pictorial diagram for measuring the resistance of a lamp. The procedure for testing a lamp follows:

1. Place test leads into the test jacks. Usually, a red lead goes to the + jack and a black lead goes to the − jack.
2. Set the function/range switch to the 1 Ω setting.
3. Place the ends of the test leads together. This will cause the pointer to move on the meter face. Adjust the zero adjust control until the pointer is over the 0 mark on the ohms scale. Any time the range switch is changed, the pointer should be adjusted to zero.
4. Place one test lead on the side or case of the lamp and the other test lead on the electrical end. This completes the circuit and the pointer should move and stop at the indicated resistance on the ohms scale. If the pointer does not move and indicates infinite resistance, the lamp is open, or in other words, burned out.

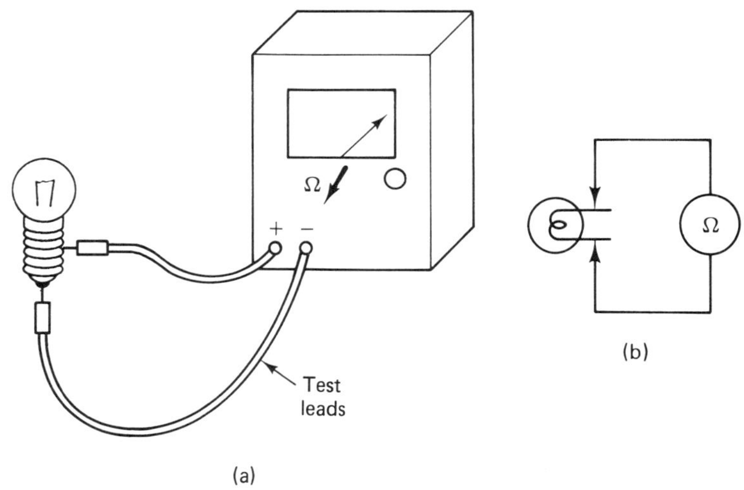

Figure 2-4 Measuring resistance of a lamp: (a) pictorial diagram; (b) schematic diagram.

Important! The ohmmeter has its own internal power source, which develops a current through the component being tested. For this reason, when you test a component in a circuit, first remove power from the circuit so as not to destroy the meter. Next, remove or unsolder one end of the component from the circuit, so that other components of the circuit do not interfere and give a false meter reading. Figure 2-4b shows the schematic diagram for this test.

2-1c.1b Testing a Switch

Figure 2-5a shows the pictorial diagram for testing a switch. The procedure for setting up the ohmmeter is the same as for the lamp. Place each test lead on one of the switch terminals. With the switch open, the meter should read infinite ohms, or the pointer should not move at all. Figure 2-5b shows the schematic representation for a normally operating open switch. If the pointer moves to 0 on the ohms scale, the switch is shorted and has a faulty mechanism. With the switch closed, the meter should indicate 0 ohms. Figure 2-5c shows the schematic representa-

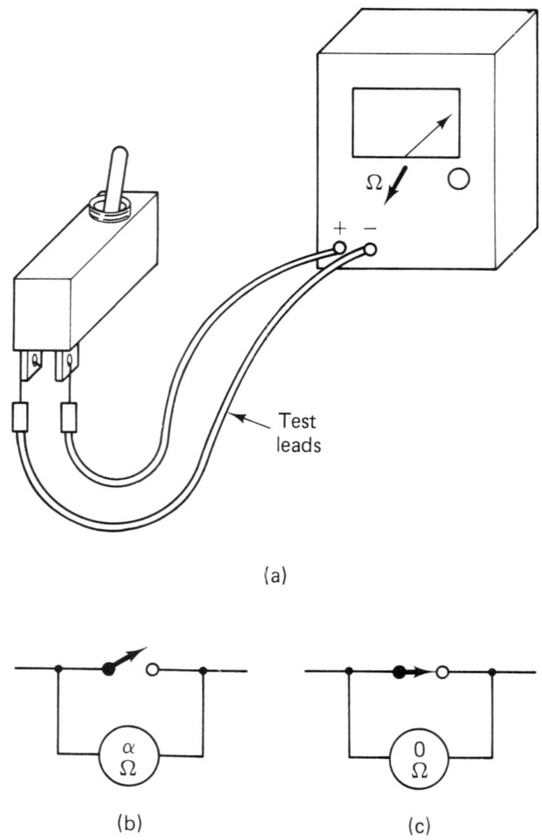

Figure 2-5 Testing a switch: (a) pictorial diagram; (b) open switch = infinite ohms; (c) closed switch = zero ohms.

tion for a normally operating closed switch. If this meter reading is infinite, the switch is open and has a faulty mechanism. A switch can be faulty with an open or shorted condition.

In the process of measuring the resistance of a component, the + test lead of the ohmmeter is placed on one end of the component and the − test lead placed on the other end. The ohmmeter is said to be in *parallel* with the component.

2-1c.2 Measuring Voltage

Reading a voltmeter scale is similar to reading an ohms scale, but with a few variations. The dc voltage scale is the second one down from the top in Figure 2-6. At this time we are concerned only with this scale. Voltage scales usually read from left to right, with zero on the left side. The numbers ascend progressing to the right until what is called a *full-scale deflection* is reached. For example, if the range switch is set on 10 DCV and the voltage being read is 10 V, the pointer will move all the way across the scale and stop at 10, which is full-scale deflection. With the range switch set on 10 DCV, each subdivision is worth 0.2 V. Therefore, the actual voltage being shown on the meter in Figure 2-6 is 9.2 V dc. If the range switch was set on 100, the meter would indicate 92 V dc. In this case the same scale is used, but a 0 is added to the numbers and each subdivision is equal to 2 V.

When the range switch is set to 250, the numbers above the scale are used and each subdivision is worth 5 V. The meter would indicate a voltage reading of 230 V dc. If the range switch is set to 25, a 0 is subtracted from the numbers and each subdivision is equal to 0.5 V. The meter reading would indicate 23 V dc at this time.

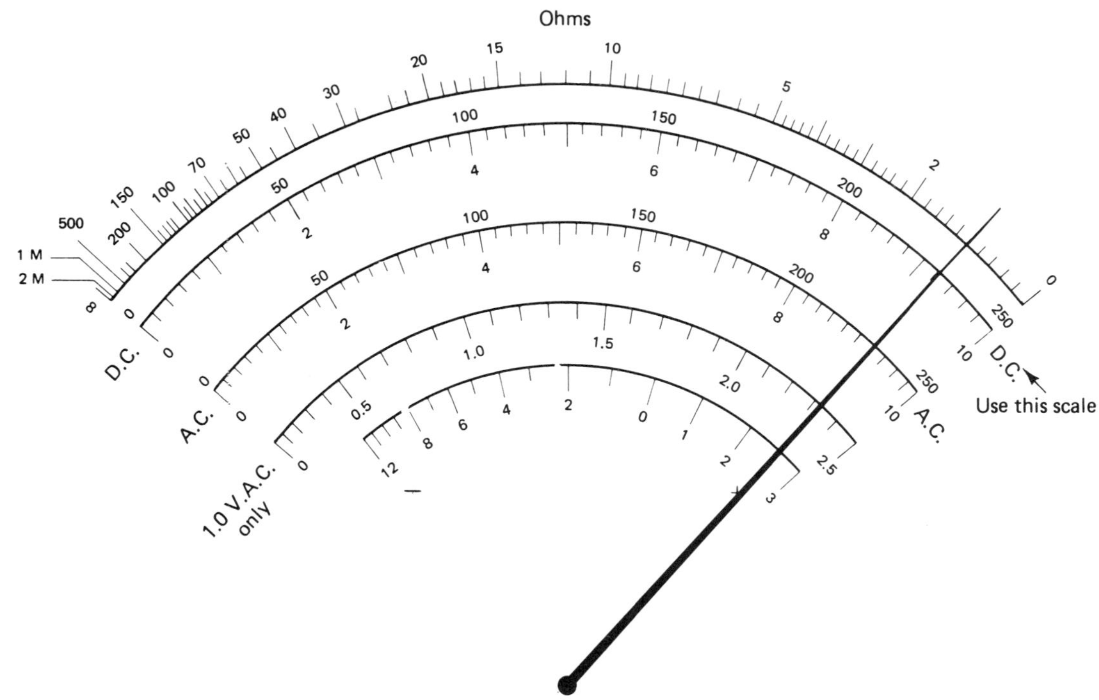

Figure 2-6 Reading dc voltage (range switch set at 10 DCV, pointer indicates 9.2 V).

When alternating-current (ac) voltage is being measured, the third and fourth scales from the top are used. The fourth scale is used only for voltages under 2.5 V ac. The same procedure is used to read ac voltage as with dc voltage, except that the range switch is set to the ACV ranges. With the range switch at 250 ACV, the pointer on the meter face shown in Figure 2-7 indicates 155 V ac. The bottom scale on the meter face is used for reading decibels and is discussed later in the book.

2-1c.2a Testing Voltage of a Circuit

There are two important procedures to remember when testing voltage with a voltmeter.

1. Set the range switch to a higher value than the voltage to be tested. If the range switch is set on a lower value than the voltage to be measured, the pointer will "race across" the meter face and possibly be bent or the meter may be damaged permanently.
2. Observe polarity of the test leads. The + lead must be placed on the most positive point of the part of the circuit being measured and the − lead placed on the less positive or negative point of the circuit.

To test a battery, simply place the + lead of the voltmeter on the positive terminal and the − lead on the case or opposite end (negative terminal) of the battery. This is not always a true indication of a good battery. A battery may test as good with this procedure, but when it is placed under load in a circuit, the voltage at the terminals may be much

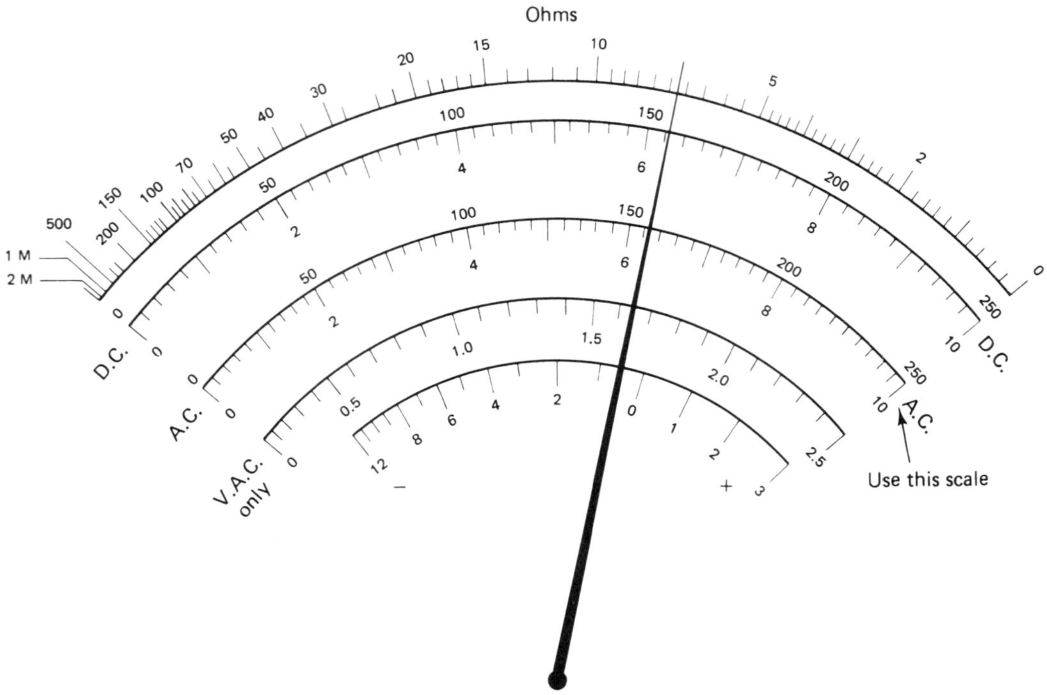

Figure 2-7 Reading ac voltage (range switch set at 250 ACV, pointer indicates 155 V).

less. Therefore, it is always best to test a battery under load, as shown in Figure 2-8a.

Notice that the + lead of the voltmeter is attached to the wire conductor connected to the positive terminal of the battery. The − lead is attached to the wire conductor, that is, connected through the switch to the negative terminal of the battery. The voltmeter is in parallel with the light and will indicate the voltage across the lamp. To measure voltage, the voltmeter is always placed in parallel with a component to measure the voltage across the component. Figure 2-8b shows the schematic representation of this procedure.

2-1c.2b Measuring Voltage across a Switch

Measuring voltage across a switch is helpful in understanding two important troubleshooting concepts in electrical circuits. Assume that the switch to be measured is the one shown in Figure 2-8b. Figure 2-9a shows a voltmeter placed across the switch in the open position. There is no current flow in the circuit because an open circuit has infinite resistance. However, the left side of the switch is connected to the negative terminal of the battery and the right side is connected to the positive terminal via (by way of) the lamp. The electrostatic charge is present on each side of the switch and there exists a difference of potential; therefore, the meter will read the 1.5 V of the battery.

When the switch is closed, current flows in the circuit. The closed switch has zero ohms and no potential difference is present; therefore, the meter will read 0 V. Electrically speaking, the meter leads are at the same potential, hence no voltage reading. Another way to look at this situation is that the meter leads are shorted together by the closed switch.

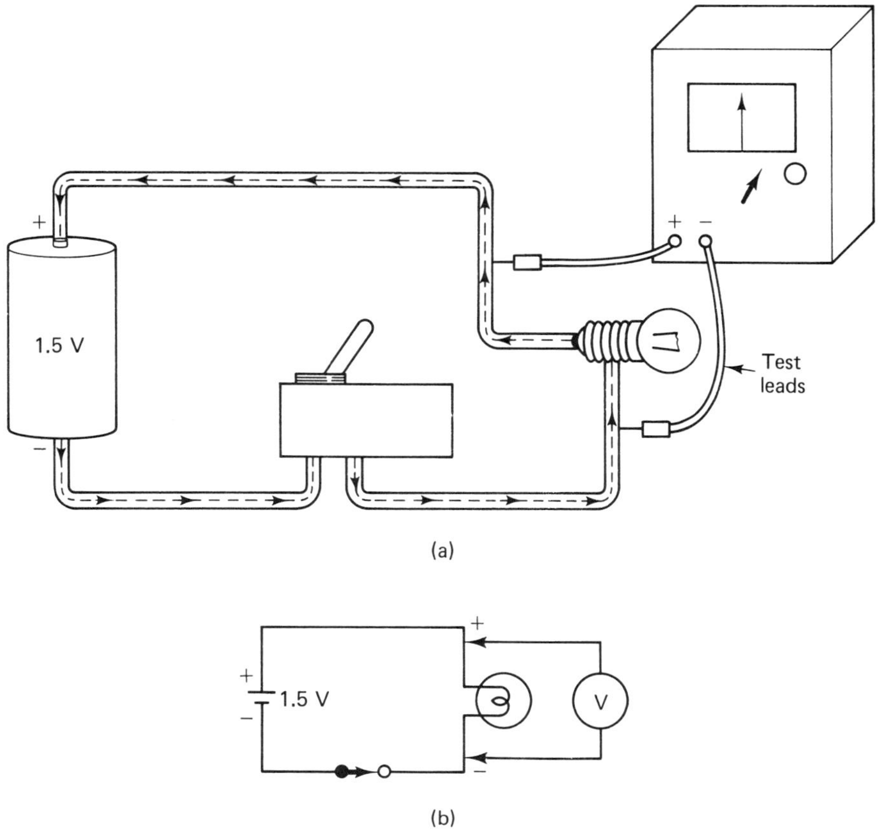

Figure 2-8 Measuring voltage: (a) pictorial diagram; (b) schematic diagram.

Summarizing these two conditions:

- An *open circuit* has infinite resistance and does not allow current to flow; therefore, voltage will be present across the open.
- A *short circuit* has zero resistance and allows current to flow; therefore, no voltage is present across the short.

2-1c.3 Measuring Current in a Circuit

When measuring dc current with a VOM, the same scale is used as with measuring dc voltage. Figure 2-10 shows a dc current reading. The ranges that are read on the scale 0 to 10 are 1.0, 10.0, and 100 mA. The ranges that are read on the scale 0 to 250 are 25 and 250 mA. The procedure for reading dc current and dc voltage is the same. With the range switch set to 25 DCMA, the pointer shown in Figure 2-10 indicates 18 mA.

Figure 2-9 Measuring voltage of a switch: (a) open switch = 1.5 V; (b) closed switch = 0 V.

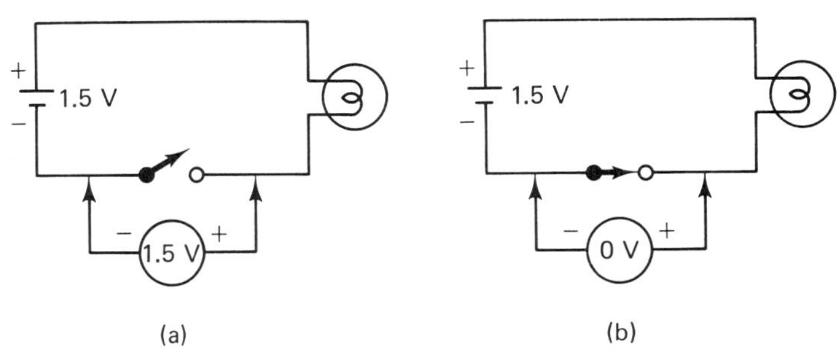

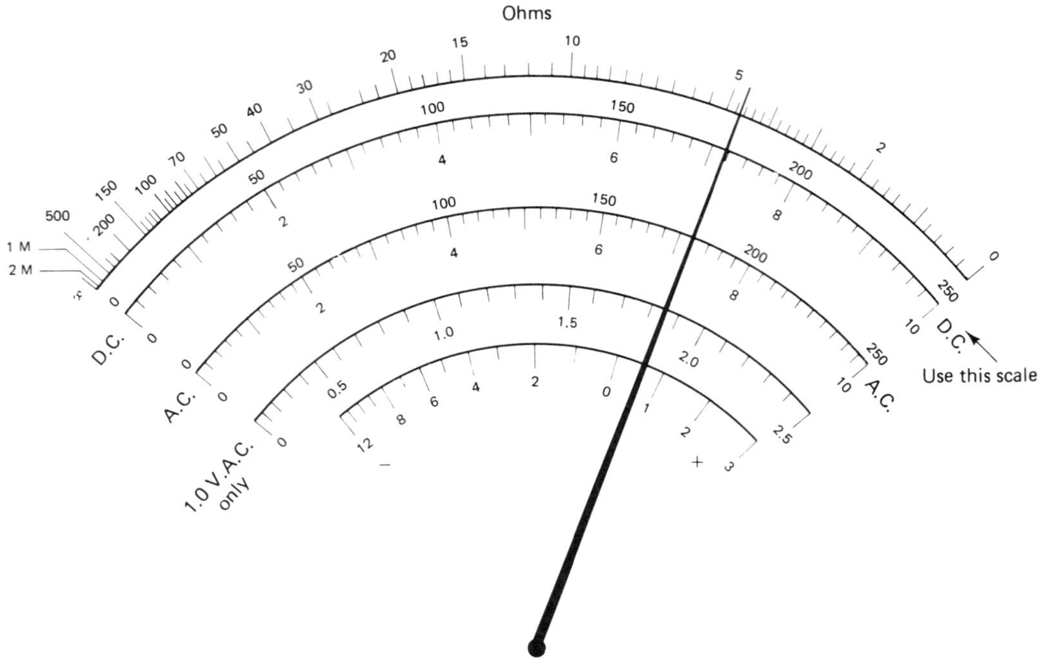

Figure 2-10 Reading dc current (range switch set at 25 DCMA, pointer indicates 18 mA).

It is important to remember that when measuring current, the ammeter must be placed "in series" with the circuit, as shown in Figure 2-11a. The procedure for connecting an ammeter follows:

1. Remove power source from the circuit (open a wire from one end of the battery).
2. Open the circuit where the meter is to be placed. Observe polarity when connecting meter, making sure that the + lead is at the more positive point in the circuit and the − lead is at the less positive point.

Figure 2-11 Measuring current in a circuit: (a) pictorial diagram; (b) schematic diagram.

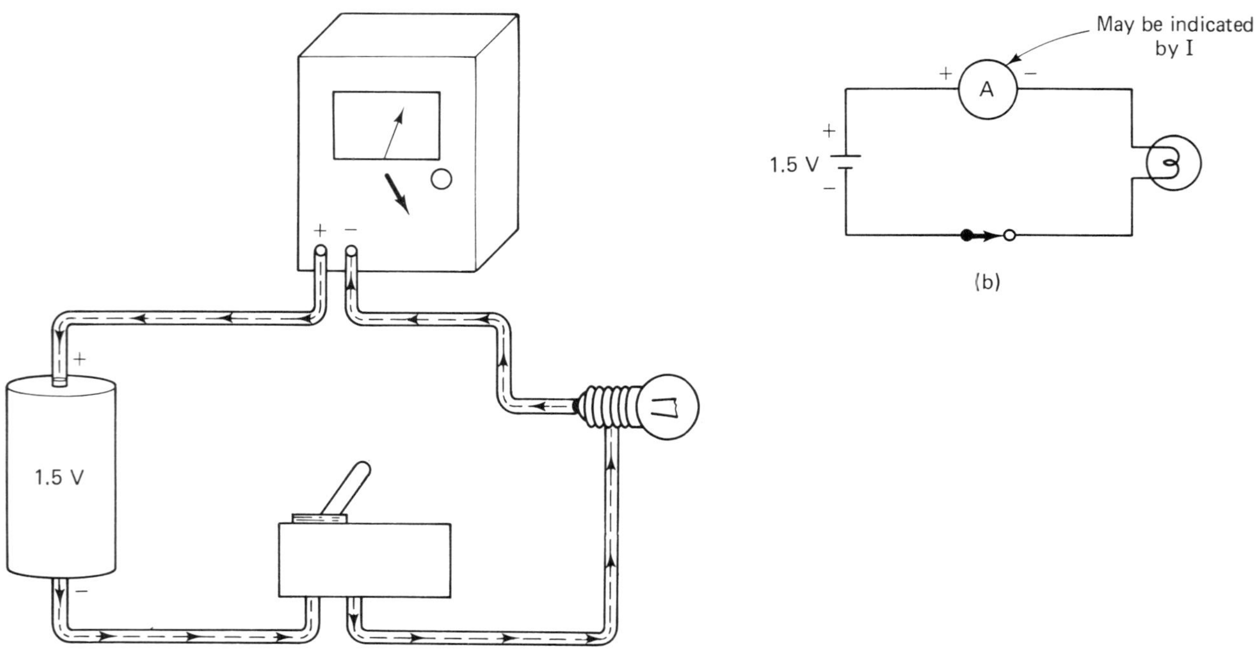

3. Set the range switch to the highest DCMA range.
4. Connect the power source to the circuit and then adjust the range switch for a proper reading.
5. Remove the power source from the circuit when disconnecting the ammeter.

Figure 2-11b shows the schematic representation for an ammeter connected in a circuit. The meter may be identified with an A for ammeter, or an I, the symbol used for current in electrical circuits.

Important! Before making a circuit measurement, always observe the function/range switch to be sure that it is in the proper place. Never place an ammeter in parallel with a circuit component (do not measure voltage with an ammeter) because the meter could be damaged.

2-1d THE BASIC ELECTRICAL PARALLEL CIRCUIT

The basic parallel circuit uses components similar to those in the basic series circuit, except that there is another path or branch for current flow. Figure 2-12a shows a pictorial diagram of the basic parallel circuit, which uses two lamps and two control switches.

When switch SWA is closed, electrons will flow through this branch and lamp 1 will light. Similarly, when switch SWB is closed, electrons will flow through this branch and lamp 2 will light. Notice that with both switches closed, there are two separate branch currents flowing. What actually happens is that electrons leave the negative terminal of the battery and flow toward the bottom junction of the two branches. At this point the current divides proportionately in relation to the resistance of each branch (in this case, evenly) and flows through the branches. At the upper junction of the two branches, the branch currents recombine and flow toward the positive terminal of the battery. This action is shown in the schematic diagram of Figure 2-12b. While this description illustrates a point-by-point analogy, remember that the electrons in the circuit all move at the same time the instant a switch is closed. In other words, when it does flow the current is nearly instantaneous and at maximum.

Figure 2-12. Basic parallel circuit: (a) pictorial diagram; (b) schematic diagram.

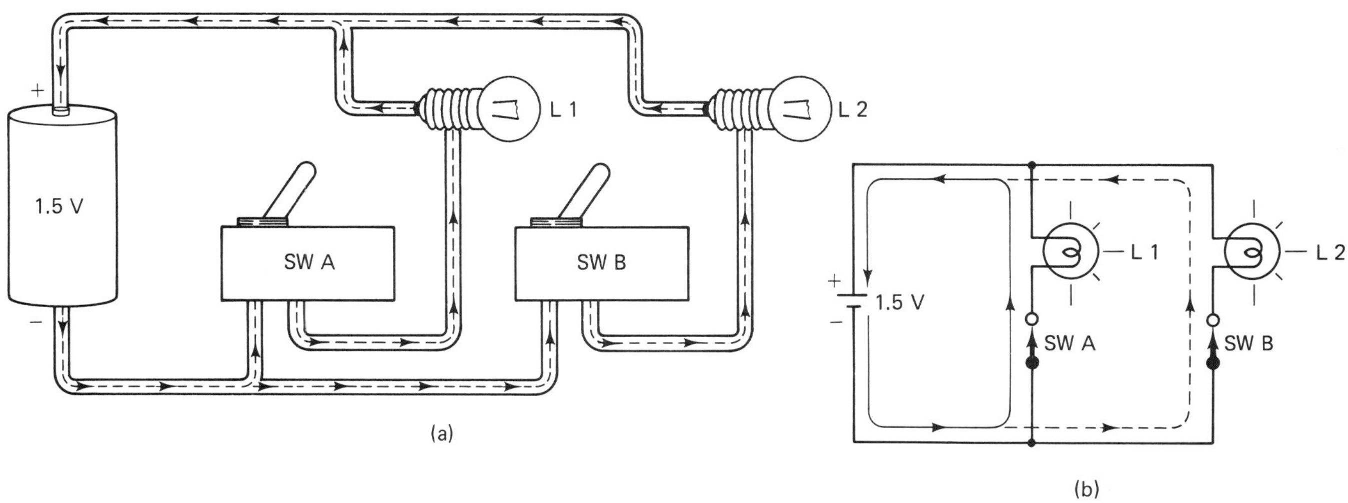

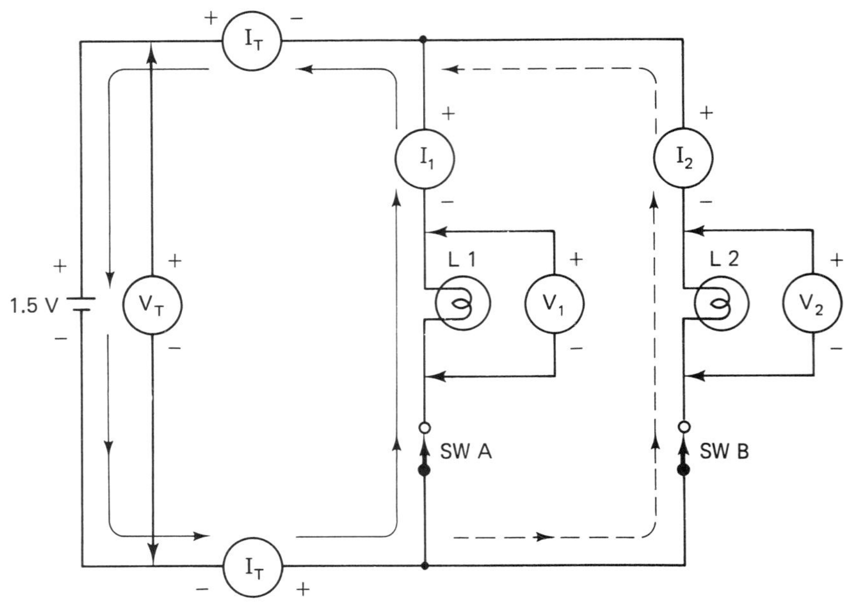

Figure 2-13 Voltage and current nomenclature and measurement in a parallel circuit.

$$V_T = V_1 = V_2$$
$$I_T = I_1 + I_2$$

With two lamps connected in parallel, twice as much current must be "drawn" or taken from the battery. For example, the total current (I_T) will be equal to the sum of the current in branch 1 (I_1) and the current in branch 2 (I_2) therefore; $I_T = I_1 + I_2$. However, since each branch is connected in parallel with the battery (V_T), the voltage across each branch (V_1 and V_2) is the same and can be expressed as $V_T = V_1 = V_2$. Figure 2-13 shows this relationship and also illustrates the voltage and current nomenclature of the components of the circuit. The round circles represent meters and how they are positioned, in terms of polarity, for making proper measurements.

2-1e BASIC CIRCUIT COMPONENTS

The basic circuit components used in this unit are the switch and the lamp. These components are available in different sizes and types for various functions. Figure 2-14 shows the physical appearance of some switches. A basic switch has a stationary contact and a movable contact referred to as the *common contact,* C. The distinguishing feature among switches is the type of C contact actuator used. A *toggle-type switch* (a) has a small round lever that moves back and forth. A *paddle-type switch* (b) is like the toggle, except that the lever is wider. A *rocker-type switch* (c) rocks back and forth. A *slide-type switch* (d) moves laterally from one side to the other. A *pushbutton-type switch* (e) moves down and up (in or out). A *rotary-type switch* (f) moves in a circular motion. *Subminiature (or micro) switches* (g) are tiny switches that operate with a small plunger or actuator arm, which may have a small roller. *Switch units used on PC boards* (h) are about the size of an IC and fit into a dual-in-line package (DIP) mounting arrangement. They usually have eight separate slide-type switches. *Proximity switches* (i) have thin metal reed contacts, which operate when an external magnet is brought close to a contact. This type of switch is often used in intrusion (burglar) alarm systems. A *mercury switch* (j) is a small container (often, glass) with a

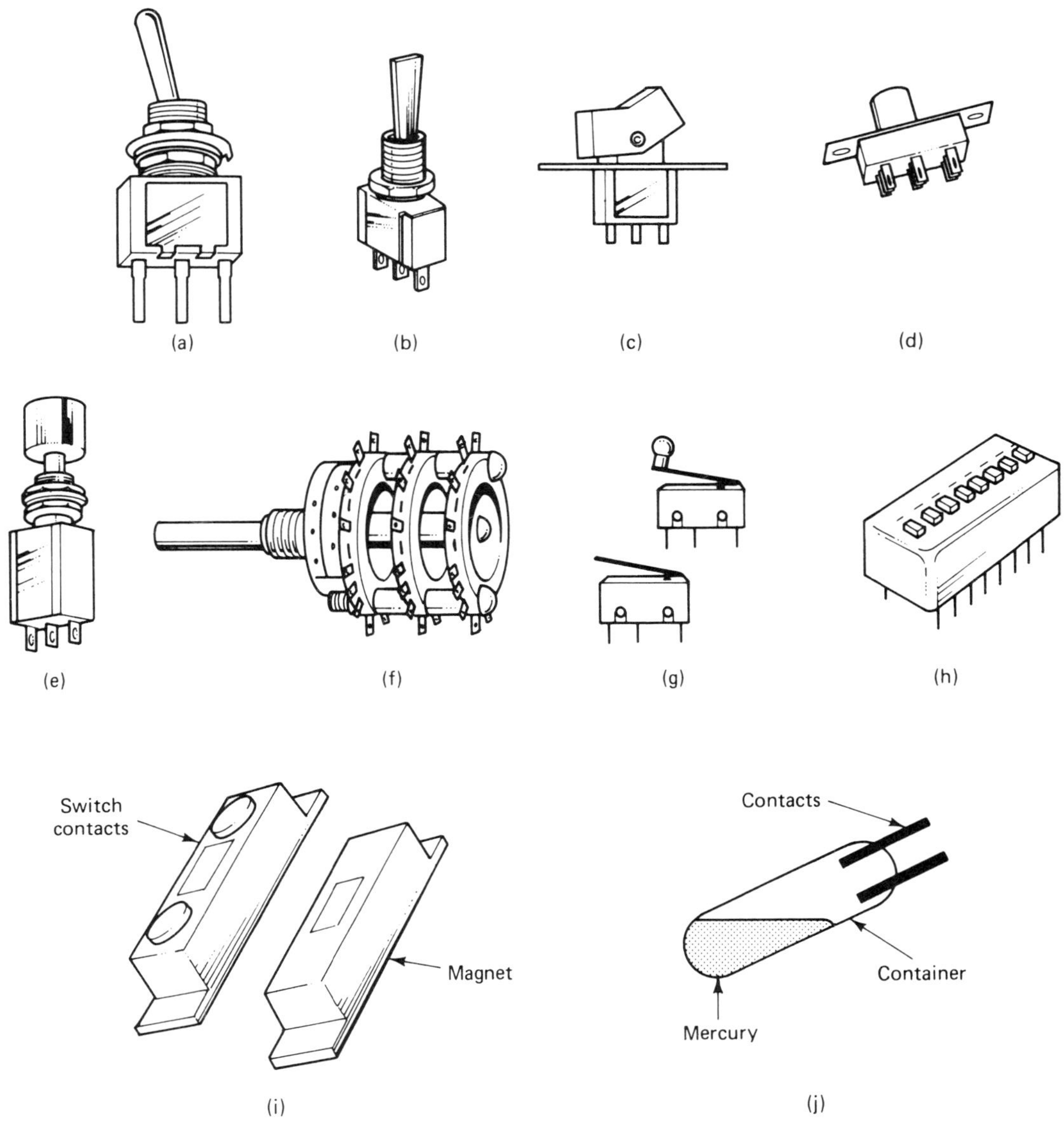

Figure 2-14 Types of switches: (a) toggle; (b) paddle; (c) rocker; (d) slide; (e) pushbutton; (f) multiwafer rotary; (g) subminiature (micro); (h) PC board eight-position; (i) proximity (magnetic); (j) mercury.

portion of a liquid mercury sealed inside. The contacts at one end protrude through the glass for external connections. When the container is tilted, the mercury flows around the contacts, which completes the electrical circuit. This type of switch is used to close or open circuits resulting from physical movement of mechanical parts, such as an interlock switch on a lid or cover leading to a dangerous piece of equipment.

Switches are used to open or close an electrical circuit. Their function is to enable or disable a particular circuit operation or transfer operation from one point to another point. Figure 2-15 shows switch classification according to their schematic symbols. The simplest switch is a *single-pole single-throw,* SPST (a), which closes a circuit from a common contact, C, to a *normally open,* NO, *contact.* A *single-pole double-throw switch,* SPDT (b), transfers the C contact from a *normally closed,*

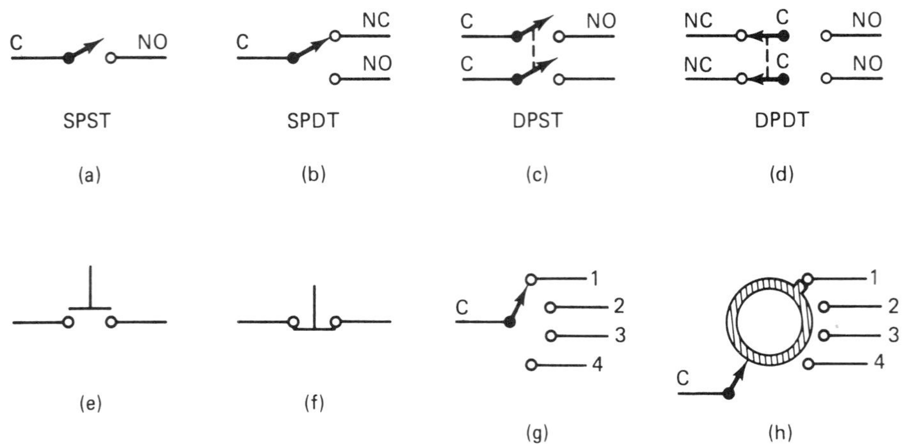

Figure 2-15 Switch schematic symbols: (a) single-pole single-throw; (b) single-pole double-throw; (c) double-pole single-throw; (d) double-pole double-throw; (e) normally open pushbutton; (f) normally closed pushbutton; (g) multiposition; (h) rotary.

NC, *contact* to a normally open, NO contact. A *double-pole single-throw switch,* DPST (c), operates similarly to a SPST switch, except that the two C contacts are "ganged" (tied) together as indicated by the dashed line and operate simultaneously. The *double-pole double-throw switch,* DPDT (d), has the C contacts "ganged" together and operates two NC and NO contacts simultaneously. A *normally open pushbutton switch* (e) closes a circuit when the button is depressed, whereas a *normally closed switch* (f) opens the circuit. The switch may keep the circuit latched in the desired condition when the button is released, or it may be spring loaded and return the contacts to the original condition. A pushbutton switch may operate several NO or NC contacts or a combination of both simultaneously. In a multiposition switch (g and h), the C contact can be set to many position contacts and is referred to as a *rotary switch.*

Miniature and subminiature electrical lamps are used primarily in flashlights and as pilot (on/off indicator) lights for electronic equipment. Figure 2-16 shows various types of lamps and lamp sockets. Incandescent lamps (a–d) have very thin filament wires that are heated when current flows through them. The wires, in turn, glow and produce light. There are two major considerations in the use of these lamps: (1) that the rated voltage of the lamp not be exceeded, and (2) that when replacing a specific lamp with a substitute, to be aware of a possible high current draw, which could activate the fuse in a piece of equipment. It is wise to know the voltage and current ratings of any lamps being used in circuits.

There are three standard bases for incandescent lamps: *screw-type* (b), *bayonet-type* (c), and *flange-type* (d). Each type of base has to match a corresponding type of socket for proper use.

The *neon lamp* (e) is a nonheated type of device, with two electrodes, that contains an inert gas. At a given voltage, the gas ionizes and current flows, causing the electrodes to glow. If used with dc voltage, one electrode glows, whereas both electrodes glow with ac use. This lamp requires a series resistor to limit the current through it.

The *light-emitting diode,* LED (f), is rugged, requires little power, come in various colors (red, yellow, green), and is relatively inexpen-

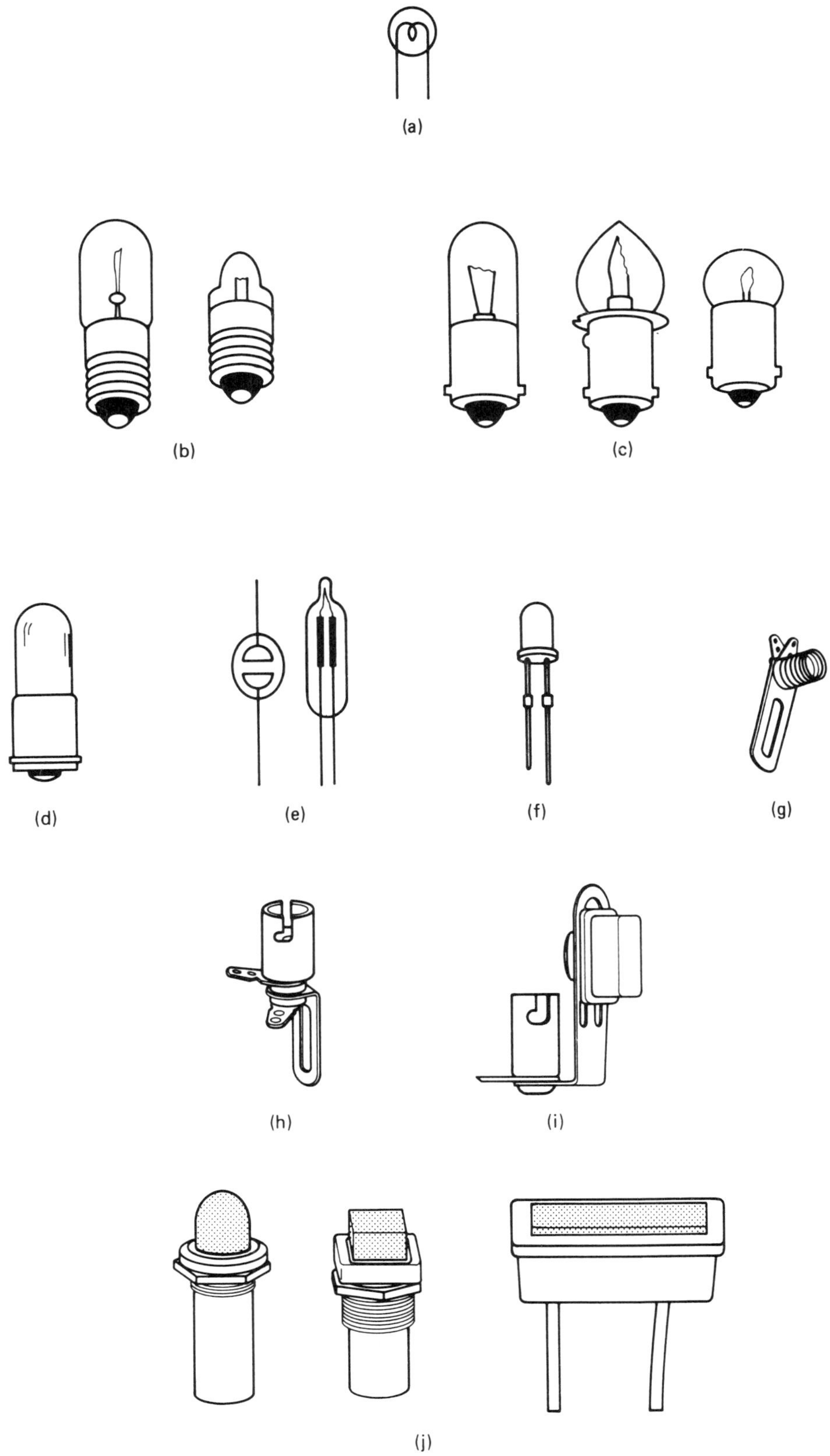

Figure 2-16 Electrical indicator components: (a) lamp electrical symbol; (b) screw-type base lamps; (c) bayonet-type base lamps; (d) flange-base type lamp; (e) neon lamp; (f) LED; (g) screw-type socket; (h) bayonet-type socket; (i) socket with colored lens; (j) through-panel pilot light indicators.

sive. In operation, the LED requires a series current-limiting resistor. Its main disadvantage is that it may not produce enough light for some specific applications.

Lamp sockets and indicators (g–j) are available with different-colored lens and various package styles. Some indicators come complete with lamp, limiting resistor (if needed), and even a switch.

SECTION 2-2 DEFINITION EXERCISES

Write a brief description for each of the following terms.

1. Electrical circuit ______________________________

____

2. Volt ______________________________

____

3. Ohm ______________________________

____

4. Ampere ______________________________

____

5. Dc ______________________________

____

6. Power source ______________________________

____

7. Load ______________________________

____

8. Switch ______________________________

____

9. Series circuit ______________________________

____

10. Parallel circuit ______________________________

____

11. Open circuit ______________________________

____

12. Closed circuit ______________________________

____

13. Voltmeter ______________________________

____

14. Ohmmeter ______________________________

____

15. Ammeter ______________________________

____

16. Multimeter ______________________________

____

17. Analog VOM ______________________________

____

18. DVM ______________________________

____

19. SPST switch ______________________________

____

20. SPDT switch ______________________________

____

21. DPST switch ______________________________

____

22. DPDT switch ______________________________

____

23. Incandescent lamp ______________________________

____

24. Neon lamp ____

____

25. LED ____

____

SECTION 2-3 EXERCISES AND PROBLEMS

Complete this section before beginning the next section.

1. Draw the schematic diagram of a basic series circuit and indicate electron flow.

2. Draw the schematic diagram of a basic parallel circuit and indicate electron flow.

3. Draw a schematic diagram of how to test a lamp with an ohmmeter.

4. Draw a schematic diagram of how to test an open switch and a closed switch with an ohmmeter.

5. Match the description in column A with the proper term in column B.

Column A

____ **a.** The force that causes electrons to move in a conductor

____ **b.** The opposition to current flow

____ **c.** A complete electrical path

Column B

1. open circuit
2. resistance
3. ohmmeter
4. schematic diagram
5. voltage

____ d. A device that consumes power while performing a useful function

____ e. An electrical drawing in which symbols are used for electrical components

____ f. A condition where no current flows in a circuit

____ g. A meter used for measuring resistance

____ h. A meter used for measuring voltage

____ i. A meter used for measuring electrical current flow.

____ j. An electrical component used in controlling current flow

6. voltmeter
7. circuit
8. switch
9. ammeter
10. load

6. Fill in the blanks in the following statements.

   a. The first thing to do when using and ohmmeter is to ____________ adjust the meter reading.

   b. To read the resistance of a component accurately, the ohmmeter must be set to the proper ____________ .

   c. The resistance of an open switch is ____________ Ω.

   d. The resistance of a closed switch is ____________ Ω.

7. In Figure 2–17, place a V in the circles that represent a voltmeter and an A in the circles that represent an ammeter.

Figure 2-17

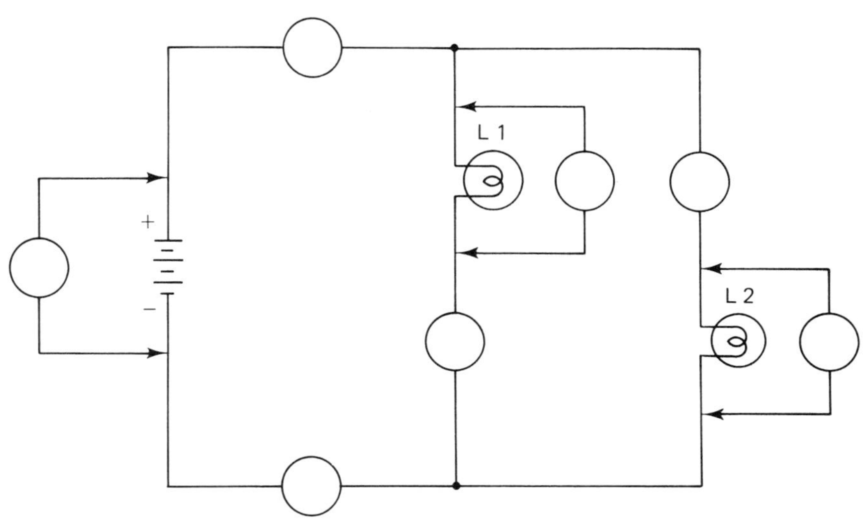

8. Fill in the blanks in the following statements.

a. To measure voltage, a voltmeter is placed ____________ the component to be measured.

b. When measuring dc voltage, ____________ must be observed.

c. To measure current, an ammeter is placed in ____________ with the component to be measured.

d. When measuring dc current, ____________ must be observed.

e. When measuring an unknown voltage or current, the meter should initially be set to the ____________ range.

9. Draw the schematic symbols for the following switches.

a. SPST **b.** SPDT **c.** DPST **d.** DPDT

10. Using Figure 2-15 as a guide, match the switches in column A with their proper use in Column B.

Column A	*Column B*
____ **a.** SPST	**1.** simultaneously, closes two circuits
____ **b.** SPDT	**2.** switches one line to many lines
____ **c.** DPST	**3.** closes one circuit
____ **d.** DPDT	**4.** simultaneously, switches two normally open lines
____ **e.** Multiposition	**5.** switches a normally open line

SECTION 2-4 EXPERIMENTS

EXPERIMENT 1. SERIES CIRCUIT

Objective: To demonstrate that the current is the same value throughout a series circuit and to develop skills in the use of a multimeter.

Introduction: To perform experiments successfully, the components used must be in good condition and proper working order. Whenever possible, test components before beginning an experiment or project and always be sure to observe the safety rules given in the preliminary safety unit, Unit A.

Materials Needed:

1 1.5-V D-cell battery
1 1.5-V lamp (Radio Shack 272-1139 or equivalent)
1 SPST switch (Radio Shack 275-701 or equivalent)
1 Breadboard for constructing circuit and appropriate wire or leads

Equipment Required:

1 Multimeter or DVM

Procedure:

1. Referring to Section 2-1c.1a and Figure 2-4, test the lamp.
2. Referring to Section 2-1c.1b and Figure 2-5, test the switch.
3. Construct the circuit shown in Figure 2-18 on the breadboard.

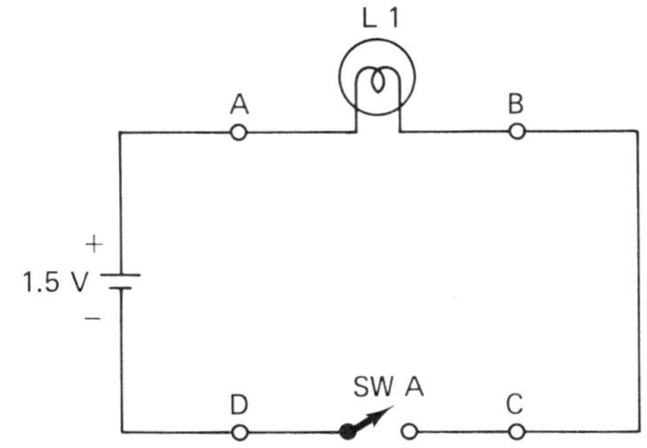

Figure 2-18 Series test circuit.

4. Set the meter to the proper voltage range. With the switch open, measure and record the voltage between the following points. Remember to observe polarity.

Points	*Measured Voltage*	*Comment*
A–D	______	Battery voltage = 1.5 V
A–B	______	Voltage across lamp = 0 V
B–C	______	Should be 0 V, electrically the same point
C–D	______	Voltage across switch = 1.5 V

5. With the switch closed, measure and record the voltage between the following points.

Points	*Measured Voltage*	*Comment*
A–D	______	Battery voltage = 1.5 V
A–B	______	Voltage across lamp = 1.5 V
B–C	______	Should be 0 V, electrically the same point
C–D	______	Voltage across switch = 0 V

6. Disconnect the battery from the circuit. Set the meter to read current on the highest mA range. Open the circuit at point A and insert the meter leads in series, observing polarity.
7. Reconnect the battery. Adjust the current range switch on the meter for a proper reading. Record the current value here. Current at point A = ____________ .
8. Repeat steps 6 and 7 for measuring and recording the current at points B through D.

Point	*Measured Current*
B	______
C	______
D	______

9. Disconnect the circuit and answer the following questions.

Fill-In Questions:

1. If the resistance of a lamp is very low, the lamp is considered to be in ____________ condition.

2. If the lamp is burned open, the resistance will be ____________ Ω.

3. The resistance of an open switch is ____________ Ω.

4. The resistance of a closed switch is ____________ Ω.

5. The voltage across the open switch is the same as voltage of the ____________ .

6. The voltage across the closed switch is ____________ V.

7. The current in a series circuit is the ____________ throughout the entire circuit.

8. The voltage of the battery decreased, when the switch was ____________

9. When a load is placed on a battery, the voltage at its terminals decreases, because of the ____________ resistance of the battery.

10. When making measurements, a voltmeter is placed in parallel or ____________ a circuit component and an ammeter is placed in ____________ with the component.

EXPERIMENT 2. PARALLEL CIRCUIT

Objective: To show how branch currents add up to the total current in a parallel circuit and that the voltage across each branch is the same.

Introduction: In this experiment you will measure the voltage across each branch, the current in each branch, and the total current being supplied by the battery.

Materials Needed:
1 1.5-V D-cell battery
2 1.5-V lamps (Radio Shack 272-1139 or equivalent)
2 SPST switches (Radio Shack 275-701 or equivalent)
1 Breadboard for constructing circuit and appropriate wire or leads

Experiment Required:
1 Multimeter or DVM

Procedure:

1. Test the lamps and switches as performed in Experiment 1.
2. Construct the circuit shown in Figure 2-19.

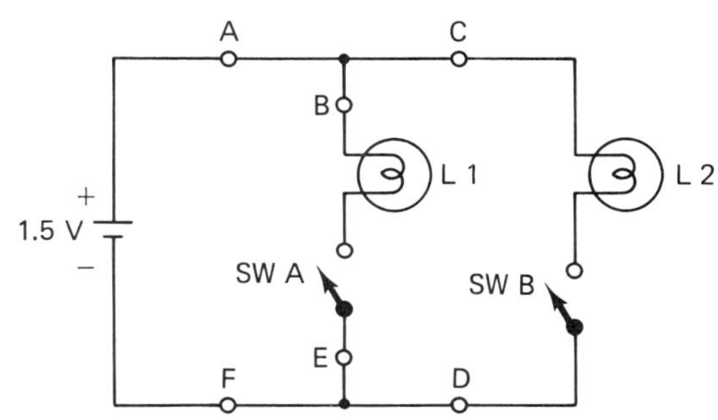

Figure 2-19 Parallel test circuit.

3. Close switches SW_A SW_B.
4. Set the meter to the proper voltage range. Measure and record the voltage between the following points.

Points	*Measured Voltage*	*Comment*
A–F	______	Battery voltage = 1.5 V
B–E	______	Voltage across L1 = 1.5 V
C–D	______	Voltage across L2 = 1.5 V

Notice that points A, B, and C are electrically the same point. Also, points D, E, and F are electrically the same.

5. Disconnect the battery. Set the meter to read current on the highest range. Open the circuit at point A and insert the meter leads in series, observing polarity.
6. Reconnect the battery. Adjust the current range switch on the meter for a proper reading. Record the current value here. Current at point A = ____________ .

7a. Open switch SW_A. Did the current value reading increase or decrease? ____________ Why?

7b. Open switch SW_B. What is the current value reading now? ____________ Why?

Close switches SW_A and SW_B.

8. Repeat steps 5 and 6 for measuring and recording the current at points B through F.

Point	*Measured Current*
B	______
C	______
D	______
E	______
F	______

9. Disconnect the circuit and answer the following questions.

Fill-In Questions:

1. The voltage across each branch in a parallel circuit is the ____________.

2. The total current in a parallel circuit is equal to the ____________ of all the branch currents.

3. When switch SW_A was opened the total current decreased to one-half because one branch was an ____________ circuit.

4. When switch SW_A and switch SW_B were both opened the total current decreased to ____________ because both branches were open circuited.

5. To measure voltage, the voltmeter is connected ____________ or in ____________ with the desired component.

6. To measure current, the circuit must be "broken" and the ammeter connected in ____________ with the component.

SECTION 2-5 TROUBLESHOOTING APPLICATION: SERIES AND PARALLEL CIRCUITS

Electronic troubleshooting procedures mostly involve making voltage measurements. With these two circuits, you will measure the correct voltages for normal circuits. You will then introduce problems and observe how the voltage readings change. All voltage measurements are entered into a table to aid in the circuit problem analysis.

2-5a THE SERIES CIRCUIT

Construct the circuit shown in Figure 2-20, using the components from the previous experiments. Notice that the power source can be two 1.5-V batteries, in series. There are also two lamps in series. Since the total power source is 3 V, each lamp will "drop" or have 1.5 V across it. Make

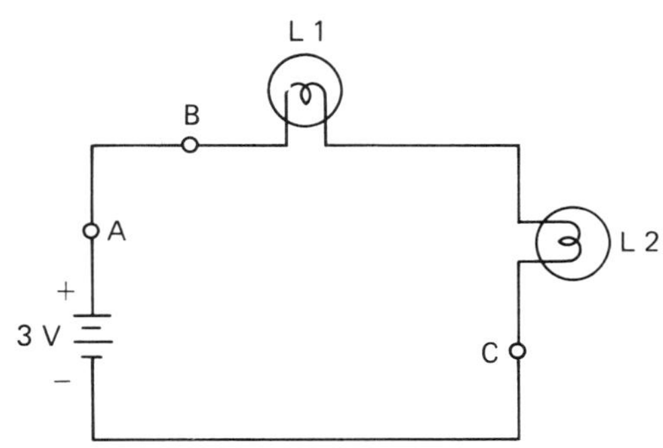

Figure 2-20

voltage measurements at the places indicated in the following table for a normal circuit and record the results in the table. Indicate if L1 and L2 are on or off. Proceed to open the circuit at the points given by the table and record the voltage measurements and condition of L1 and L2. Work on one problem at a time.

	Measure Voltage across:				*On/Off Condition of:*		
Problem	*Battery*	*L1*	*L2*	*Points B–C*	*L1*	*L2*	*Comment*
None	______	______	______	______	______	______	Circuit normal
Remove L1	______	______	______	______	______	______	Open circuit
Remove L2	______	______	______	______	______	______	Open circuit
Open point A	______	______	______	______	______	______	No voltage to L1 and L2

In most instances a circuit problem is a burned-out lamp (an open circuit). With a series circuit, the total current will stop flowing and all lamps will go out. The voltage across an open circuit (in this case L1 or L2) will be equal to the supply voltage (3 V). If points A, B, or C become open, there will be no voltage applied to the lamps.

2-5b THE PARALLEL CIRCUIT

Construct the circuit shown in Figure 2-21. The batteries are placed in parallel to provide sufficient current to the lamps. Perform the measurements and problems as indicated in the following table. Record the measurements in the table.

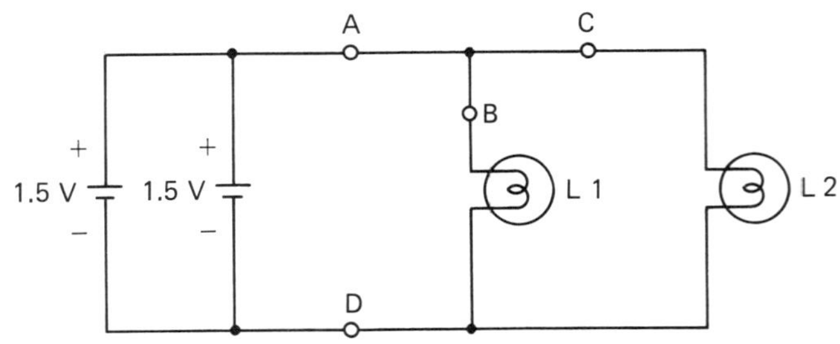

Figure 2-21

Problem	*Measure Voltage across:* Battery	L1	L2	*On/Off Condition of:* L1	L2	Comment
None	______	______	______	______	______	Circuit normal
Remove L1	______	______	______	______	______	Branch 1 open
Remove L2	______	______	______	______	______	Branch 2 open
Open point A	______	______	______	______	______	No voltage to L1 and L2
Open point B	______	______	______	______	______	No voltage to L1

If either lamp burns out, the voltage will increase slightly across the entire circuit. This occurs because there is less loading on the power supply. The lamp that is out will usually indicate that it is the one that is "open." A resistance or continuity test should be made on any lamp suspected of being open. If point A or D is open, no voltage will be applied to the lamps. If point B or C is open, the respective lamp will be off.

A piece of wire or metal can fall across the terminals of a lamp and short it out. The lamp will go out because the current will take the path of least resistance and flow through the short. Short circuits can cause fires or destroy power supplies. *Do not attempt to short-circuit any device.*

SECTION 2-6 INSTANT REVIEW

- An electrical circuit is a complete electrical path between two or more devices.

- The unit of voltage is the volt and is defined as the force required to produce 1 ampere of current through a resistance of 1 ohm.
- The unit of resistance is the ohm and is defined as the value of electrical resistance that will allow 1 ampere of current to flow through a material when 1 volt is applied.
- The symbol for ohm is Ω.
- The unit of current is the ampere and is defined as the current flow produced when 1 volt is applied across 1 ohm of resistance.
- The power source supplies electrical energy to a circuit.
- A load is the device or circuit resistance that consumes power from the power source in performing a useful function.
- A switch is a device that controls (on or off) the current flow in an electrical circuit.
- The total current in a series circuit flows through all the components in the circuit. $I_T = I_1 = I_2 = \cdots$.
- The total current in a parallel circuit is the sum of all the branch currents. $I_T = I_1 + I_2 + \cdots$.
- The voltage across a parallel circuit is the same for each branch. $V_T = V_1 = V_2 = \cdots$.
- A voltmeter measures voltage.
- An ohmmeter measures resistance.
- An ammeter measures current.
- A multimeter measures voltage, resistance, and current.
- An analog meter uses a needle indicator on a meter scale.
- A digital meter uses seven-segment numerical displays.
- A voltmeter is always placed across a component to read voltage.
- An ammeter is placed in series with a component to read current.
- Remember to remove power from the circuit while an ammeter is being connected.
- An ohmmeter supplies its own power.
- The power to a circuit must be removed before an ohmmeter is used to take resistance measurements.
- Observe polarity when connecting a meter to a dc circuit.
- A dc voltage circuit allows current to flow in one direction only.
- An open circuit has infinite resistance.
- An open circuit does not allow current to flow.
- A closed circuit allows current to flow.
- A short circuit has zero resistance.

SECTION 2-7 SELF-CHECKING QUIZ

Circle the best answer for each question.

1. The force that produces current flow in a circuit is called:

a. ampere **b.** voltage

c. ohm **d.** coulomb

2. When 6.25×10^{18} electrons pass a given point in one second, it is called:

a. an ohm **b.** a volt

c. an ampere **d.** none of the above

3. The ohm is the unit for electrical:

a. resistance
b. voltage
c. current
d. energy

4. Two lamps are connected in a series circuit. If one lamp burns out:

a. the other lamp will remain on
b. the other lamp will go out

5. A particular lamp has a resistance reading of infinite ohms. The lamp is:

a. okay
b. partially all right
c. not good, because it is open
d. not good, because it is shorted

6. The voltage measured across a closed switch will be:

a. the total power source voltage
b. one-half the power source voltage
c. zero volts
d. not enough information is given

Questions 7 through 13 refer to Figure 2-22.

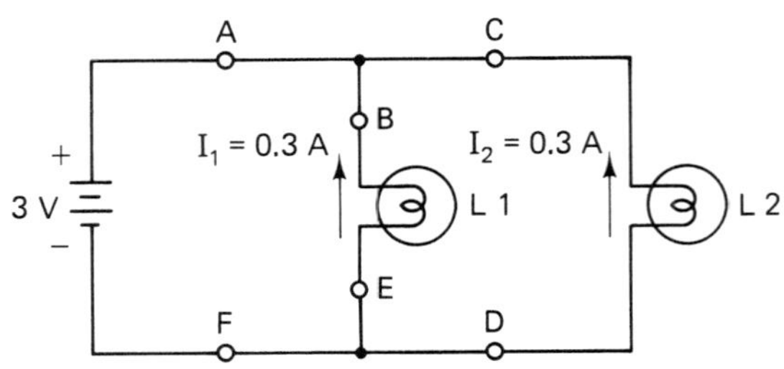

Figure 2-22

7. The total current for the circuit is:

a. 0.15 A
b. 0.3 A
c. 0.6 A
d. none of the above

8. The voltage across L2 (V_2) is:

a. 1.5 V
b. 3 V
c. 6 V
d. none of the above

9. To read the voltage across L1, you would:

a. open the circuit at point C and place the positive lead of the voltmeter at point A and the negative lead at L2
b. open point B and place the positive lead of the voltmeter at point A and the negative lead at point F
c. place the positive lead of the voltmeter at point F and the negative lead at point D.
d. place the positive lead of the voltmeter at point B and the negative lead at point E

10. To read the current through L1, (I_1) you would:

a. open the circuit at point B and place the negative lead of the ammeter at point A and the positive lead at L1
b. open the circuit at point C and place the positive lead of the ammeter at point B and the negative lead at L2
c. open the circuit at point E and place the negative lead of the ammeter at point F and the positive lead at L1
d. none of the above

11. To read the total current through L2 (I_2) you would:

a. open the circuit at point A and place the positive lead of the ammeter at the + terminal of the battery and the negative lead at point C

12. To read the total current (I_T) in the circuit, you would:

a. open the circuit at point D and place the negative lead of the ammeter at point F and the positive lead at point E

b. open the circuit at point C and place the positive lead of the ammeter at point B and the negative lead at L2

c. open the circuit at point D and place the positive lead of the ammeter at point E and the negative lead at L2

d. place the positive lead of the ammeter at point C and the negative lead at point D

b. open the circuit at point A and place the positive lead of the ammeter at point C and the negative lead at point B

c. open the circuit at point A and place the positive lead of the ammeter at point C and the negative lead at point B

d. open the circuit at point F and place the negative lead of the ammeter at the − terminal of the battery and the positive lead at point E

e. none of the above

13. If lamp L1 burns out:

a. lamp L2 will go out

b. the current through L2 will increase two times

c. the voltage across L2 will decrease

d. the total current will decrease

Questions 14 through 16 refer to Figure 2-23.

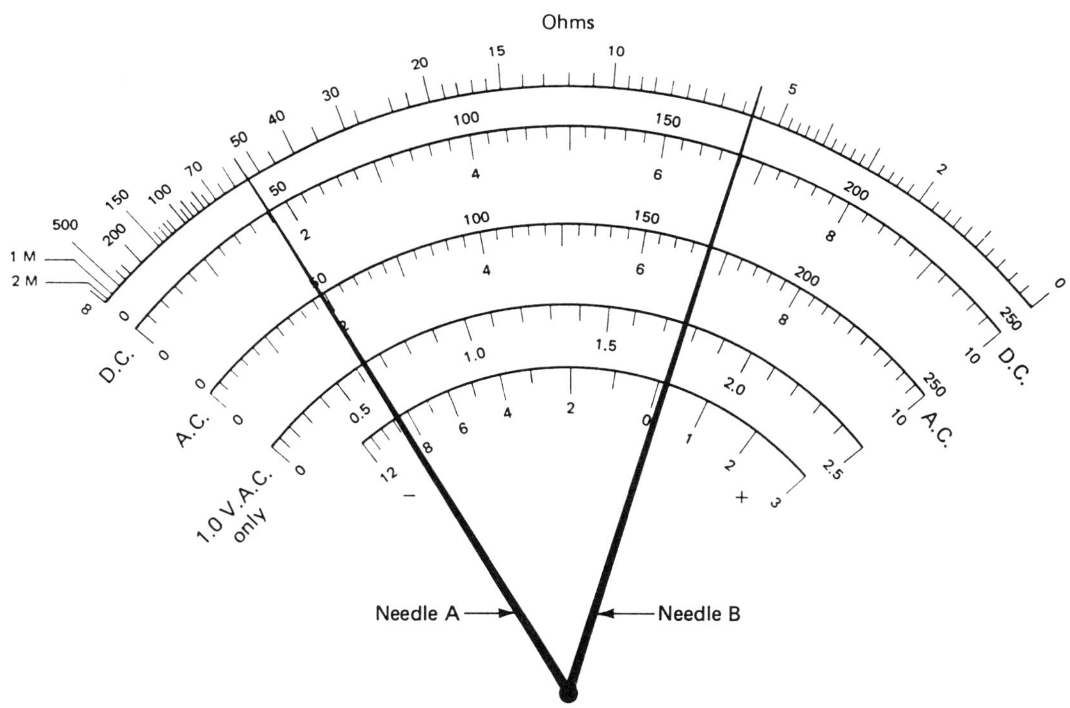

Figure 2-23

14. If the range switch is set on R = 10 k, needle A would indicate an ohmmeter reading of:

a. 55 Ω

b. 550 Ω

c. 55 kΩ

d. 550 kΩ

15. If the range switch is set at 25 DCV, needle A would indicate a voltage reading of:

a. 4.5 V

b. 1.8 V

c. 18 V

d. 45 V

16. If the range switch is set at 250 DCMA, needle B would indicate a current reading of:

a. 68 mA **b.** 150.8 mA

c. 170 mA **d.** none of the above

17. The formula $I_T = I_1 = I_2$ identifies and explains the current in a:

a. series circuit **b.** parallel circuit

18. A load is a:

a. force that causes current flow

b. component used in controlling current flow

c. device that uses voltage to perform a useful function

d. all of the above

19. The formula $V_T = V_1 = V_2$ identifies and explains the voltage across a:

a. series circuit **b.** parallel circuit

20. The resistance of an open circuit is:

a. zero ohms **b.** infinite ohms

c. cannot be determined **d.** none of the above

ANSWERS TO EXPERIMENTS AND QUIZ

Experiment 1: **(1)** good **(2)** infinite **(3)** infinite **(4)** zero **(5)** battery **(6)** zero **(7)** same **(8)** closed **(9)** internal **(10)** across, series

Experiment 2: **(1)** same **(2)** sum **(3)** open **(4)** sum **(5)** across, parallel **(6)** series

Self-Checking Quiz: **(1)** b **(2)** c **(3)** a **(4)** b **(5)** c **(6)** c **(7)** c **(8)** b **(9)** d **(10)** c **(11)** b **(12)** d **(13)** d **(14)** d **(15)** a **(16)** c **(17)** a **(18)** c **(19)** b **(20)** b

Unit 3

Conductors, Insulators, and Resistors

INTRODUCTION

The flow of electrical current depends upon conductors, resistance, and insulators. In some cases there must be conductors large enough to safely carry the amount of current in a circuit. Insulators are used to protect persons and devices from destructive electrical currents. Resistors are used to control the electrical current in circuits in order to perform specific tasks.

UNIT OBJECTIVES

Upon completion of this unit, you will be able to:

1. State what constitutes a good electrical conductor.
2. List materials that make good conductors.
3. State what constitutes a good electrical insulator.
4. List materials that make good insulators.
5. Explain positive and negative temperature coefficients in terms of material resistance.
6. Describe the effect of voltage breakdown.
7. Calculate the resistance of a specific conductor.
8. Show the differences in wire sizes and gauge numbers.
9. Explain the different types of wire and their use.
10. List basic electrical wire and cable connectors, such as phono, pin, banana, and coaxial.
11. Describe the composition of resistors.
12. Calculate the value of carbon-type resistors using the resistor color code.
13. Explain the other classifications of resistors in terms of power, type of material, tolerance, and failure rate.
14. Describe variable and adjustable resistors.
15. Identify the terms *potentiometer, "pot," rheostat, trimmer, resistance taper,* and *helipot.*
16. Explain nonlinear resistors such as thermistors, varistors, and photoresistors.
17. Define *conductance* and how to calculate it.
18. Test resistors and find the unknown value of a resistor in a circuit.

SECTION 3-1 FUNDAMENTAL CONCEPTS

3-1a CONDUCTORS AND WIRE

An *electrical conductor* is a material that has a large amount of valence electrons that very easily can become free electrons. A good conductor will allow a large number of electrons to flow through it and is usually made of metal.

3-1a.1 Types of Conducting Materials

Silver is the best known practical electrical conductor. Table 3-1 shows a comparison of some conductive materials with a diameter of 0.001 in. (1 mil) and a length of 1 ft at a constant temperature of 20 °C. Some metals are more expensive and cannot be found in large quantities. Copper, second in the list of conductivity values, is readily available and used primarily in the manufacture of electrical wire.

TABLE 3-1
Comparison of Conductor Materials

Material	*ohms/mil, foot*	*Relative Conductivity*
Silver	9.8	1 (high)
Copper	10.4	2
Gold	14.7	3
Aluminum	17.0	4
Tungsten	33.2	5
Nichrome	660.0	6 (low)

3-1a.2 Resistance of a Conductor

The principal factors determining the resistance of a conductor are the type of material it is made of, represented by the Greek lowercase letter rho (ϱ), its cross-sectional area, and its length. Figure 3-1 shows the relationship of these factors. The resistance of a conductor varies directly proportional with its length (l) and inversely proportional with its cross-sectional area. Stated in a formula, we have

$$R = \varrho\frac{l}{A}$$

where R = resistance of the material
l = length of the conductor (feet)
A = cross-sectional area (circular mils), which is equal to the area of a circle 0.001 in. (1 mil) in diameter (also, $A = d^2$)

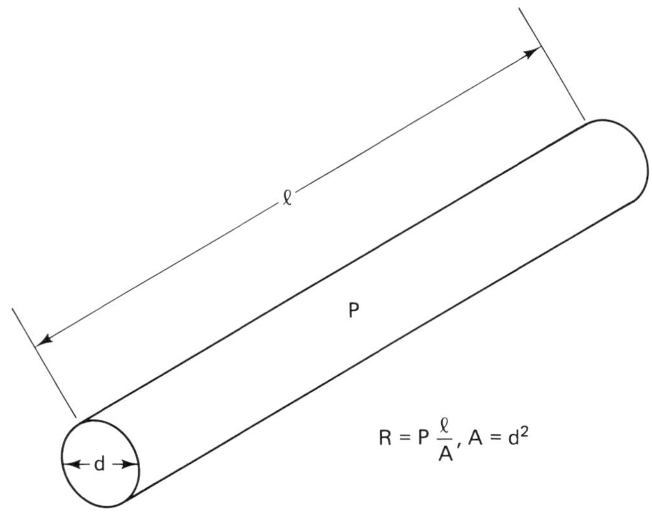

Figure 3-1 Resistance of a conductor.

As an example, No. 22 gauge copper wire commonly used in electronic wiring has a diameter of about 25.3 mils. The resistance of a 100-ft length of this wire can be calculated thus:

$$R = \varrho \frac{l}{A} \qquad A = d^2 = (25.3)^2 = 640$$
$$= 10.4 \left(\frac{100}{640}\right) = \frac{1040}{640} = 1.625\ \Omega$$

The resistance of a conductor varies inversely proportional to its cross-sectional area. In other words, if the cross-sectional area decreases, the resistance increases, and vice versa. As an example, if the diameter of the 100-ft copper wire were decreased by one-half (diameter 25.3 mils divided by 2 = 12.65 mils), its resistance would increase thus:

$$A = d^2 = (12.65)^2 = 160$$
$$R = 10.4 \left(\frac{100}{160}\right) = \left(\frac{1040}{160}\right) = 6.5$$

Notice that the resistance has increased four times (6.5 ÷ 1.625 = 4), even though the diameter of the wire was decreased only twofold.

3-1a.3 Wire Sizes

A comparison of the cross-sectional area of several commonly used wires is shown in Figure 3-2. Notice the difference in size between house wiring and electronic wiring. "AWG" stands for "American wire

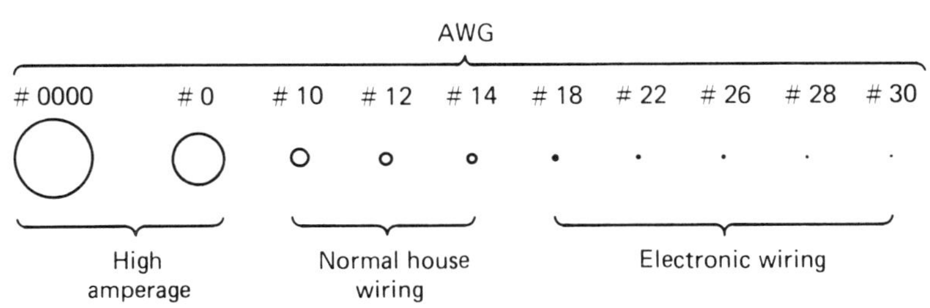

Figure 3-2 Approximate cross-sectional area comparison of several wire sizes.

TABLE 3-2
Comparison of Wire Sizes

Gauge	*Diameter (mils)*	*Resistance/100 ft*	*Current Capacity (A)*
0000	460.2	0.0049	225
0	325.0	0.0098	125
10	101.9	0.099	25
12	80.8	0.159	20
14	64.0	0.252	15
18	40.3	0.639	3
22	25.4	1.614	<1
26	15.9	4.08	<1
28	12.6	6.49	<1
30	10.0	10.30	<1

gauge," a standard used for wire sizes. Table 3-2 compares these wires with diameter, resistance, and safe current-carrying capacity.

3-1a.4 Effects of Temperature

Temperature affects most materials used in electrical and electronic devices. When the temperature of the material changes, its resistance also changes. When an increase in temperature causes an increase in the resistance of a material, it is said to have a *positive temperature coefficient.* If an increase in temperature causes the resistance of a material to decrease, it is called a *negative temperature coefficient.* Copper has a positive temperature coefficient, whereas carbon that is used in making resistors has a negative temperature coefficient.

Temperature has little effect on most basic circuits and is usually ignored; however, temperature is a critical factor with circuits using solid-state components. Special design considerations and devices must be used with these circuits.

3-1a.5 Types of Wire

Various types of wire are used in electrical and electronic circuits and devices. Figure 3-3 shows a few of these types and the following is a brief description of their use.

Solid wire is easy to handle but cannot withstand much flexing. Bare solid wire is used for buses, a common connection line for components and other wires. Solid wire with varnish or paint insulation (a) is used for windings of coils, transformers, generators, and motors. Insulated (rubber or plastic) solid wire (b) has advantages in some circuits. Small (AWG No. 30) insulated solid wire is best to use for integrated circuits (ICs) and connections that are very close.

Flexible wire consists of several strands of twisted wire covered by insulation (rubber, paper, plastic, Teflon, etc.) and is referred to as *stranded wire* (c). Stranded wire is used extensively in electronic circuits. Two stranded wires may be joined together to form a *two-conductor wire* (d), such as that used in electrical appliances.

Shielded wire (e) consists of a stranded inner conductor covered with insulation and then a braid of woven segments of stranded wire. This shield is used as a conductor and also to block electrical interfer-

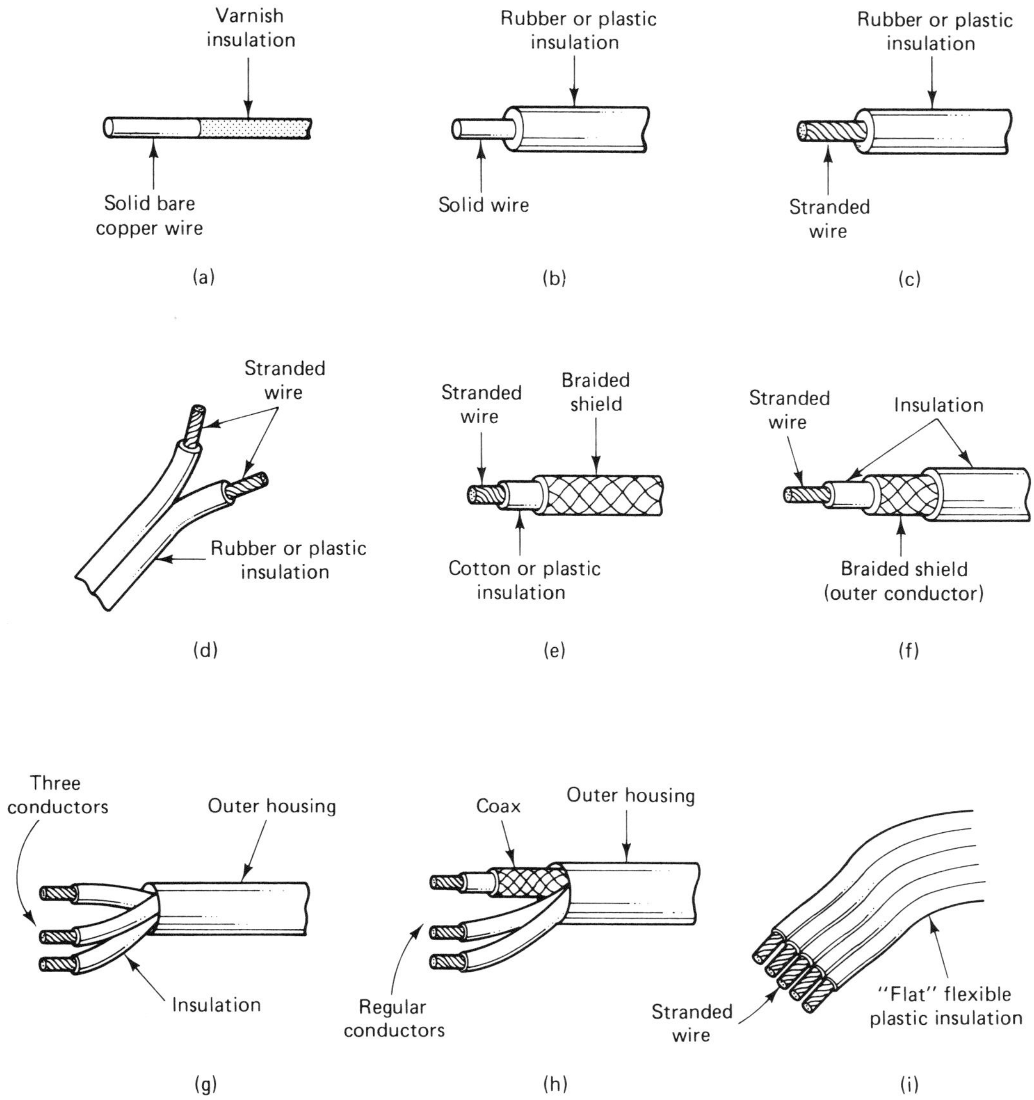

Figure 3-3 Types of wire: (a) solid strand with/without insulation; (b) insulated solid strand; (c) insulated stranded wire; (d) two-conductor wire; (e) shielded wire; (f) coaxial cable; (g) interconnecting (three-wire) cable; (h) special-purpose interconnecting cable; (i) "flat" ribbon cable.

ence from the inner conductor. *Coaxial cable* (f) is similar to shielded wire, except that an insulated covering is used around the shield.

Several wires can be put into a larger plastic tube interconnecting cable. Some of these cables may contain shielded or coaxial cable and other special arrangements (g, h). The newer type of flat "ribbon" cable has stranded wire sizes of AWG Nos. 22 to 30, which are connected side by side, with up to 64 conductors in a single cable. Ribbon cables are used primarily in computer installations.

3-1b INSULATORS

An *electrical insulator* is a material that has very few loosely bound electrons, which can become free electrons. In other words, an insulator has very few free electrons and opposes the current flow through the material. Insulators and insulating material are used with wire and components to protect people from getting shocked or electrocuted and to prevent short circuits from destroying circuits and causing fires. Wires have rubber, plastic, or Teflon covering them for insulation. Rubber gloves are worn by utility employees working on high-voltage power lines. The power lines themselves are insulated from the power poles by glass and/or porcelain insulators. Printed circuit boards are made from plastic, Bakelite, or fiberglass to insulate electronic components from each other.

Even though insulators oppose the flow of electron current, sufficient voltage can be placed on a material to cause current flow. The voltage is extremely high and a point is reached where electrons will begin to flow. The insulating material is usually destroyed and in many cases will burn, which creates a carbon residue and allows more electrons to flow. The high-voltage point at which electrons begin to flow in an insulator is called the *breakdown voltage.* Table 3-3 shows a comparison of different types of insulating materials.

TABLE 3-3
Comparison of Insulating Materials

Material	*Resistance (Ω/cm)*	*Relative Insulating Ability*	*Breakdown Voltage (kV/cm)*
Glass	10^{14}	1 (high)	1200
Teflon	10^{14}	2	600
Rubber	10^{13}	3	270
Bakelite	10^{12}	4	150
Waxed paper	10^{9}	5	500
Air (dry)	N/A	6 (low)	30

3-1c PLUGS, JACKS, AND CABLE CONNECTORS

Plugs, jacks, and cable connectors facilitate wire connections between various electronic circuits and equipment. *Plugs* are usually attached to the wire, while their receptacle or *jack* is rigidly mounted to a chassis or some equipment. A plug fits into a jack. Plugs are shown in Figure 3-4.

Phone plugs (a) are two or more wire connectors that are used in telephone central offices, radio–television broadcasting, and professional entertainment (musical instrument amplifiers and microphones). The smaller *phono plug* (b) is a two-wire connector used in audio systems (i.e., from turntable to amplifier, from tape deck to amplifier, etc.). Two-wire conductors and shielded and coaxial cables are used with phone and phono plugs. A *pin plug* (c) is a single-wire connector used where circuit changes are sometimes required. The *banana plug* (d) has the same function as the pin plug, except that it is larger and the spring connectors on the tip can be adjusted to form a better electrical connection.

The *phone jack* is a connector of two or more wires that matches the *phone plug.* It may also have contacts that open or close circuits when the plug is inserted. A *phono jack* is a two-wire connector that accepts the phono plug. The pin and banana jacks are single-wire connectors that match their respective plug types and usually are mounted through a panel or chassis. Jacks are shown in Figure 3-5.

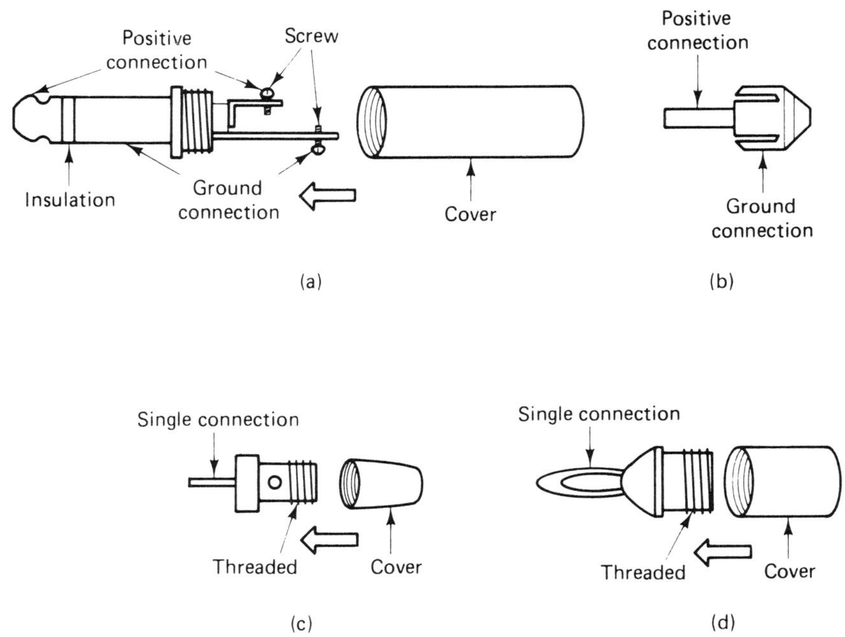

Figure 3-4 Plugs: (a) phone; (b) phono; (c) pin; (d) banana.

Coaxial-cable connectors are used on high-frequency equipment and test equipment. There are both *screw-type* and *bayonet-type connectors,* which provide good electrical connections. *Pin-contact connectors* require a *male* half and a *female* half. A keyway and key is used on these connectors to align the pins and socket holes properly. *Rack-and-panel connectors* are used with subassemblies or subchassis that fit into a large mounting rack. This connector serves as a quick-disconnect for

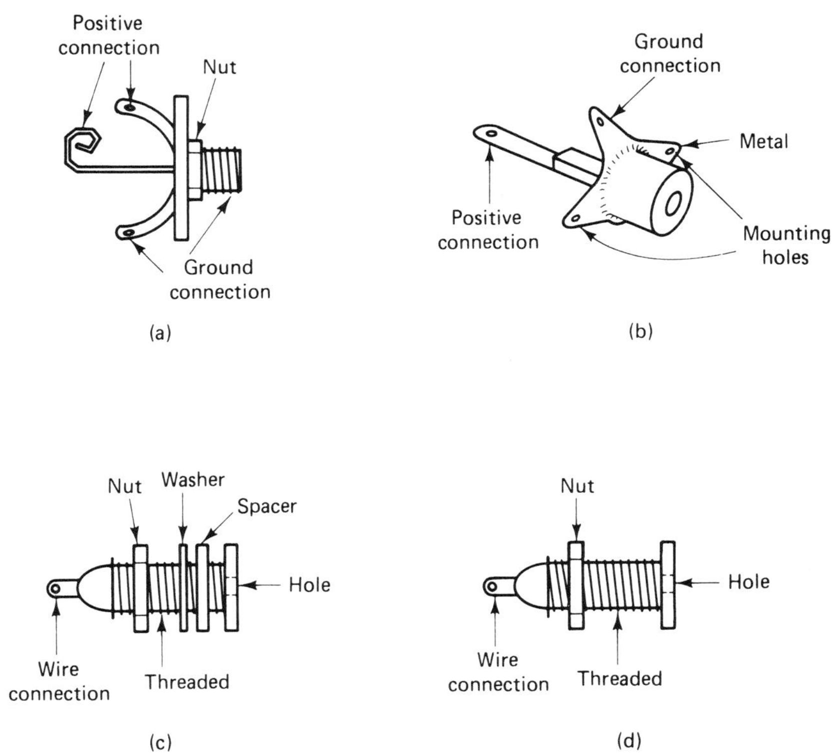

Figure 3-5 Jacks: (a) phone; (b) phono; (c) pin; (d) banana.

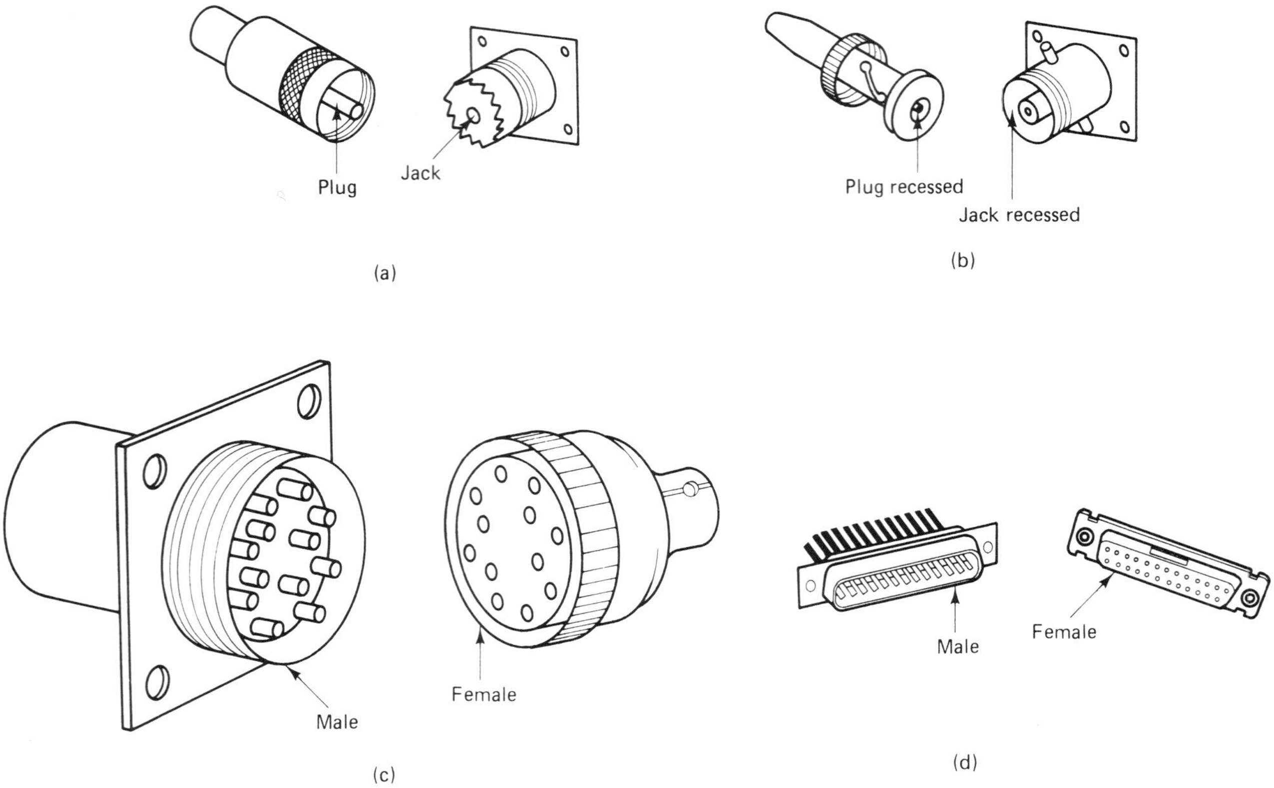

Figure 3-6 Cable connectors: (a) coaxial connector (screw type); (b) coaxial connector (bayonet type); (c) pin-contact-type connector; (d) rack-and-panel connector.

rapid test or replacement of subassemblies. Cable connectors are shown in Figure 3-6.

3-1d LINEAR FIXED RESISTORS

All materials have some resistance. The previous sections explained how conductors such as silver and copper have low resistance and offer very little opposition to current; also, that insulators such as glass and rubber have very high resistance and offer a large opposition to current flow. A resistor is a specially designed and manufactured electronic component that presents a specifically desired amount of resistance into a circuit.

3-1d.1 The Carbon Resistor

The most commonly used resistor is made of carbon. Carbon is neither a good conductor nor a good insulator, but rather is in between these two extremes, in the category called *semiconductor*. A composition of carbon granules, an insulating material, and a binder substance are mixed together and shaped into a small solid rod. Wire leads are attached to each end of the rod and then it is covered with a nonconductive solid coating. The carbon resistor can have ohmic values of less than 1 Ω to over 20 MΩ.

3-1d.2 Resistor Color Code

Most resistor values are indicated by color bands painted around the resistor, which are referenced to a color code. Figure 3-7 shows the schematic symbol for a resistor and the color code. The colors represent numbers 0 through 9, powers of 10, tolerance percentages, and failure rates.

The color bands of a resistor are grouped toward one end. To "read" a resistor and calculate its value, the end with the bands is placed to the left. Reading from left to right, the bands indicate:

First color band: first significant number
Second color band: second significant number
Third color band: multiplier
Fourth color band (if any): tolerance
Fifth color band (if any): failure rate

Figure 3-7 Resistor: (a) schematic symbol; (b) color code chart.

(a)

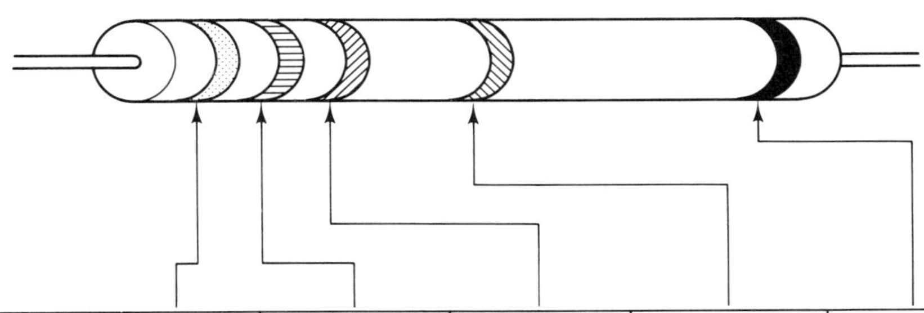

	First color band	Second color band	Third color band	Fourth color band	Fifth color band
Color	First significant number	Second significant number	Multiplier	Tolerance	Failure rating
Black	0	0	10^0		
Brown	1	1	10^1		1%
Red	2	2	10^2		0.1%
Orange	3	3	10^3		0.01%
Yellow	4	4	10^4		0.001%
Green	5	5	10^5		
Blue	6	6	10^6		
Violet	7	7	10^7		
Gray	8	8	10^8		
White	9	9	10^9		
Gold			0.1	±5%	
Silver			0.01	±10%	
No color				±20%	

(b)

For example, if a resistor had the following color bands:

first band = yellow = 4
second band = violet = 7
third band = red = 2 or $\times 10^2$
fourth band = no color = $\pm 20\%$

the resistor would read

$$47 \times 10^2 = 4700\ \Omega \text{ at } \pm 20\% = 4.7\ \text{k}\Omega \text{ at } \pm 20\%$$

Similarly, if a resistor had the following color bands:

first band = green = 5
second band = blue = 6
third band = gold = 0.1
fourth band = gold = $\pm 5\%$

the resistor would read

$$56 \times 0.1 = 5.6\ \Omega \text{ at } \pm 5\%$$

It is very difficult and expensive to manufacture carbon resistors to an exact value. Most circuits do not require absolute precision; therefore, these types of resistors have a tolerance or range in which they are usable. For example, a 1-kΩ resistor may have a ±10% tolerance. Ten percent of 1000 is 100; therefore, the actual value of the resistor can vary from 900 to 1100 Ω (1000 − 100 = 900 and 1000 + 100 = 1100).

The failure rating, as indicated by the fifth color band on a resistor, is a relatively new specification. It shows the reliability of a resistor in terms of the maximum failure rate in percent per 1000 hours. These resistors are normally required for specialized circuits in military and aerospace applications.

3-1d.3 Resistor Sizes and Power Ratings

From the discussion on conductors, it was shown how resistance was determined by the length, the cross-sectional area, and the resistivity of the material used. These factors are not necessarily true with resistors, since their resistance is determined by a mixture and not the physical size.

The different physical sizes of resistors relates to their power-handling, capability. This is referred to as the wattage rating or the amount of heat that a resistor can dissipate without burning up or altering its internal structure. Normally, if the internal structure of a resistor changes due to overheating, its value will increase. Figure 3-8 shows a comparison of the physical size and power rating of resistors.

Carbon resistors are generally found in sizes of $\frac{1}{8}$ W up to 2 W and are usually color-coded. Resistors with higher power ratings are of different types and normally have their resistance value printed on the side of the resistor. These power ratings can go as high as 250 W. More information on power is given in Unit 4.

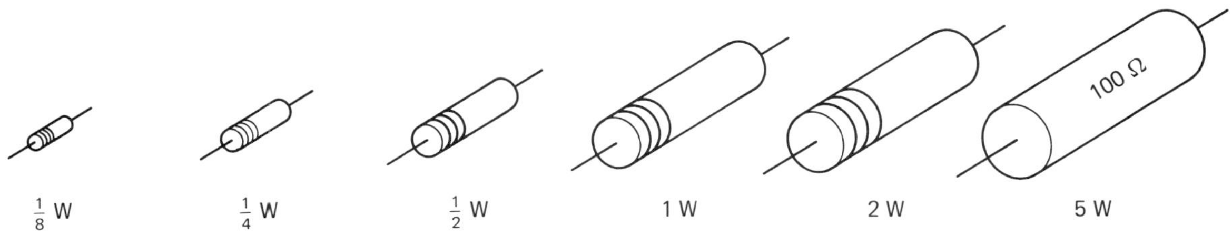

Figure 3-8 Resistor sizes and power ratings.

3-1d.4 Other Fixed Resistors

There are other types of fixed resistors manufactured for specific purposes. The resistance tolerances are usually more precise, on the order of 1% or less; however, they tend to be more expensive.

Wire-wound resistors are copper, nichrome, or other metal alloy wire that is wound around a ceramic core. Leads are attached to each end and the entire unit is sealed in an insulating material. Higher-power-rated resistors are usually wire-wound types. Their main disadvantage is the property of coiled wire and its interaction with the flow of electron current. This problem is related to inductive reactance, which is covered later in this book.

Deposited-film resistors consist of nichrome, aluminum-oxide, tin-oxide, or other metal alloys that are heated to high temperatures and a thin film of the material deposited on insulated substrate (support or foundation) material. With some resistors this substrate is cut into a spiraling pattern and placed on a ceramic core. *Thin-film resistors* are classified as having a thickness on the order of 10^{-6} in., whereas *thick-film resistors* are greater than 10^{-6} in.

Some resistors are made into an *integrated circuit* (IC) form and placed into IC packages containing eight or more pins. They are low-power types but take up less space on PC boards.

3-1e VARIABLE RESISTORS

Variable resistors have some means of mechanically adjusting their resistance. Figure 3-9 shows some common types of variable resistors. A normal resistor has two leads or terminals. A variable resistor has a third terminal, which is movable and indicated by an arrow, as shown, on the schematic symbol of Figure 3-9a.

The *wire-wound adjustable resistor* shown in Figure 3-9b has a section removed from its outer insulating material, thereby exposing the wire. A metal slider can be moved along the wire, which varies the resistance from that variable terminal to the end fixed terminals. This type of resistor is normally of the higher-power type and its resistance is not changed very often. A locking screw is also provided on the variable terminal for securing that terminal to a desired resistance setting.

The *trimmer resistor,* shown in Figure 3-9c, is usually very small and mounts on PC boards. It is normally adjusted by a screwdriver. Trimmer resistors are used to "fine adjust" or set specific resistances to control other circuit functions. They are not adjusted too often.

The *potentiometer,* commonly referred to as a "pot," shown in Figure 3-9d, is the most used variable resistor. The resistance material is usually carbon or copper wire wound on a circular form. A fixed termi-

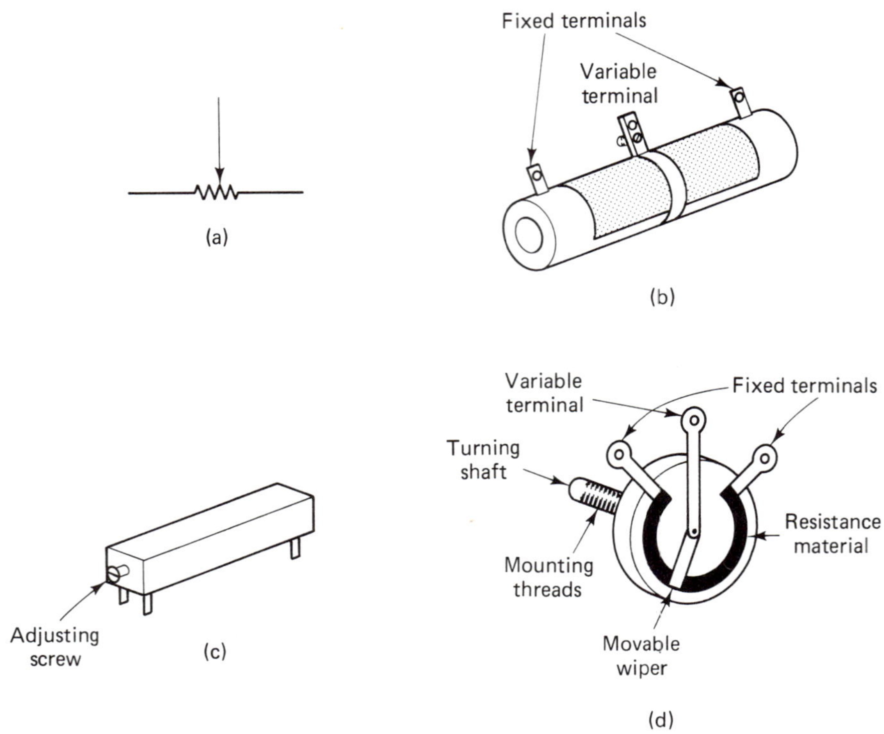

Figure 3-9 Variable resistors: (a) schematic symbol; (b) wire-wound adjustable type; (c) trimmer type; (d) potentiometer.

nal is attached to each end of the resistance material. A variable terminal, also called a *wiper* since it "wipes" across the resistance material, is connected to a shaft that comes out the front of the unit. When the shaft is turned, the resistance changes at the variable terminal in relation to the fixed terminals. The potentiometer is used in a circuit whenever the resistance must be varied frequently. You have used potentiometers countless times. They are used as volume controls, tone controls, balancing controls, and adjusting controls for graphic equalizers in such equipment as stereo music systems, radios, and television receivers.

3-1e.1 Potentiometer Operation

To better understand the operation of a potentiometer, refer to Figure 3-10. If an ohmmeter is placed across the fixed terminals, the total resistance of the pot can be read. If the ohmmeter is placed across the vari-

Figure 3-10 Operation of a potentiometer.

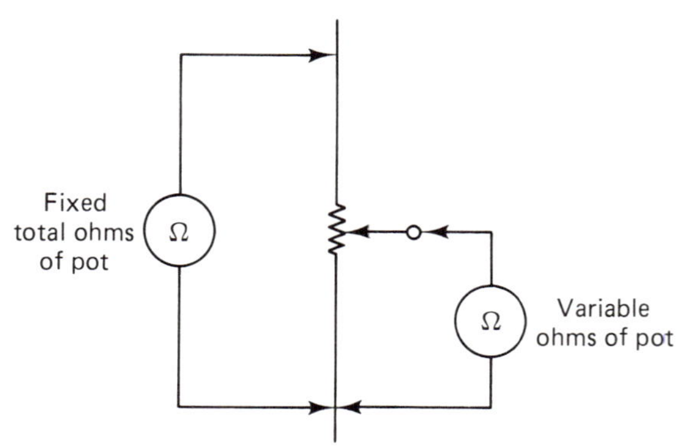

able terminal and one fixed terminal, the resistance reading will vary when the shaft is rotated. When the wiper moves up, the resistance will increase. When the wiper moves down, the resistance will decrease. If the wiper is placed all the way to the bottom, the ohmmeter will indicate zero ohms. In other words, the leads of the ohmmeter are effectively shorted together. Potentiometers are usually connected where a clockwise motion increases the resistance and a counterclockwise motion decreases the resistance.

The manner in which the resistance is distributed between the fixed terminals is referred to as the *taper* of the pot. A *linear taper* means that the resistance change at the variable terminal is directly proportional to the movement of the wiper. With an *audio or logarithmic taper* the resistance changes at a logarithmic rate of base 10 in proportion to the movement of the wiper. This type of pot is used as the volume control in audio circuits because of the nature of the human ear to hear various power levels. With an *S taper,* the resistance increases proportionally with the wiper up to a point, and then as the wiper is increased, the resistance decreases.

A *helical potentiometer,* sometimes called a *helipot,* is a precision potentiometer which requires several turns of the shaft to move the wiper from one fixed terminal to the other. This pot is used in extremely fine adjustment control circuits.

There are many variations with pots. Several mounted on the same shaft are referred to as being "ganged together" since the rotation of the shaft varies the resistance of each. Some pots have switches mounted behind them and serve as on/off-volume controls for different equipment.

A *rheostat* is a two-terminal variable resistor: one fixed terminal and one variable terminal. A pot can be wired as a rheostat simply by not connecting one of the fixed terminals.

3-1f NONLINEAR RESISTORS

Nonlinear resistors are devices whose resistance varies on a nonlinear basis when circuit conditions change or with external environmental changes. The majority of these devices have a negative temperature coefficient, although some with a positive temperature coefficient are used in specific circuit applications. Three of these devices are shown in Figure 3-11.

The *thermistor* or *thermal resistor* responds to changes in temperature. It can control current flow in a circuit, depending on the temperature of the circuit. Thermistors also find applications in meters where temperature changes are indicated as voltage or current changes.

The *voltage-dependent resistor,* sometimes called a *varistor,* changes resistance with an increase in voltage. In some circuits, this type of resistor is used to oppose voltage surges and protect other components. It may be called a "surgistor" for this application.

The *photoresistor* is made of a semiconductor material: cadmium sulfide (CdS) or cadmium selenide (CdSe). This material is deposited on a ceramic substrate with two attached metal leads and mounted in a case with a window. Its resistance reponds to light. When light falls through the window onto the material, electron–hole carriers are released. The current has increased; therefore, it is said that the resistance has decreased. The minimum resistance for a given amount of

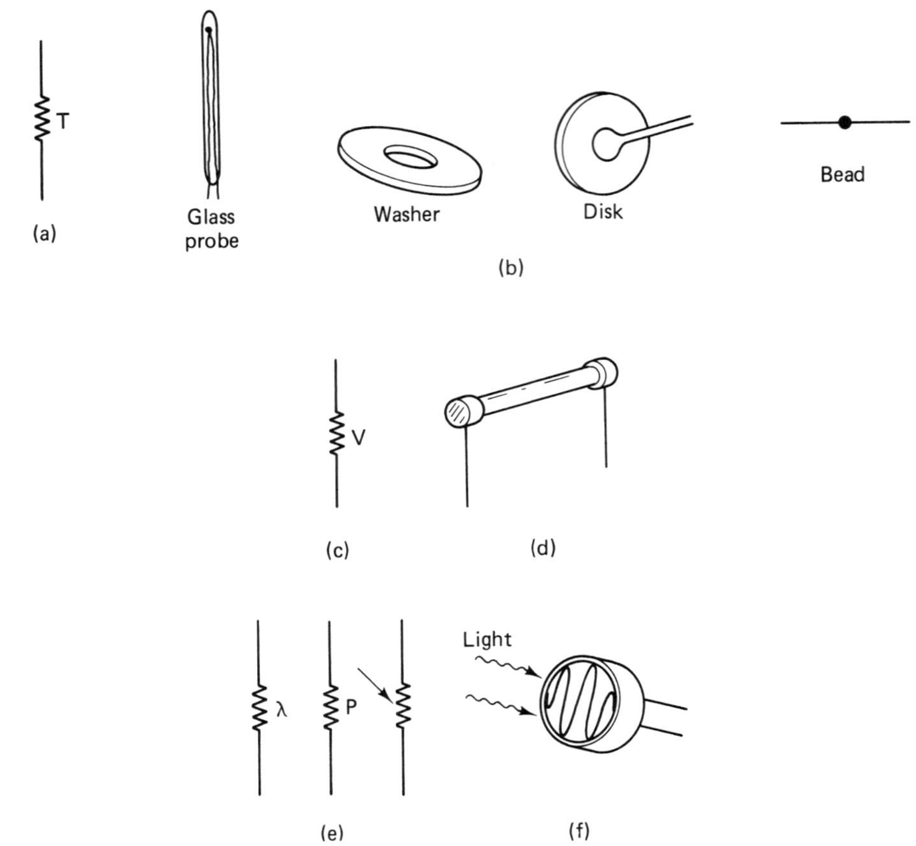

Figure 3-11 Nonlinear resistors: (a) thermistor schematic symbol and (b) physical types; (c) voltage dependant resistor and (d) physical type; (e) photoresistor schematic symbols and (f) physical type.

light is called the *light resistance* (R_{light}) and the maximum resistance, when no light is present, is called the *dark resistance* (R_{dark}). The photoresistor is usually classified as an optoelectronic device. One of its more popular applications is to control the automatic turning on of a lamp when there is no sunlight.

3-1g CONDUCTANCE

In some cases it is easier to understand the operation of an electronic component or electrical circuit in terms of how well it conducts electron current rather than opposing it. The opposite of resistance is called *conductance* and it is defined as the ease with which a component or material passes current flow. Mathematically stated, conductance is the reciprocal of resistance. Written as a formula, it appears as

$$\text{conductance} = \frac{1}{\text{resistance}}$$

$$\text{or} \quad G = \frac{1}{R}$$

where G is the conductance and R resistance.

The original unit of conductance is the mho (pronounced "moe"), which is "ohm" spelled backwards. The newer unit of conductance is the siemens (S). A resistance of 1 ohm equals the conductance of 1 mho or 1 siemens:

$$G = \frac{1}{R} = \frac{1}{1} = 1\ \text{S}$$

However, larger values of resistance are equal to smaller values of conductance. For example:

$$2\ \Omega\text{:}\quad G = \frac{1}{R} = \frac{1}{2} = 0.5\ \text{S}$$

$$10\ \Omega\text{:}\quad G = \frac{1}{R} = \frac{1}{10} = 0.1\ \text{S}$$

$$1\ \text{k}\Omega\text{:}\quad G = \frac{1}{R} = \frac{1}{1 \times 10^3} = 0.001\ \text{S} = 1\ \text{mS}$$

$$2\ \text{M}\Omega\text{:}\quad G = \frac{1}{R} = \frac{1}{2 \times 10^6} = 0.0000005\ \text{S} = 0.5\ \mu\text{S}$$

SECTION 3-2 DEFINITION EXERCISES

Write a brief description for each of the following terms.

1. Electrical conductor ______________________________

____

2. Rho ______________________________

____

3. Mil ______________________________

____

4. Positive temperature coefficient ______________________________

____

5. Negative temperature coefficient ______________________________

____

6. Solid wire ______________________________

____

7. Stranded wire ______________________________

____

8. Braided shield ______________________________

____

9. Coaxial cable ______________________________

____

10. Ribbon cable ______________________________

____

11. Electrical insulator ______________________________

____

12. Breakdown voltage ______________________________

____

13. Plugs ______________________________

____

14. Jacks ______________________________

____

15. Coaxial cable connectors ______________________________

____

16. Pin-contact connectors ______________________________

____

17. Rack-and-panel connectors ______________________________

____

18. Semiconductor ______________________________

____

19. Carbon resistor ______________________________

____

20. Color code ______________________________

____

21. Tolerance

22. Wattage rating

23. Power rating

24. Wire-wound resistors

25. Thin-film resistors

26. Thick-film resistors

27. IC resistors

28. Wire-wound adjustable resistor

29. Trimmer resistor

30. Potentiometer

31. Rheostat

32. Resistance taper

33. Nonlinear resistor

34. Thermistor ____

____

35. Voltage-dependent resistor ____

____

36. Photoresistor ____

____

37. Conductance ____

____

38. Mho ____

____

39. Siemens ____

____

SECTION 3-3 EXERCISES AND PROBLEMS

Complete this section before beginning the next section.

1. List three features that explain a conductor.

a.

b.

c.

2. Draw an example of a conductor that shows how to find its resistance (refer to Figure 3-1).

3. List the principal factors that determine the resistance of a conductor.

a.

b.

c.

4. Solve the following problems for a No. 18 gauge copper wire, where its $\varrho = 10.4$, diameter = 40.3 cmil, and length = 1000 ft.

a. resistance of No. 18 gauge wire at 1000 ft = ____________

b. resistance if its length is doubled = ____________

c. resistance if its length is halved = ____________

d. resistance if its diameter is doubled = ____________

e. resistance if its diameter is halved = ____________

5. List three features that explain an insulator

a.

b.

c.

6. Identify the current flow properties of the following materials by placing the appropriate letter next to them. Let C = conductor, S = semiconductor, and I = insulator.

____ **a.** rubber	____ **b.** Bakelite	____ **c.** silver	____ **d.** carbon
____ **e.** gold	____ **f.** aluminum	____ **g.** nichrome	____ **h.** Teflon
____ **i.** copper	____ **j.** glass	____ **k.** air	____ **l.** tungsten

7. Draw the schematic symbol for the following devices.

a. fixed resistor

b. potentiometer

8. List five ways in which fixed resistors are classified or rated.

a.

b.

c.

d.

e.

9. Match the description in column A with the appropriate term in column B.

Column A	*Column B*
____ a. Determines a resistor's physical size	1. stranded wire
____ b. Device used as a volume control	2. ribbon cable
____ c. One thousand ohms	3. conductance
____ d. A device that fits into a jack	4. 6.8 MΩ
____ e. Flexible wire	5. solid wire
____ f. Wire used in motors	6. 1 kΩ
____ g. Blocks electrical interference from inner conductor	7. potentiometer
____ h. Multiconductor wire used in computer installations	8. plug
____ i. 6,800,000 Ω	9. shielded wire
____ j. The opposite of resistance	10. wattage rating

10. Calculate the resistance value and tolerance for each resistor. Express the resistance in kilohms or megohms where applicable: for example, 2.2 kΩ, 22 kΩ, 220 kΩ, 3.3 MΩ, and so on.

	Band Color					
Resistor	*First*	*Second*	*Third*	*Fourth*	Ω	*at %*
a	Red	Red	Brown	Gold	______	______
b	Yellow	Violet	Orange	Silver	______	______
c	Blue	Gray	Yellow	None	______	______
d	Orange	Orange	Red	Silver	______	______
e	Green	Blue	Gold	Gold	______	______
f	Brown	Black	Blue	None	______	______
g	Red	Violet	Brown	Silver	______	______
h	Brown	Black	Yellow	Gold	______	______
i	Orange	White	Brown	Gold	______	______

Resistor	Band Color First	Second	Third	Fourth	Ω	at %
j	Brown	Black	Orange	Silver	_______	_______
k	Brown	Black	Gold	Gold	_______	_______
l	Yellow	Violet	Silver	Gold	_______	_______
m	Brown	Green	Red	Silver	_______	_______
n	Gray	Red	Gold	Gold	_______	_______
o	Brown	Black	Red	None	_______	_______

11. List the color of each band for the resistors given.

Resistor	Band Color First	Second	Third	Fourth
a. 2.2 kΩ at 20%	_______	_______	_______	_______
b. 4.7 Ω at 5%	_______	_______	_______	_______
c. 560 Ω at 10%	_______	_______	_______	_______
d. 180 kΩ at 20%	_______	_______	_______	_______
e. 12 kΩ at 5%	_______	_______	_______	_______
f. 0.33 Ω at 5%	_______	_______	_______	_______
g. 2.7 Ω at 5%	_______	_______	_______	_______
h. 10 kΩ at 10%	_______	_______	_______	_______
i. 470 kΩ at 20%	_______	_______	_______	_______
j. 680 Ω at 20%	_______	_______	_______	_______
k. 15 kΩ at 10%	_______	_______	_______	_______
l. 750 kΩ at 5%	_______	_______	_______	_______
m. 5.6 MΩ at 10%	_______	_______	_______	_______
n. 39 kΩ at 20%	_______	_______	_______	_______
o. 10 MΩ at 5%	_______	_______	_______	_______

12. Indicate which of the following statements are True (T) or False (F)

_____ **a.** An insulator has many free electrons.

_____ **b.** If the diameter of a conductor is doubled, its resistance is halved.

_____ **c.** A No. 22 gauge wire can carry a larger current than can be carried by a No. 18 gauge wire.

_____ **d.** With a positive temperature coefficient, when the temperature increases the resistance of the conductor increases.

_____ **e.** If the breakdown voltage of an insulator is exceeded, current will flow.

_____ **f.** Bakelite has a higher resistance per centimeter than that of glass.

_____ **g.** The resistance measured between the two fixed terminals of a potentiometer is its total resistance.

_____ **h.** The resistance measured between one fixed terminal and the wiper terminal of a potentiometer is variable.

_____ **i.** If a particular resistor is 22 kΩ, its conductance is 45.4 μS.

_____ **j.** A potentiometer can be connected as a rheostat.

_____ **k.** High-power resistors are usually wire-wound types.

_____ **l.** An audio taper of a pot is linear.

_____ **m.** A helipot requires several complete turns of the shaft to vary the resistance from one end to the other.

_____ **n.** The resistance of a thermistor varies with temperature.

_____ **o.** The resistance of a photoresistor varies with light.

_____ **p.** A voltage-dependent resistor is not affected by voltage surges.

SECTION 3-4 EXPERIMENTS

EXPERIMENT 1. MEASURING RESISTOR VALUES

Objective: To show how to measure the value of resistors using an ohmmeter.

Introduction: In Unit 2 you learned how to use a multimeter. In this experiment you will use the ohmmeter to test fixed resistors. Review Section 2-1c.1.

Materials Needed: Six carbon-composition resistors, suggested values:

1	47-Ω	*1*	22 kΩ
1	220 Ω	*1*	470 kΩ
1	1 kΩ	*1*	1 MΩ

Equipment Required: *1* Multimeter or DVM

Procedure:

1. Take each resistor and read its color code.
2. List the colors in the following table as indicated. You may want to use the following abbreviations for the colors:

Black	= Blk	Brown	= Brn
Red	= Red	Orange	= Org
Yellow	= Yel	Green	= Grn
Blue	= Blu	Violet	= Vio
Gray	= Gra	White	= Wht
Silver	= Sil	Gold	= Gol
None	= No		

Resistor	*Color band*				*Calculated Ohms (%)*	*Calculated Tolerance (± Ω)*	*Measured Ohms*	*Within Tolerance?*
	First	*Second*	*Third*	*Fourth*				
a	______	______	______	______	______	______	______	______
b	______	______	______	______	______	______	______	______
c	______	______	______	______	______	______	______	______
d	______	______	______	______	______	______	______	______
e	______	______	______	______	______	______	______	______
f	______	______	______	______	______	______	______	______

3. Calculate the resistance value and tolerance and enter them in the table at the appropriate places.
4. Calculate the tolerance of the resistor in terms of ± ohms and list in the table. As an example:

$$4.7\ \text{k}\Omega \text{ at } \pm 10\% = 4700 \times 0.1 = \pm 470\ \Omega$$

5. Measure the resistor with an ohmmeter, making sure to zero-adjust the meter each time where applicable. Enter this value into the table at the appropriate place.
6. Compare the measured resistance value with the calculated resistance value and the ± tolerance. Indicate by "yes" or "no" in the table if the resistor is within tolerance.

   For example:

$$\text{Measured resistance} = 4850\ \Omega$$

$$\begin{aligned} \text{Calculated resistance} &= \ 4700\ \Omega \\ \pm \text{ tolerance} &= +\ \underline{470\ \Omega} \\ & \ 5170\ \Omega \end{aligned}$$

Is 4850 Ω within or below 5170 Ω? Yes.

Note: If the measured resistance value is greater than the calculated resistance value, add the tolerance value; however, if it is less, subtract it.

Fill-In questions:

1. When using an ohmmeter, the first thing to do is to ____________ adjust the meter.
2. When measuring the resistance of a resistor, be sure to set the ohmmeter to the proper ____________ .
3. A resistor to be used in a circuit should be within the ____________ indicated on the resistor.
4. The physical size of resistors is determined by their ____________ rating.

EXPERIMENT 2. TESTING A POTENTIOMETER

Objective: To demonstrate the operation of a potentiometer and how to test it.

Introduction: Potentiometers used as controls in circuits can become faulty and need to be tested. They can also become dirty, resulting in erratic operation of a device. This is particularly evident in audio circuits when a scratchy sound is heard as the volume control is turned. Special cleaning fluids can be used in some cases to correct a problem such as this, but in other instances the entire pot must be replaced.

Materials Needed: *1* Potentiometer

Equipment Required: *1* Multimeter or DVM

Procedure:

1. Draw Figure 3-10 in the space below.

2. Connect the ohmmeter leads across the fixed terminals of the pot (refer to Figure 3-9d). Read the meter and list the resistance measurement here: total resistance of pot = ____________ Ω.

3. Connect the ohmmeter leads from one fixed terminal to the variable terminal. Rotate the shaft of the pot and notice how the resistance reading changes.
4. Set the shaft of the pot halfway between the two extremes of its movement. Read the meter and list the resistance measurement here: resistance of wiper at halfway point = ____________ Ω.
5. Does this pot seem to have a linear taper? ____________ . (If it does, the resistance at this point will be about half of the total resistance.)

Fill-In Questions:

1. The total resistance of a pot is measured between the ________ ____________ terminals.
2. To make sure that the pot is variable, a test should be made between one fixed terminal and the ____________ terminal.
3. Potentiometers can be cleaned with special cleaning ____________ .
4. Refer to Figure 3-10. If the wiper moves up, the resistance between the variable terminal and the bottom terminal will ____________ .
5. Refer to Figure 3-10. If the wiper moves down, the resistance at the variable terminal will ____________ .

EXPERIMENT 3. NONLINEAR RESISTORS

Objective: To demonstrate the resistance characteristics of nonlinear resistive devices.

Introduction: In this experiment you will measure the resistance of a thermistor at normal room temperature. Heat is then applied to the thermistor and you will observe the change in resistance. You will also measure the resistance of a photoresistor with and without light being present.

Materials Needed:

1 Photoresistor (Radio Shack 276-1657 or equivalent)

1 Thermistor with 150-Ω cold resistance or less

Equipment Required:

1 Multimeter or DVM

1 50-W or less soldering iron

Procedure:

1. Plug the soldering iron into an ac outlet so that it can be heating up. *Be careful* not to touch the tip of the soldering iron or let it come in contact with anything. Allow heating for at least 5 minutes.

2. Draw the schematic symbols for a thermistor and photoresistor (refer to Figure 3-11) in the space below. Label each symbol.

3. Set the ohmmeter on the R = 1 scale and zero-adjust it. Measure the resistance of the thermistor and list the reading here: thermistor cold resistance = ____________ Ω.
4. Touch the tip of the soldering iron to the edge of the thermistor and observe the ohmmeter. What happens to the ohmmeter reading?

5. When the ohmmeter reading becomes stable, list it here and remove the soldering iron. Unplug the soldering iron. Thermistor hot resistance = ____________ Ω.
6. Does the thermistor have a positive or negative temperature coefficient? ____________
7. Set the ohmmeter to a middle range, zero-adjust it, and then connect its leads across the leads of the photoresistor.
8. Face the window side of the photoresistor toward the light. Observe the ohmmeter reading and list it here: R_{light} = ____________ Ω.
9. Place a thick dark-colored object such as cardboard or a book against the window of the photoresistor to block out all light. Observe the ohmmeter reading and list it here: R_{dark} = ____________ Ω.
10. Does the photoresistor have a positive or a negative resistance coefficient? ____________
11. Remove the object from the photoresistor and place your finger over it. What is the reading of the ohmmeter now? ____________ Ω.
12. Move your hand various distances over the window of the photoresistor and observe the reading of the ohmmeter.

Fill-In Questions:

1. When the temperature of the thermistor increased, its resistance ____________ .

2. The thermistor has a ____________ temperature coefficient.

3. The minimum resistance a photoresistor has when light strikes it is called the ____________ resistance.

4. The maximum resistance a photoresistor has when all light is blocked from it is called ____________ resistance.

5. The photoresistor has a ____________ resistance coefficient.

6. The photoresistor test proved that some substances, such as the ____________, allow more light to pass through than is expected.

SECTION 3-5 TROUBLESHOOTING APPLICATION: FINDING THE UNKNOWN VALUE OF A RESISTOR

A resistor may burn open and be so discolored that it is impossible to read its color code. If there is an indication of the voltage that is supposed to be dropped by or across the resistor, its resistive value can be found.

Very simply, a pot is connected across the open resistor. A voltmeter is then placed across the pot and it is adjusted to the indicated voltage. Next, the pot is removed from the circuit and then an ohmmeter reads the adjusted resistance. A fixed resistor of the value indicated by the pot is then placed into the circuit.

Construct the circuit shown in Figure 3-12a. Leave an open where resistor R_2 is located. Connect a pot across the open as shown, and then connect a voltmeter across the pot as shown. Adjust the pot for a reading of 1 V on the voltmeter.

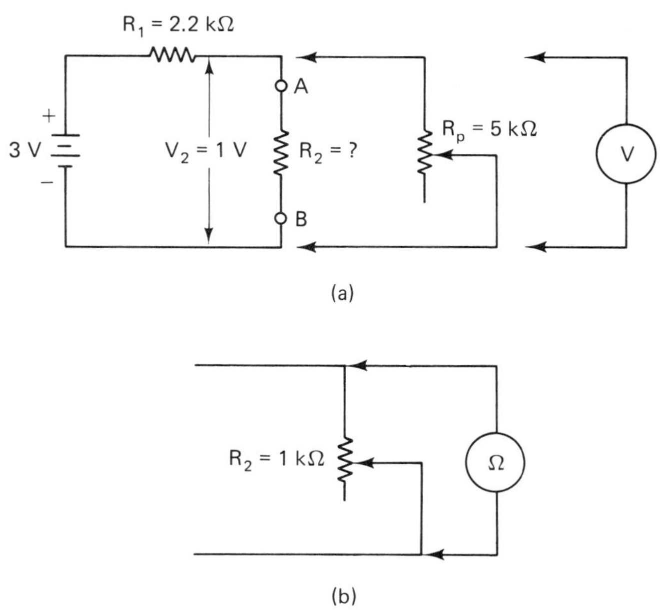

Figure 3-12 Finding the unknown value of a resistor: (a) adjusting the voltage; (b) measuring the resistance.

Disconnect the voltmeter and pot, but be sure not to touch or move the setting of the pot shaft. With an ohmmeter, measure the resistance between the terminal of the pot that was connected to point A and the wiper terminal shown in Figure 3-12b. This resistance should be about 1 kΩ. A 1-kΩ fixed resistor could now be placed into the circuit.

SECTION 3-6 INSTANT REVIEW

- An electrical conductor (1) has many free electrons, (2) has low resistance, (3) allows current to flow, and (4) is usually made of metal.
- An electrical insulator (1) has few free electrons, (2) has high or infinite resistance, (3) blocks current flow, and (4) is non-metallic.
- A positive temperature coefficient means an increase in resistance with an increase in temperature.
- A negative temperature coefficient means a decrease in resistance with an increase in temperature.
- Breakdown voltage is a point where the voltage becomes so great that current will flow in an insulator.
- The resistance of a conductor is directly proportional to its resistivity (ϱ) and length and inversely proportional to its cross-sectional area. (Temperature also affects resistance.) $R = \varrho\,(l/A)$.
- If the diameter of a conductor is doubled, its resistance will decrease fourfold.
- The smaller the wire gauge number, the larger its cross-sectional area.
- The smaller the wire gauge number, the greater its current capability.
- A semiconductor material exhibits properties that are between those of a conductor and an insulator.
- With the resistor color code, the

first color band = first significant number
second color band = second significant number
third color band = multiplier
fourth color band = tolerance
fifth color band = failure rate

- Resistors are made of carbon composition, wire-wound, thin-film, thick-film, and integrated circuit materials.
- Resistors are rated or classified by resistance value, power, type of material, tolerance, and failure rate.
- Variable resistors are adjustable types, potentiometers, rheostats, and trimmers.
- The taper of a pot is determined by the manner in which the resistance is distributed between its fixed terminals.
- A helipot requires several complete turns of its shaft to vary the complete resistance.

- Nonlinear resistors include thermistors, varistors, and photoresistors.
- Conductance is the reciprocal of resistance.
- The unit of conductance is the mho or siemens.

SECTION 3-7 SELF-CHECKING QUIZ

Circle the most correct answer for each question.

1. A good conductor has:

a. few free electrons **b.** low resistance

c. a breakdown voltage **d.** none of the above

2. A conductor has a ρ value of 14.7 Ω/mil, ft, a diameter of 2 cmil, and a length of 500 ft. Its resistance is about:

a. 1837 Ω **b.** 3675 Ω

c. 7350 Ω **d.** 14,700 Ω

3. A good insulator:

a. has many free electrons

b. allows current to flow

c. is usually made of metal

d. none of the above

4. The resistance of a conductor is:

a. inversely proportional to both its length and cross-sectional area

b. directly proportional to its cross-sectional area and inversely proportional to its length

c. directly proportional to its length and cross-sectional area

d. directly proportional to its length and inversely proportional to its cross-sectional area

5. A wire in a circuit is carrying 8 A of current. If this wire is replaced by another wire with half the diameter, the amount of current flow would be:

a. 8 A **b.** 4 A

c. 2 A **d.** 16 A

6. The best insulator is the material:

a. carbon **b.** rubber

c. glass **d.** Bakelite

7. The best type of wire to use as a line cord to electrical/electronic equipment such as fans, radios and TV sets is:

a. stranded wire **b.** coaxial cable

c. solid strand wire **d.** ribbon cable

8. Devices used to facilitate wiring between various electrical/electronic equipment are:

a. conductors, resistors, and thermistors

b. plugs, jacks, and cable connectors

c. photoresistors, varistors, and potentiometers

d. all of the above

9. The point where current flows in an insulator is called the:

a. breakdown voltage
b. ionization point
c. resistance breakdown
d. none of the above

10. The resistance of a particular material decreases when its temperature increases. The material has a:

a. positive temperature coefficient
b. negative temperature coefficient
c. defect in the material
d. neutralizing effect, making it more stable

11. A fixed resistor has the following colors, reading left to right: blue, gray, orange, and silver. Its resistance is:

a. 6.8 kΩ ± 10%
b. 790 kΩ ± 10%
c. 68 kΩ ± 10%
d. 79 kΩ ± 5%

12. A 2.2-kΩ resistor has a fourth color band of gold. Its acceptable resistance range is:

a. 1100–3300 Ω
b. 1980–2420 Ω
c. 2090–2310 Ω
d. 2178–2222 Ω

13. The physical size of a resistor is determined by its:

a. resistance value
b. tolerance value
c. type of material
d. wattage rating

14. Variable resistors include:

a. potentiometers, trimmers, and photoresistors
b. adjustable types, potentiometers, and trimmers
c. trimmers, thermistors, and potentiometers
d. potentiometers, trimmers, and varistors

15. A particular pot has a total resistance of 100 kΩ. When the wiper is set halfway between the fixed terminals, the variable terminal is at 50 kΩ. The pot is said to have:

a. a linear taper
b. an audio taper
c. no taper
d. a negative coefficient

16. A 270-Ω ± 5% resistor would have a color code of:

a. brown, blue, black, and gold
b. red, blue, brown, and silver
c. red, violet, brown, and gold
d. brown, red, brown, and no color

17. A potentiometer that requires several complete turns to cover its total resistance is called a:

a. varistor
b. helipot
c. slidistor
d. varipot

18. The resistance of a photoresistor is varied by:

a. sound
b. voltage
c. light
d. temperature

19. The conductance of a 1.2-kΩ resistor is:

a. 833.3 μS
b. 0.8333 S
c. 1200 S
d. 34.6 S

20. The resistance of a thermistor varies with changes in:

a. sound
b. voltage
c. light
d. temperature

ANSWERS TO EXPERIMENTS AND SELF-CHECKING QUIZ

Experiment 1: **(1)** zero **(2)** range **(3)** tolerance **(4)** wattage

Experiment 2: **(1)** fixed **(2)** variable **(3)** fluids **(4)** increase **(5)** decrease

Experiment 3: **(1)** decreased **(2)** negative **(3)** light **(4)** dark **(5)** negative **(6)** human body

Self-Checking Quiz: **(1)** b **(2)** a **(3)** d **(4)** d **(5)** c **(6)** c **(7)** a **(8)** b **(9)** a **(10)** b **(11)** c **(12)** c **(13)** d **(14)** b **(15)** a **(16)** c **(17)** b **(18)** c **(19)** a **(20)** d

Unit 4

Ohm's Law and Power

INTRODUCTION

Specific laws are used in calculating the parameters for electrical circuits. There are relationships between voltage, resistance, and current to produce usable circuits. Other laws are used to analyze complicated circuits and select the proper size components for the circuit.

UNIT OBJECTIVES

Upon completion of this unit, you will be able to:

1. State Ohm's law.
2. Calculate the current in a circuit using Ohm's law.
3. Find the voltage or resistance of a circuit or resistor using Ohm's law.
4. Define *power.*
5. Calculate power from voltage and current.
6. Determine other circuit variables from the power formula.
7. Apply Ohm's law to a series circuit.
8. Describe Kirchhoff's voltage law.
9. Apply Ohm's law to a parallel circuit.
10. Describe Kirchhoff's current law.
11. Solve for total resistance in a series circuit and a parallel circuit.
12. Solve for circuit unknowns using Ohm's law and Kirchhoff's law.
13. Operate a power supply.
14. Explain the use of circuit protectors.
15. Define the terms *node, joule, work, power, watt, voltage drop, loading effect, overload, slow-blow fuse,* and *ground.*

SECTION 4-1 FUNDAMENTAL CONCEPTS

4-1a OHM'S LAW

In 1827 a German mathematics and physics professor, Georg Simon Ohm, published a paper entitled "The Galvanic Circuit Investigated Mathematically." Although debated and ridiculed initially, this work later became known as Ohm's law and is the basis of all electrical and electronic circuits.

Ohm's Law states:

> The current flowing in a circuit is directly proportional to the voltage (or electromotive force) across the circuit and inversely proportional to the resistance of the circuit.

As you know, there are three quantities associated with an electrical circuit: voltage, represented by V or E; resistance, indicated by R; and current, represented by I. This relationship is shown in Figure 4-1a. Ohm's law can then be expressed as a formula, as shown in Figure 4-1b. If any two quantities are known, the unknown quantity can be found by rearranging the basic formula. However, the diagram shown in Figure 4-1c can be used as an aid to finding the unknown quantity. Simply cover the unknown quantity with your finger and the arrangement of the two known quantities indicates the appropriate formula. For example, to find current, cover I and the result is V over R, or in other words, V divided by R.

Figure 4–1 Ohm's law: (a) basic circuit; (b) basic formula; (c) formula triangle diagram.

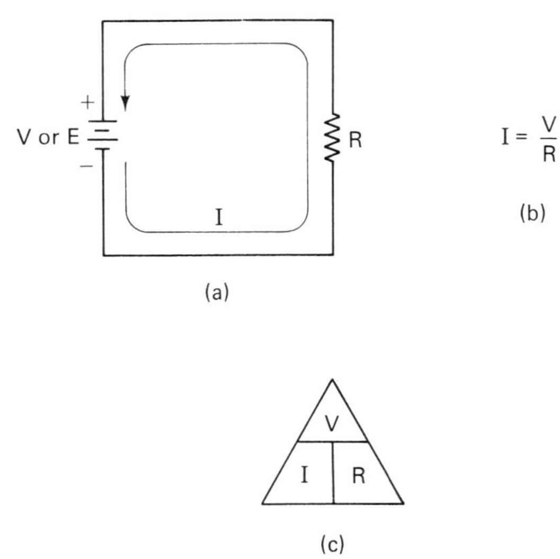

4-1a.1 Finding Current

To find the current flow in a basic circuit as shown in Figure 4-1a, let $V = 6$ V and $R = 2\ \Omega$. Solving for I using the Ohm's law formula gives

$$I = \frac{V}{R} = \frac{6\text{ V}}{2\ \Omega} = 3\text{ A}$$

Suppose that the voltage is doubled to 12 V; then

$$I = \frac{V}{R} = \frac{12\text{ V}}{2\ \Omega} = 6\text{ A}$$

which confirms Ohm's law that the current is directly proportional to voltage. In other words, if the voltage increases, the current increases. Similarly, if the voltage was halved in the initial circuit to 3 V, then

$$I = \frac{V}{R} = \frac{3\text{ V}}{2\ \Omega} = 1.5\text{ A}$$

This shows that if the voltage decreases, the current decreases.

If the resistance is doubled to 4 Ω, then

$$I = \frac{V}{R} = \frac{6\text{ V}}{4\ \Omega} = 1.5\text{ A}$$

which proves the other part of Ohm's law—that the current is inversely proportional to the resistance. This means that if resistance increases, the current decreases. Similarly, if the resistance were halved to 1 Ω, then

$$I = \frac{V}{R} = \frac{6\text{ V}}{1\ \Omega} = 6\text{ A}$$

This shows that if the resistance decreases, the current increases.

4-1a.2 Finding Voltage

With some situations, you will know the resistance of a component and the current flowing through it. With these two quantities known, you can calculate the unknown voltage across the component. Using the formula triangle diagram shown in Figure 4-1c, cover the V and the result is I next to R, or I multiplied by R. In other words,

$$V = IR$$

Even though this is an easy method for finding V, it is important to know how to transpose terms in a formula or equation to be able to solve any of its unknown variables. The basic rule for transposing terms states that whatever is performed mathematically on one side of the equation must also be performed on the other side. To find V in the basic formula,

$$I = \frac{V}{R}$$

we want to have V on one side of the equation by itself. To be able to cancel R, we have to multiply both sides of the equation by R:

$$I \times \frac{R}{1} = \frac{V}{R} \times \frac{R}{1}$$

Combine: $$IR = \frac{VR}{R}$$

Cancel: $$IR = \frac{V\cancel{R}}{\cancel{R}}$$

Thus $$IR = V$$

To better understand the formula, it is reversed:

$$V = IR$$

To prove the new equation, let $I = 3$ A and $R = 2\ \Omega$; therefore,

$$V = IR = 3\text{ A} \times 2\ \Omega = 6\text{ V}$$

Notice the initial values used in the formula for finding current in the preceding section. Solving for V then verifies the original answer for 3 A of current.

4-1a.3 Finding Resistance

In some cases you will know the current flowing through a component and the voltage across it. With these two quantities known, you can calculate the unknown resistance of the component. Using the formula triangle diagram shown in Figure 4-1c, cover the R and the result is V over I, or in other words,

$$R = \frac{V}{I}$$

We can also find the resistance by transposing terms in the equation. To find R in the basic formula

$$I = \frac{V}{R}$$

we want to have R on one side of the equation by itself. To accomplish this we use the following procedure: Multiply both sides of equation by R:

$$R \times I = \frac{V}{R} \times \frac{R}{1}$$

The R's on the right-hand side of the equation cancel each other:

$$R \times I = \frac{V}{R} \times \frac{R}{1}$$

Therefore,

$$R \times I = V$$

Divide both sides of the equation by I:

$$\frac{R \times I}{I} = \frac{V}{I}$$

The I's on the left-hand side of the equation cancel each other:

$$\frac{R \times I}{I} = \frac{V}{I}$$

The final result is

$$R = \frac{V}{I}$$

As an example, let $V = 6$ V and $I = 3$ A; therefore,

$$R = \frac{V}{I} = \frac{6\text{ V}}{3\text{ A}} = 2\ \Omega$$

Again, notice the initial values used in the formula for finding current in the preceding section. Solving for R also verifies the original answer for 3 A of current.

4-1b POWER

In Unit 1 it was shown that energy is the invisible force that provides the capacity to do work. Remember, energy cannot be created or destroyed. It can only be transformed or changed from one form to another.

> Work is usually defined as a force applied to a specific weight of material to move it a specific distance. Electrical work is the voltage applied to a circuit to cause electrons to flow.

Work is done in the process of converting energy from one form to another; therefore, energy is stored work, such as the stored charge of a battery.

> The joule is a standard unit of work. One electrical joule of work is accomplished when 1 coulomb of electrons is moved through a potential difference of 1 volt.

Power is the rate of doing work. Electrical power is the rate at which electrical energy is expended. Since 1 coulomb is equal to 1 ampere of current, a formula can be found for calculating the power of an electrical circuit. This formula states that power is equal to voltage multiplied by current:

$$P = VI$$

The unit of electrical power is the watt, named after James Watt, a Scottish engineer and inventor who made improvements on the steam engine, which resulted in making devices more efficient to reduce power. From the example in Section 4-1a.1, where $V = 6$ V and $R = 2$ Ω, resulting in $I = 3$ A, the power can be calculated thus:

$$P = VI = 6\text{ V} \times 3\text{ A} = 18\text{ W}$$

When electrical current flows through a device, heat is produced, which is released into the surrounding air. Here is an example of the transformation of energy (i.e., from electrical energy to heat energy). However, the heat energy cannot be returned to the electrical source, and some of this energy has been consumed. The rate at which the heat energy is released is also referred to as power.

4-1b.1 Finding Current from Power

Similar to Ohm's law, if two of the quantities for power are known, the third unknown quantity can be calculated. If the power and voltage of a circuit are known, the power formula can be transposed to find current, thus:

$$P = VI$$

We want to get I on one side of the equation by itself. In this case the operation between V and I is multiplication. To cancel V, we have to divide both sides of the equation by V:

$$\frac{P}{V} = \frac{VI}{V}$$

Cancel:

$$\frac{P}{V} = \frac{\cancel{V}I}{\cancel{V}}$$

Thus

$$\frac{P}{V} = I$$

Rearrange:

$$I = \frac{P}{V}$$

As an example, if $P = 18$ W and $V = 6$ V, then

$$I = \frac{P}{V} = \frac{18\text{ W}}{6\text{ V}} = 3\text{ A}$$

4-1b.2 Finding Voltage from Power

To find voltage using the power formula, we must transpose I from the formula

$$P = VI$$

To transpose I, we must divide both sides of the equation by I:

$$\frac{P}{I} = \frac{VI}{I}$$

Cancel: $$\frac{P}{I} = \frac{V\cancel{I}}{\cancel{I}}$$

Thus $$\frac{P}{I} = V$$

Rearrange: $$V = \frac{P}{I}$$

As an example, if $P = 18$ W and $I = 3$ A, then

$$V = \frac{P}{I} = \frac{18\text{ W}}{3\text{ A}} = 6\text{ V}$$

4-1b.3 Finding Resistance from Power

If the power of a circuit or device is known and at least one other quantity (V or I) is known, the resistance can be calculated by using Ohm's law. For example,

$$I = \frac{P}{V} \quad \text{and} \quad V = \frac{P}{I}$$

and from Section 4-1a.3, to find resistance, you use the formula

$$R = \frac{V}{I}$$

If $P = 18$ W and $I = 3$ A, then by the formulas

$$V = \frac{P}{I} = \frac{18\text{ W}}{3\text{ A}} = 6\text{ V} \quad \text{and} \quad R = \frac{V}{I} = \frac{6\text{ V}}{3\text{ A}} = 2\ \Omega$$

4-1b.4 Other Power Formulas

The basic power formula is $P = VI$; however, if V or I are not known but R is known, substituting Ohm's law can be used to solve for power. For example,

$$P = VI$$

and Ohm's law states that $V = IR$. Then substituting V in the power formula results in

$$P = IR \times I \qquad \text{or} \qquad P = I^2R$$

If $I = 3$ A and $R = 2\ \Omega$, then

$$P = I^2R = (3)^2 \times 2 = 9 \times 2 = 18\ \text{W}$$

Similarly,

$$P = VI$$

and Ohm's law states that $I = \frac{V}{R}$. Then substituting I in the power formula results in

$$P = V \times \frac{V}{R} \qquad \text{or} \qquad P = \frac{V^2}{R}$$

If $V = 6$ V and $R = 2\ \Omega$, then

$$P = \frac{V^2}{R} = \frac{(6\ \text{V})^2}{2\ \Omega} = \frac{36\ \text{V}}{2\ \Omega} = 18\ \text{W}$$

There are four basic quantities inherent in every electrical circuit: voltage, resistance, current, and power. If any two of these quantities are known, transposing the power formula and Ohm's law formula can find the other two unknown quantities. There are other formulas that can be derived which make the process of calculation easier, such as

$$V = \sqrt{PR} \qquad R = \frac{P}{I^2} \qquad \text{and} \qquad I = \sqrt{\frac{P}{R}}$$

Rather than try and remember all the formulas or derive them from other basic formulas, the formula wheel diagram shown in Figure 4-2 can be used. The hub of the wheel is divided into four center sec-

Figure 4–2 Formula wheel diagram for finding *V, I, R,* and *P.*

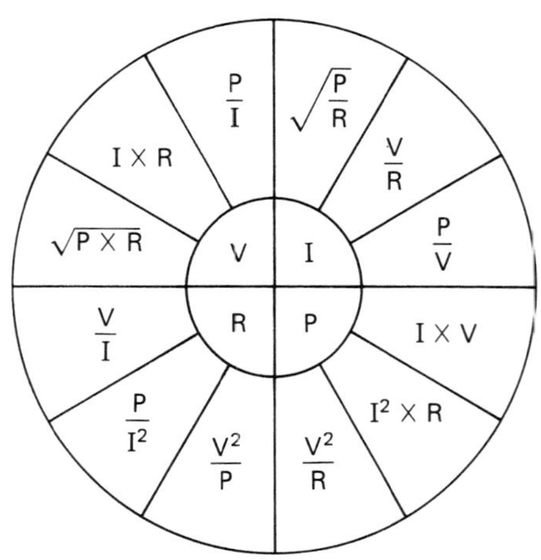

tions, which represent the four basic quantities of a circuit. There are three more sections that radiate outward (in a spoke-like fashion) from each center section that contain the formula for finding that particular unknown. For example, if you wanted to find the current in a circuit and you knew the resistance and power, you would look at the I center section. From the three sections radiating from the I center section, you would locate the formula that contained R and P: in this case, the square root of power divided by resistance:

$$I = \sqrt{\frac{P}{R}}$$

4-1c OHM'S LAW AND THE SERIES CIRCUIT

Ohm's law is applied to circuits to determine various parameters or working conditions of the circuits. Figure 4-3a shows a simple three-resistor series circuit. The total current in the circuit is found by the Ohm's law formula:

$$I_T = \frac{V_T}{R_T}$$

where I_T = total current
V_T = total voltage or applied voltage (V_A)
R_T = total circuit resistance

Figure 4–3 Series circuit using Ohm's law: (a) schematic diagram; (b) I_T versus V_T graph.

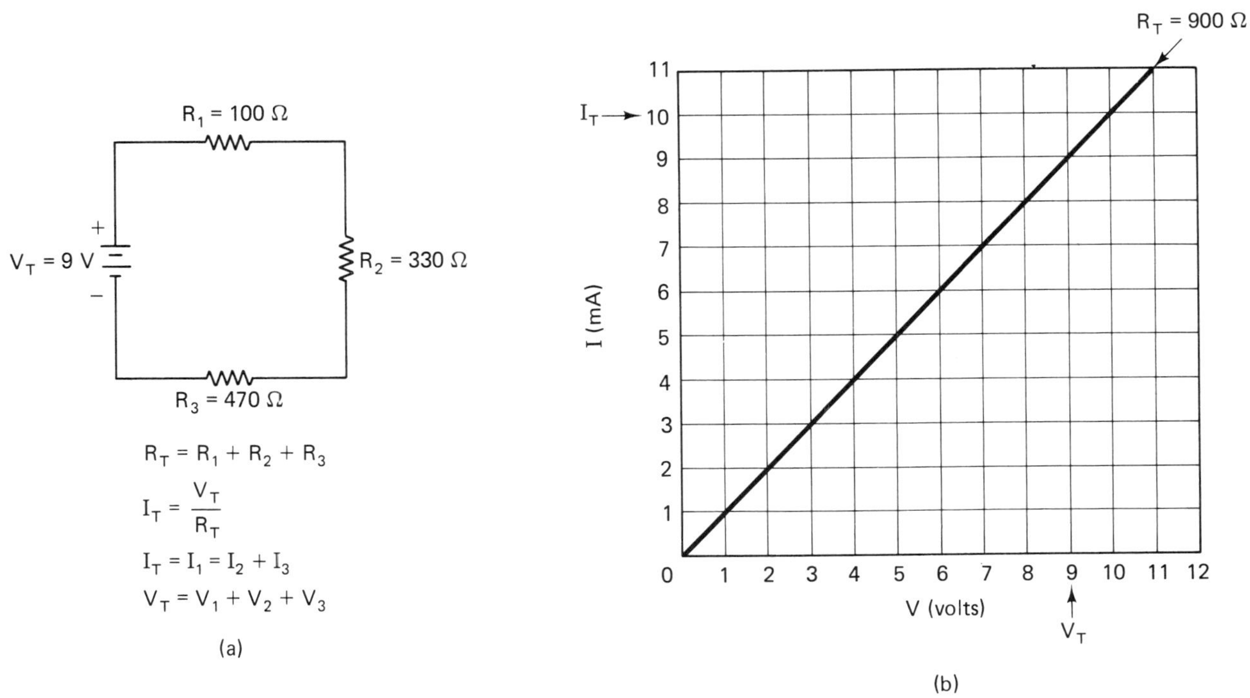

The total resistance (R_T) for a series circuit is found by adding up the individual resistances using the formula

$$R_T = R_1 + R_2 + R_3$$

Therefore,

$$R_T = 100\ \Omega + 330\ \Omega + 470\ \Omega = 900\ \Omega$$

The total current can now be found using the formula

$$I_T = \frac{V_T}{R_T} = \frac{9\text{ V}}{900\ \Omega} = 0.01\text{ A} = 10 \times 10^{-3}\text{ A} = 10\text{ mA}$$

Remember that the total current is the same throughout a series circuit, as expressed by the formula

$$I_T = I_1 = I_2 = I_3$$
$$10\text{ mA} = 10\text{ mA} = 10\text{ mA} = 10\text{ mA}$$

Figure 4-3b shows a linear graph representing current I versus voltage V for a resistance of 900 Ω. Notice where I_T (10 mA) and V_T (9 V) intersect on the graph. The line was drawn from this point to zero to establish the relationship of I_T to V_T for a 900-Ω resistance. Referring to the graph, if V_T were decreased to 4.5 V, I_T would be at 5 mA. From the graph you can see how I_T changes if V_T is changed. A circuit with a different resistance would have a different slope to the line, but the same relationship between V_T and I_T would exist.

Remember that the current in a series circuit is the same throughout the circuit and is indicated by the formula

$$I_T = I_1 = I_2 = I_3$$

The applied voltage or (V_T) to a circuit is referred to as a *voltage rise* since it is the source of electrical energy used by the circuit. Each resistive component in a circuit uses some of this applied energy. As the current flows through the component, a voltage is developed across it which is called a *voltage drop.* All the voltage drops in a series circuit add up to equal the applied voltage, which is indicated by the formula

$$V_T = V_1 + V_2 + V_3$$

To prove this for the circuit shown in Figure 4-3a, the voltage drop across each resistor has to be calculated.

$$V_1 = I_1R_1 = 10\text{ mA} \times 100\ \Omega = 10 \times 10^{-3} \times 100 = 1000 \times 10^{-3} = 1\text{ V}$$
$$V_2 = I_2R_2 = 10\text{ mA} \times 330\ \Omega = 10 \times 10^{-3} \times 330$$
$$= 3300 \times 10^{-3} = 3.3\text{ V}$$
$$V_3 = I_3R_3 = 10\text{ mA} \times 470\ \Omega = 10 \times 10^{-3} \times 470$$
$$= 4700 \times 10^{-3} = 4.7\text{ V}$$

Finally,

$$V_T = V_1 + V_2 + V_3$$
$$= 1\text{ V} = 3.3\text{ V} = 4.7\text{ V}$$
$$= 9.0\text{ V}$$

The total power (P_T) of the circuit is the sum of all the individual powers consumed by the components and is expressed by the formula

$$P_T = P_1 + P_2 + P_3$$

Therefore, the power of each resistor of the circuit in Figure 4-3a must be calculated.

$$P_1 = V_1 I_1 = 1\text{ V} \times 10\text{ mA} = 1 \times 10 \times 10^{-3} = 10^{-3}\text{ W} = 10\text{ mW}$$
$$P_2 = V_2 I_2 = 3.3\text{ V} \times 10\text{ mA} = 3.3 \times 10 \times 10^{-3} = 33 \times 10^{-3}\text{ W} = 33\text{ mW}$$
$$P_3 = V_3 I_3 = 4.7\text{ V} \times 10\text{ mA} = 4.7 \times 10 \times 10^{-3} = 47 \times 10^{-3}\text{ W} = 47\text{ mW}$$

Finally,

$$P_T = P_1 + P_2 + P_3 = 10\text{ mW} + 33\text{ mW} + 47\text{ mW} = 90\text{ mW}$$

Of course, the total power can be found or checked with the basic power formula:

$$P_T = V_T I_T = 9\text{ V} \times 10\text{ mA} = 9 \times 10 \times 10^{-3}\text{ W} = 90\text{ mW}$$

4-1c.1 Kirchhoff's Voltage Law

Gustave Robert Kirchhoff, a German physicist who lived during the mid-nineteenth century, contributed to the world's electrical knowledge by developing two laws. These laws are used in analyzing circuit conditions. His first law applies to a series circuit, as shown in Figure 4-4. Kirchhoff's voltage law states:

> The algebraic sum of all voltages around a closed loop is equal to zero.

This can be represented by the formula

$$V_T + V_1 + V_2 + V_3 = 0$$

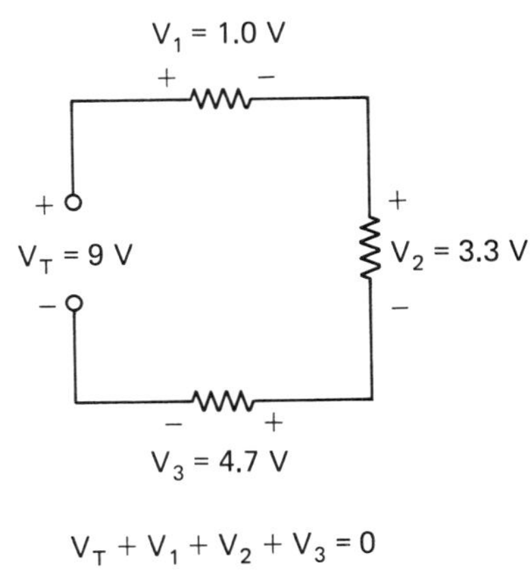

Figure 4–4 Kirchhoff's voltage law.

As current flows through a component, the voltage drop that develops across it has a polarity. There are more electrons entering the component than leaving it, so that the side that is entered is more negative than the side that is exited.

To apply Kirchhoff's voltage law, a few procedures must be established:

1. Choose the direction of current flow that you prefer to trace on the circuit (clockwise or counterclockwise).
2. Label the polarities of the circuit voltage drops.
3. Select a specific starting point in the circuit (usually, one side of the voltage source).
4. The first polarity indication of each component that you come to as you trace around the circuit is the one to use in the calculation.

For example, in Figure 4-4, let's select the negative terminal of the voltage source as the starting point. Going counterclockwise, the first component we come to is R_3. The first polarity encountered is negative; therefore, $V_3 = -4.7$ V. Continuing around the circuit, we find that $V_2 = -3.3$ V, $V_1 = -1.0$ V, and the voltage source $V_T = +9$ V. Algebraically summing these values, we have

$$V_3 + V_2 + V_1 + V_T = 0$$
$$(-4.7\text{ V}) + (-3.3\text{ V}) + (-1\text{ V}) + 9\text{ V} = 0$$
$$(-9\text{ V}) + 9\text{ V} = 0$$

If we had started on the positive side of the voltage source and traced the circuit clockwise, the result would have been

$$V_1 + V_2 + V_3 + V_T = 0$$
$$1\text{ V} + 3.3\text{ V} + 4.7\text{ V} + (-9\text{ V}) = 0$$

As you can see, Kirchhoff's voltage law is correct. The sum of the voltages around a closed-loop circuit is zero. Understanding this principle will aid you in analyzing more complex circuits.

4-1d OHM'S LAW AND THE PARALLEL CIRCUIT

Ohm's law can be applied to a parallel circuit as shown in Figure 4-5. The total current formula is the same as with the series circuit,

$$I_T = \frac{V_T}{R_T}$$

However, the total resistance is calculated differently. When more than two resistors of unequal value are in parallel, the reciprocal formula is used:

$$R_T = \frac{1}{\dfrac{1}{R_1} + \dfrac{1}{R_2} + \dfrac{1}{R_3}}$$

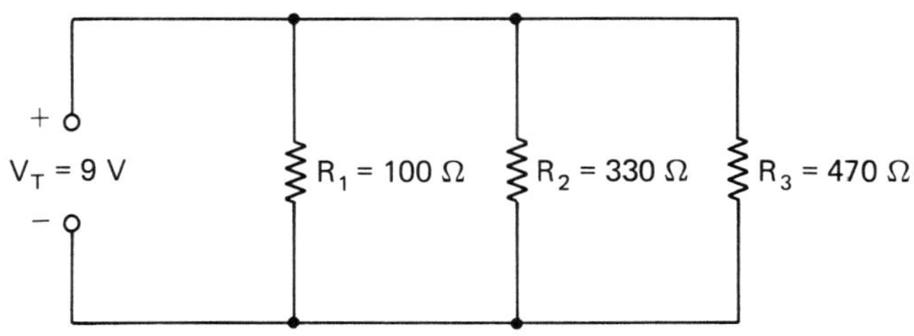

$$R_T = \frac{1}{\frac{1}{R_1} + \frac{1}{R_2} + \frac{1}{R_3}}$$

$$I_T = \frac{V_T}{R_T}$$

$$I_T = I_1 + I_2 + I_3$$

$$V_T = V_1 = V_2 = V_3$$

Figure 4–5 Parallel circuit using Ohm's law.

Then we fill in the values:

$$R_T = \frac{1}{\frac{1}{100\,\Omega} + \frac{1}{330\,\Omega} + \frac{1}{470\,\Omega}}$$

The first thing to do is to find the least common denominator by multiplying the resistor values together:

$$100 \times 330 \times 470 = 15{,}510{,}000$$

This number is used as the lowest denominator:

$$R_T = \frac{1}{\frac{X_1 + X_2 + X_3}{15{,}510{,}000}}$$

Now each of the resistor values is divided into this number to find its corresponding factor, which is placed where the X's are shown:

$$15{,}510{,}000 \div R_1\,(100) = 155{,}100$$
$$15{,}510{,}000 \div R_2\,(330) = 47{,}000$$
$$15{,}510{,}000 \div R_3\,(470) = 33{,}000$$

These numbers can now be placed into the formula:

$$R_T = \frac{1}{\frac{155{,}100 + 47{,}000 + 33{,}000}{15{,}510{,}000}}$$

To simplify the equation, you could use powers of 10, but remember to keep the power the same when adding:

$$R_T = \frac{1}{\frac{1551 \times 10^2 + 470 \times 10^2 + 330 \times 10^2}{1551 \times 10^4}}$$

Adding produces

$$R_T = \frac{1}{\dfrac{2351 \times 10^2}{1551 \times 10^4}}$$

Next divide 2351×10^2 by 1551×10^4, which gives

$$R_T = \frac{1}{1.5158 \times 10^{-2}}$$

Notice that the power of 10 is now negative. The last step is to take the reciprocal of 1.5158×10^{-2} by dividing it into 1; thus

$$R_T = 0.6597 \times 10^2 = 66\ \Omega$$

Notice that this time, the power of 10 becomes positive. The total resistance is approximately 66 Ω. This proves a very important rule for resistances in parallel. *The total resistance of resistors in parallel will always be less than the smallest resistor value.*

The total current can now be calculated:

$$I_T = \frac{V_T}{R_T} = \frac{9\ \text{V}}{66\ \Omega} = 0.13636 = 136.4 \times 10^{-3} = 136.4\ \text{mA}$$

The numbers are rounded off and expressed in milliamperes.

Remember that the voltage across a parallel circuit is the same across each branch,

$$\begin{aligned} V_T &= V_1 = V_2 = V_3 \\ 9\ \text{V} &= 9\ \text{V} = 9\ \text{V} = 9\ \text{V} \end{aligned}$$

The current through each branch can now be calculated:

$$I_1 = \frac{V_1}{R_1} = \frac{9\ \text{V}}{100\ \Omega} = 0.09\ \text{A} = 90 \times 10^{-3}\ \text{A} = 90\ \text{mA}$$

$$I_2 = \frac{V_2}{R_2} = \frac{9\ \text{V}}{330\ \Omega} = 0.02727\ \text{A} = 27.3 \times 10^{-3}\ \text{A} = 27.3\ \text{mA}$$

$$I_3 = \frac{V_3}{R_3} = \frac{9\ \text{V}}{470\ \Omega} = 0.0191\ \text{A} = 19.1 \times 10^{-3}\ \text{A} = 19.1\ \text{mA}$$

Again, remember that the total current in a parallel circuit is the sum of all its branch currents and can be found by the formula

$$\begin{aligned} I_T &= I_1 + I_2 + I_3 \\ &= 90\ \text{mA} + 27.3\ \text{mA} + 19.1\ \text{mA} \\ &= 136.4\ \text{mA} \end{aligned}$$

The power for a parallel circuit is calculated the same way as for a series circuit; therefore, the power dissipated by each resistor must be found:

$$P_1 = V_1 I_1 = 9\,\text{V} \times 90 \times 10^{-3}\,\text{A} = 810 \times 10^{-3}\,\text{W} = 810\,\text{mW}$$
$$P_2 = V_2 I_2 = 9\,\text{V} \times 27.3 \times 10^{-3}\,\text{A} = 245.7 \times 10^{-3}\,\text{W} = 245.7\,\text{mW}$$
$$P_3 = V_3 I_3 = 9\,\text{V} \times 19.1 \times 10^{-3}\,\text{A} = 171.9 \times 10^{-3}\,\text{W} = 171.9\,\text{mW}$$

The total power is the sum of the individual resistor powers, calculated thus:

$$\begin{aligned} P_T &= P_1 + P_2 + P_3 \\ &= 810\,\text{mW} + 245.7\,\text{mW} + 171.9\,\text{mW} \\ &= 1227.6\,\text{mW} \quad \text{or} \quad 1.2276\,\text{W} \end{aligned}$$

The total power can also be calculated from the formula

$$\begin{aligned} P_T &= V_T I_T \\ &= 9\,\text{V} \times 136.4\,\text{mA} \\ &= 1227.6\,\text{mW} \quad \text{or} \quad 1.2276\,\text{W} \end{aligned}$$

Since the same resistor values and voltage were used for each circuit, a comparison can be made. The series circuit produced less current and consumed less power, whereas the parallel circuit had more current and dissipated more power.

Parallel connections tend to produce a *loading effect,* where the total resistance decreases and the total current increases. An example of this is connecting speakers in parallel to a stereo system. Connecting too many speakers could cause an overload, where too much current flows and may damage the output stage of the amplifier.

4-1d.1 Kirchhoff's Current Law

Kirchhoff's second law relates to current flow in a circuit. Kirchhoff's current law states:

> The current flowing into a junction or node is equal to the current flowing away from that point.

Another way it can be stated is:

> The algebraic sum of the currents entering and leaving any node or junction of conductors is equal to zero. *A node is a specific reference point on a circuit.*

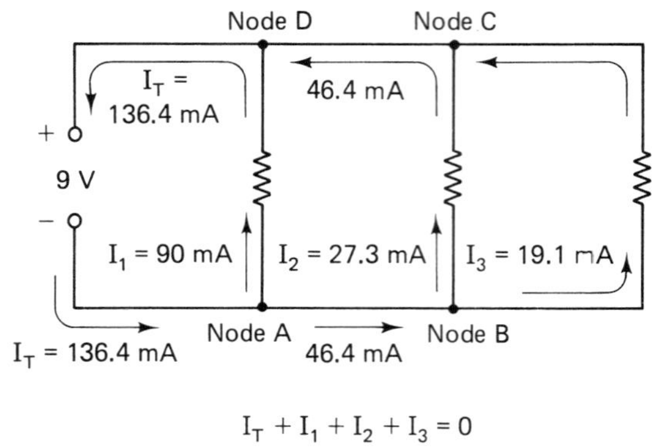

Figure 4-6 Kirchhoff's current law.

Figure 4-6 is the same circuit shown in Figure 4-5 except that it shows the current paths. In Figure 4-6 you see that 136.4 mA is flowing into node A. From this point, 90 mA is flowing up through R_1 and 46.4 mA flows to the right toward node B. Applying the current law, we have 136.4 mA = 90 mA + 46.4 mA. If we add the currents algebraically, we have to assign polarities to the currents. Let current flowing toward the node be positive and current flowing away from the node be negative; therefore,

$$+136.4 \text{ mA} + (-90 \text{ mA}) + (-46.4 \text{ mA}) = 0$$

Notice that the 46.4 mA flowing toward node B divides again, with 27.3 mA flowing through R_2 and 19.1 mA flowing through R_3. These two currents are recombined at node C to 46.4 mA and flows toward node D. At node D, the 46.4 mA is recombined with the 90 mA flowing through R_1 to equal the total current of 136.4 mA.

Applying the laws of voltage drops and current flow, will help you to understand more complex circuits given in the next unit.

4-1d.2 Same-Value Resistors in Parallel

When all the resistors in parallel have the same value, their total resistance can be found simply by dividing the total number of resistors into the value of one resistor. Given in terms of a formula, we have

$$R_T = \frac{R}{N_R}$$

where R is the value of one resistor and N_R is the total number of resistors. For example, if four 1-kΩ resistors are in parallel, the total resistance as given by the formula is

$$R_T = \frac{R}{N_R} = \frac{1 \text{ k}\Omega}{4} = \frac{1 \times 10^3}{4} = 0.25 \times 10^3 = 250\ \Omega$$

This answer can be proven by the reciprocal method formula, thus:

$$R_T = \frac{1}{\frac{1}{R_1} + \frac{1}{R_2} + \frac{1}{R_3} + \frac{1}{R_4}}$$
$$= \frac{1}{\frac{1}{1000} + \frac{1}{1000} + \frac{1}{1000} + \frac{1}{1000}}$$
$$= \frac{1}{\frac{4}{1000}}$$
$$= \frac{1}{0.004}$$
$$= 250\ \Omega$$

4-1d.3 Two Resistors in Parallel

The total resistance of two resistors in parallel can be found with a method called *product over sum.* The values of both resistors are multiplied together and then divided by sum of the two values. It is expressed as a formula, thus:

$$R_T = \frac{R_1 \times R_2}{R_1 + R_2}$$

As an example, $R_1 = 2.2\ \text{k}\Omega$ and $R_2 = 3.3\ \text{k}\Omega$ and their total resistance is

$$R_T = \frac{R_1 \times R_2}{R_1 + R_2}$$
$$= \frac{2.2\ \text{k}\Omega \times 3.3\ \text{k}\Omega}{2.2\ \text{k}\Omega + 3.3\ \text{k}\Omega}$$
$$= \frac{2.2 \times 10^3 \times 3.3 \times 10^3}{2.2 \times 10^3 + 3.3 \times 10^3}$$
$$= \frac{7.26 \times 10^6}{5.5 \times 10^3}$$
$$= 1.32 \times 10^3 = 1.32\ \text{k}\Omega$$

4-1e SOLVING FOR CIRCUIT UNKNOWNS

Sometimes it is necessary to calculate unknown circuit values based on the values that are known. The solution to this problem is to begin finding basic formulas to calculate the unknown. Normally, other unknown values are encountered and we must apply other formulas based on what we know about the circuit parameters. When the various unknown values are calculated, they are placed back into the initial formulas to solve for the original unknown. This process is more easily understood and mastered through practice.

4-1e.1 Unknown Values in a Series Circuit

Figure 4-7 shows a series circuit with the value of resistor R_2 unknown. In analyzing this problem, we ask ourselves the basic Ohm's law formula for finding R_2, which is

$$R_2 = \frac{V_2}{I_2}$$

We need to have two known values to find the third unknown value. Since this is a series circuit, I_2 is known because

$$I_T = I_1 = I_2 = I_3$$

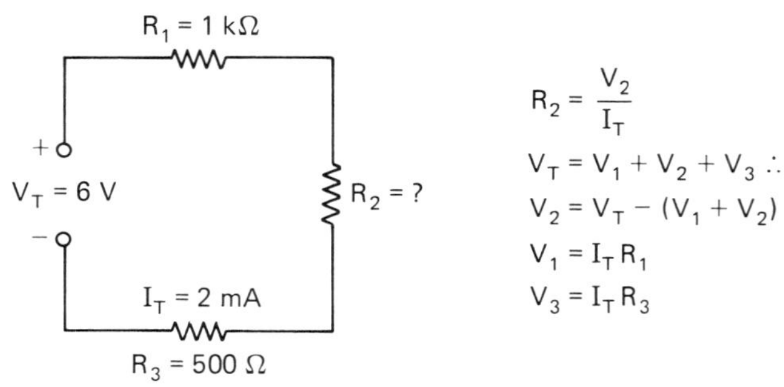

Figure 4-7 Solving for unknown values in a series circuit.

The voltage drop across $R_2 (V_2)$ is unknown but can be found. If $V_T = V_1 + V_2 + V_3$, then V_2 is equal to the sum of V_1 and V_3 subtracted from V_T as indicated by the formula

$$V_2 = V_T - (V_1 + V_3)$$

The value of V_T is given as 6 V, but V_1 and V_3 are not known. However, V_1 and V_3 can be calculated from what is given, by using the Ohm's law formula for finding voltage:

$$V_1 = I_T R_1 = 2\text{ mA} \times 1\text{ k}\Omega = 2 \times 10^{-3} \times 1 \times 10^3 = 2\text{ V}$$
$$V_3 = I_T R_3 = 2\text{ mA} \times 500 = 2 \times 10^{-3} \times 500 = 1000 \times 10^{-3} = 1\text{ V}$$

The voltage drop across R_2 can now be found:

$$\begin{aligned} V_2 &= V_T - (V_1 + V_3) \\ &= 6\text{ V} - (2\text{ V} + 1\text{ V}) \\ &= 6\text{ V} - 3\text{ V} \\ &= 3\text{ V} \end{aligned}$$

and the value of R_2 can be calculated:

$$R_2 = \frac{V_2}{I_2} = \frac{3\text{ V}}{2\text{ mA}} = \frac{3}{2 \times 10^{-3}} = 1.5 \times 10^3 = 1.5\text{ k}\Omega$$

4-1e.2 Unknown Values in a Parallel Circuit

With a similar approach, we can solve for the unknown value of R_2 shown in the parallel circuit in Figure 4-8. Here again, the basic Ohm's law formula for finding the value of R_2 is

$$R_2 = \frac{V_2}{I_2}$$

The voltage across R_2 (V_2) is known since the voltage of a parallel circuit is the same across each branch, as given by the formula

$$V_T = V_1 = V_2 = V_3$$

The current through R_2 (I_2) is unknown but can be found. If $I_T = I_1 + I_2 + I_3$, then I_2 is equal to the sum of I_1 and I_2 subtracted from I_T, as given by the formula

$$I_2 = I_T - (I_1 + I_3)$$

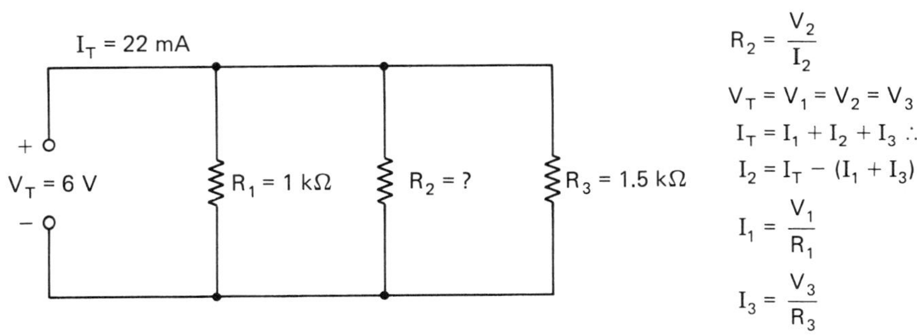

Figure 4–8 Solving for unknown values in a parallel circuit.

The value of I_T is given as 22 mA, but I_1 and I_3 are not indicated. These currents can be calculated from Ohm's law formulas, thus:

$$I_1 = \frac{V_1}{R_1} = \frac{6\text{ V}}{1\text{ k}\Omega} = \frac{6}{1 \times 10^3} = 6 \times 10^{-3}\text{ A} = 6\text{ mA}$$

$$I_3 = \frac{V_3}{R_3} = \frac{6\text{ V}}{1.5\text{ k}\Omega} = \frac{6}{1.5 \times 10^3} = 4 \times 10^{-3}\text{ A} = 4\text{ mA}$$

The current through R_2 can now be found:

$$\begin{aligned} I_2 &= I_T - (I_1 + I_3) \\ &= 22\text{ mA} - (6\text{ mA} + 4\text{ mA}) \\ &= 22\text{ mA} - 10\text{ mA} \\ &= 12\text{ mA} \end{aligned}$$

Finally, the value of R_2 can be calculated, thus:

$$R_2 = \frac{V_2}{I_2} = \frac{6\text{ V}}{12\text{ mA}} = \frac{6}{12 \times 10^{-3}} = 0.5 \times 10^3 = 500\ \Omega$$

4-1f USE OF A POWER SUPPLY

When batteries are used frequently, their energy becomes depleted and the circuits they are powering will in most cases have faulty operation. Power supplies are used to replace batteries. *A power supply is a device that converts the 120-ac line voltage to dc voltage for operating other electrical/electronic circuits.* Power supplies are more efficient and dependable than batteries. Nearly every electronic device that operates on 120 V ac has an internal power supply built into it. Separate self-contained power supplies are used for performing laboratory experiments and testing PC boards or other discrete electronic components. There are various types of power supplies, but one very popular version is shown in Figure 4-9.

The pilot light is an indication that the power supply is on or off. Even when this light is on, the output terminals should be measured to determine if voltage is present. The main output terminals are marked −V, GND, and +V or some other similar indication. The GND terminal is the common connection for most circuits. It is usually the color black with electronic circuits. Every circuit needs two connections to a power supply. The most used terminals are the GND and the +V. The +V terminal supplies a positive voltage with respect to the GND terminal. In many cases it is red in color. The +V control is a pot that adjusts the voltage level at the +V terminal. The −V terminal supplies a negative voltage with respect to the GND terminal. The −V control is a pot that adjusts the voltage at the −V terminal. We explain negative voltage in greater detail in Unit 5.

Many circuits today, such as operational amplifiers (op amps), use a + and a − voltage power supply. These types of power supplies may also have a fixed +5-V terminal and ground terminal that is used with digital circuits.

Figure 4-9 Typical ± *V* power supply.

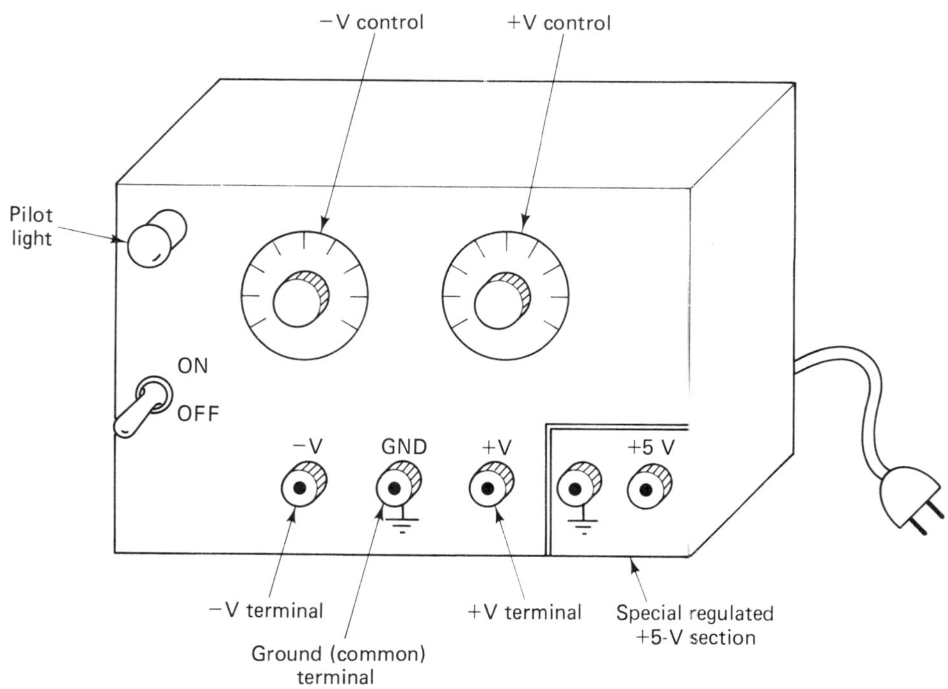

Before you use a power supply, the voltage control should be turned all the way counterclockwise, so that zero or minimum voltage is present at the output terminals. The following sequence is the proper way to use a power supply.

1. Have the correct drawing or schematic diagram of the circuit being tested.
2. Connect the power supply to the circuit.
3. Place a voltmeter across the voltage terminals connected to the circuit.
4. Slowly adjust the voltage control until the voltmeter indicates the desired voltage.
5. Turn off the power supply when you are not testing or working with the circuit.

   Caution!

   Never short-out the terminals of the power supply with a wire or metal object. This can damage the power supply or create more problems.

4-1g CIRCUIT PROTECTORS

Most electrical/electronic equipment and even the electrical system in your house has circuit protectors in the case of an overload. The most common circuit protector is the fuse, and the schematic symbol is shown in the circuit of Figure 4-10.

A fuse is normally found after the on/off switch mounted in a special holder or bracket. It is placed in series with the power source and the load or device. The fuse is made of a single strand of wire that is designed to burn open when its specific current rating is exceeded. The excessive current flowing through the fuse develops sufficient heat that the wire melts, which creates an open circuit. With this situation, the power source is removed from the circuit.

In the design of equipment, the fuse current rating selected is usually 20 to 50% greater than the total circuit current. The fuse will "operate" or burn open if one of two conditions occur. If the power source voltage increases, often referred to as *voltage surge,* the current in the

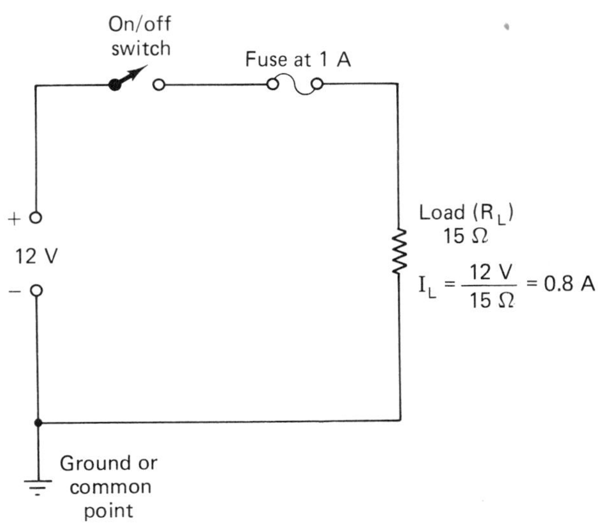

Figure 4–10 Fuse as a circuit protector.

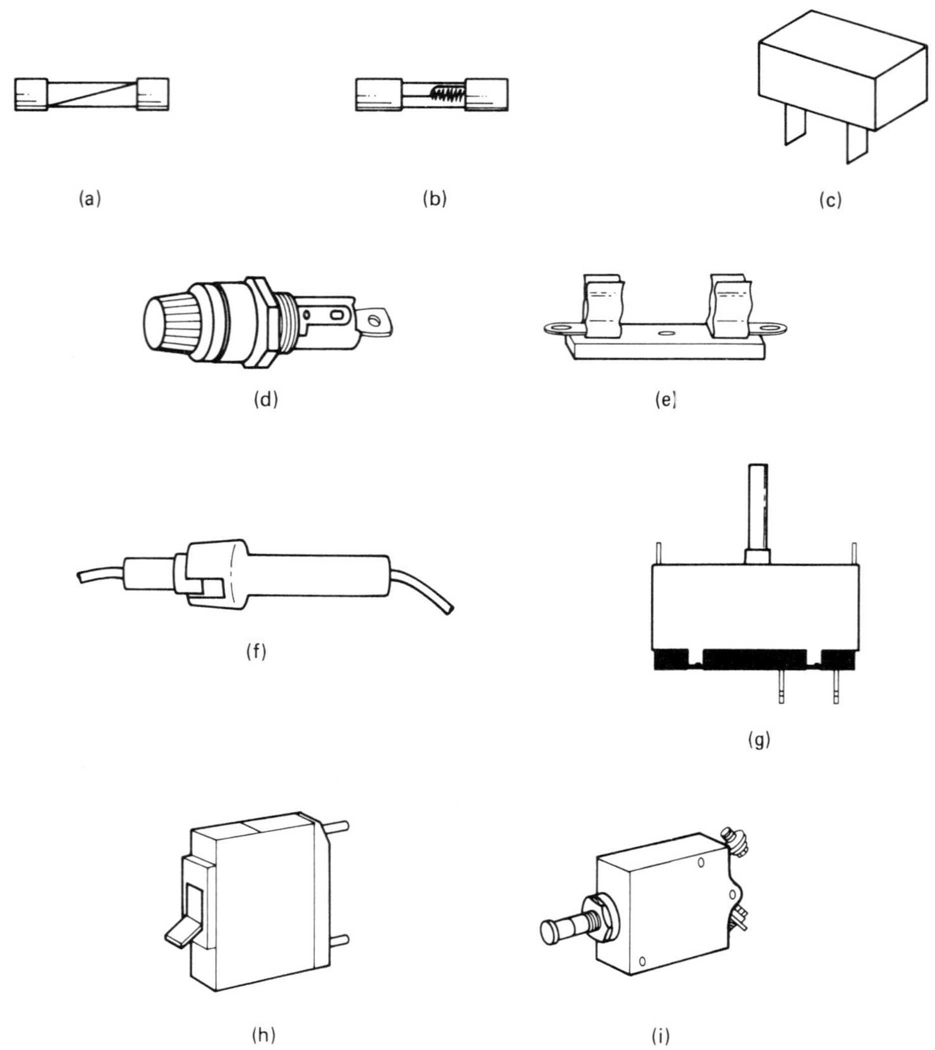

Figure 4–11 Circuit protectors: (a) regular "fast-blow" fuse; (b) slow-blow fuse; (c) bel fuse; (d) through-chassis fuse holder; (e) top-mount fuse holder; (f) in-line fuse holder (car radio); (g) television circuit breaker; (h) toggle-switch-type circuit breaker; (i) pushbutton-type circuit breaker.

circuit will increase. Voltage surges are usually of short duration and do not cause an excessive amount of current flow to damage some circuits. In this case a *slow-blow fuse* is used. It will continue to carry the excessive amount of current for a little longer, so that the voltage will probably return to normal before it burns open. In the other condition, the resistance of the load decreases or becomes shorted, which causes the circuit current to increase.

Notice the ground or common point schematic symbol in Figure 4-10. This is the circuit's reference point and the place at which the black or common lead of a voltmeter is connected. You will see this symbol very often in electronic schematic diagrams. In this circuit it is the most negative point.

Figure 4-11 shows some types of circuit protectors. A circuit breaker is a thermal device that operates a switch when an excessive current develops enough heat. The circuit breaker normally can be reset after the problem in the circuit is corrected and the temperature decreases.

SECTION 4-2 DEFINITIONS EXERCISES

Write a brief description for each of the following terms.

1. Ohm's law ______________________________

____

2. Power ______________________________

____

3. Kirchhoff's voltage law ______________________________

____

4. Kirchhoff's current law ______________________________

____

5. Node ______________________________

____

6. Power supply ______________________________

____

7. Fuse ______________________________

____

8. Circuit breaker ______________________________

____

SECTION 4-3 EXERCISES AND PROBLEMS

Complete this section before beginning the next section.

1. Draw a simple circuit of a resistor connected to a battery. Show current flow direction and label all parts (see Figure 4-1a).

2. Write the basic formula for Ohm's law.

3. Draw the Ohm's law formula triangle diagram.

4. In the space below, draw three resistors connected in series to a battery similar to the circuit shown in Figure 4-3a. Label the resistors R_1, R_2, and R_3, respectively.

Solve the following circuit values from the information given.

a. $V_T = 12$ V	Find: $R_T =$ ____	$I_3 =$ ____	$P_T =$ ____
$R_1 = 100\ \Omega$	$I_T =$ ____	$V_1 =$ ____	$P_1 =$ ____
$R_2 = 330\ \Omega$	$I_1 =$ ____	$V_2 =$ ____	$P_2 =$ ____
$R_3 = 470\ \Omega$	$I_2 =$ ____	$V_3 =$ ____	$P_3 =$ ____
b. $V_T = 12$ V	Find: $R_T =$ ____	$I_3 =$ ____	$P_T =$ ____
$R_1 = 2.2\ k\Omega$	$I_T =$ ____	$I_3 =$ ____	$P_T =$ ____
$R_2 = 1\ k\Omega$	$I_1 =$ ____	$V_2 =$ ____	$P_2 =$ ____
$R_3 = 3.3\ k\Omega$	$I_2 =$ ____	$V_3 =$ ____	$P_3 =$ ____
c. $V_T = 6$ V	Find: $R_T =$ ____	$I_3 =$ ____	$P_T =$ ____
$R_1 = 1.2\ k\Omega$	$I_T =$ ____	$V_1 =$ ____	$P_1 =$ ____
$R_2 = 1.5\ k\Omega$	$I_1 =$ ____	$V_2 =$ ____	$P_2 =$ ____
$R_3 = 470\Omega$	$I_2 =$ ____	$V_3 =$ ____	$P_3 =$ ____

5. List the three formulas to find total resistance in a parallel circuit.

6. In the space below, draw three resistors connected in parallel to a battery similar to the circuit shown in Figure 4-5. Label the resistors R_1, R_2, and R_3, respectively.

Solve the following circuit values from the information given.

a. $V_T = 12\ \text{V}$ Find: $R_T =$ ____ $I_3 =$ ____ $P_T =$ ____

$R_1 = 100\ \Omega$ $I_T =$ ____ $V_1 =$ ____ $P_1 =$ ____

$R_2 = 330\ \Omega$ $I_1 =$ ____ $V_2 =$ ____ $P_2 =$ ____

$R_3 = 470\ \Omega$ $I_2 =$ ____ $V_3 =$ ____ $P_3 =$ ____

b. $V_T = 12\ \text{V}$ Find: $R_T =$ ____ $I_3 =$ ____ $P_T =$ ____

$R_1 = 2.2\ \text{k}\Omega$ $I_T =$ ____ $V_1 =$ ____ $P_1 =$ ____

$R_2 = 1.0\ \text{k}\Omega$ $I_1 =$ ____ $V_2 =$ ____ $P_2 =$ ____

$R_3 = 3.3\ \text{k}\Omega$ $I_2 =$ ____ $V_3 =$ ____ $P_3 =$ ____

c. $V_T = 11\ \text{V}$ Find: $R_T =$ ____ $I_3 =$ ____ $P_T =$ ____

$R_1 = 330\ \Omega$ $I_T =$ ____ $V_1 =$ ____ $P_1 =$ ____

$R_2 = 330\ \Omega$ $I_1 =$ ____ $V_2 =$ ____ $P_2 =$ ____

$R_3 = 330\ \Omega$ $I_2 =$ ____ $V_3 =$ ____ $P_3 =$ ____

7. Find the unknown values from the information given for the following problems.

a. P = ______	**b.** P = 360 mW	**c.** P = 102 mW
V = 9 V	V = ______	V = 6 V
I = 20 mA	I = 30 mA	I = ______
d. P = ______	**e.** P = ______	**f.** P = 1.08 W
V = 9 V	I = 10 mA	V = ______
R = 500 Ω	R = 2.2 kΩ	R = 1.2 kΩ
g. P = 1.35 W	**h.** P = 2.24 W	**i.** P = 225 mW
V = 45 V	I = ______	I = 3 mA
R = ______	R = 5.6 kΩ	R = ______

8. Find the value of R_1 in Figure 4-12. *Hint:* $R_1 = V_1/I_1$, $I_T = I_1 = I_2 = I_3$, $I_T = V_2/R_2$, $V_3 = I_3R_3$, and $V_1 = V_T - (V_2 + V_3)$.

R_1 = ______ Ω

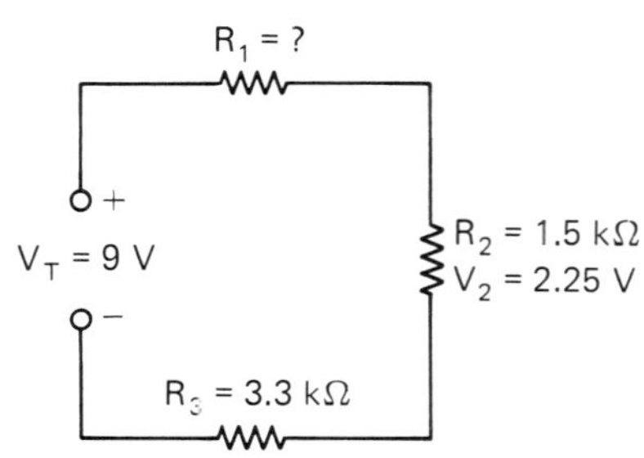

Figure 4–12

9. Find the values of V_T, I_T, and R_3 in Figure 4-13. *Hint:* $V_2 = I_2R_2$, $V_T = V_1 = V_2 = V_3$, $I_1 = V_1/R_1$, $I_T = I_1 + I_2 + I_3$, and $R_3 = V_3/I_3$.

V_T = ______

I_T = ______

R_3 = ______ Ω

Figure 4–13

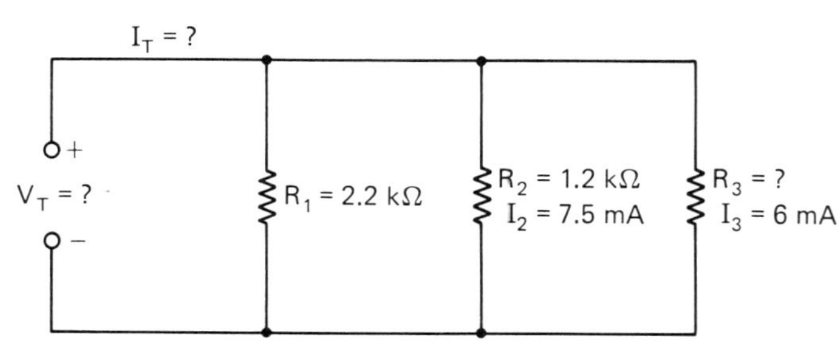

10. Draw the schematic symbol for a fuse.

11. Draw the schematic symbol for ground reference.

12. Indicate S for series circuit and P for parallel circuit next to each formula that applies. Some formulas may be used for both types of circuits.

____ **a.** $V_T = V_1 = V_2 = V_3$ ____ **b.** $R_T = R_1 + R_2 + R_3$

____ **c.** $I_T = V_T/R_T$ ____ **d.** $V_T = V_1 + V_2 + V_3$

____ **e.** $P_T = P_1 + P_2 + P_3$ ____ **f.** $I_T = I_1 = I_2 = I_3$

____ **g.** $P_T = V_T I_T$ ____ **h.** $I_T = I_1 + I_2 + I_3$

____ **i.** $R_T = \dfrac{1}{\dfrac{1}{R_1} + \dfrac{1}{R_2} + \dfrac{1}{R_3}}$

SECTION 4-4 EXPERIMENTS

EXPERIMENT 1. RESISTORS IN SERIES

Objective: To prove the parameters for a circuit with resistors in series and to verify Ohm's law.

Introduction: In this experiment, you will verify the following formulas: $R_T = R_1 + R_2 + R_3$, $I_T = V_T/R_T$, $V_T = V_1 + V_2 + V_3$, and $I_T = I_1 = I_2 = I_3$. You will also calculate the power dissipated by the resistors.

Materials Needed:
1 100-Ω ± 10% resistor at 0.5 W
1 330-Ω ± 10% resistor at 0.5 W
1 470-Ω ± 10% resistor at 0.5 W
1 1.5-kΩ ± 10% resistor at 0.5 W
1 2.2-kΩ ± 10% resistor at 0.5 W
1 3.3-kΩ ± 10% resistor at 0.5 W
1 Breadboard for constructing circuit and appropriate wires or leads

Equipment Required:
1 Variable power supply up to 30 V
1 Multimeter or DVM

Procedure:

1. In the space below, draw the circuit shown in Figure 4-3a.

2. Measure the ohmic value of all the resistors to be sure that they are within tolerance.
3. Construct the circuit using the values shown but without the battery or any power source.
4. Calculate the total resistance of the circuit.

$$R_T = R_1 + R_2 + R_3 = \underline{\hspace{6cm}}\ \Omega$$

5. Place an ohmmeter across the points where the power source will be connected. List the reading here: ______ Ω. Is the reading the same as the calculated value? ______________________________

   Why? ____

6. Remove the ohmmeter from the circuit and calculate the following formulas:

$I_T = V_T/R_T =$ ________________ $I_T = I_1 = I_2 = I_3$

$I_1 =$ ______

$I_2 =$ ______

$I_3 =$ ______

$P_T = V_T I_T =$ ______

$V_1 = I_1 R_2 =$ ______ $P_1 = V_1 I_1 =$ ______

$V_2 = I_2 R_2 =$ ____ $P_2 = V_2 I_2 =$ ______

$V_3 = I_3 R_3 =$ ______ $P_3 = V_3 I_3 =$ ______

$V_T = V_1 + V_2 + V_3 =$ ________ $P_T = P_1 + P_2 + P_3 =$ ________

7. Connect the power supply to the points on the circuit shown by the battery.
8. Place a voltmeter across the power supply output terminals. Turn on the power supply and adjust the output voltage for +9 V.
9. Measure the voltage drop across each resistor (observe proper polarity).

$V_1 =$ ______ $V_2 =$ ______ $V_3 =$ ______

10. Turn off the power supply. Open the circuit between the + V terminal of the power supply and R_1 and connect an ammeter. Turn on the power supply and read the current: $I_1 =$ ____ .
11. Turn off the power supply, remove the meter, and reconnect the original circuit.
12. Open the circuit between R_1 and R_2. Connect the ammeter. Turn on the power supply and read the current: $I_2 =$ ____ .
13. Turn off the power supply, remove the ammeter, and reconnect the original circuit.
14. Open the circuit between R_2 and R_3. Connect the ammeter, turn on the power supply, and read the current: $I_3 =$ ____ .
15. Turn off the power supply, remove the ammeter, and reconnect the original circuit. Calculate the following power formulas:

$P_1 = V_1 I_1 =$ ____ $P_2 = V_2 I_2 =$ ____ $P_3 = V_3 I_3 =$ ____

$P_T = V_1 + V_2 + V_3 =$ ____

16. Place the voltmeter across the power supply output terminals. Turn on the power supply and adjust the output voltage for +12 V.
17. Repeat steps 9 through 15, but list the values of the readings and calculations here.

$V_1 =$ ____________ $I_1 =$ ____________ $P_1 =$ ____________

$V_2 =$ ____________ $I_2 =$ ____________ $P_2 =$ ____________

$V_3 =$ ____________ $I_3 =$ ____________ $P_3 =$ ____________

$V_T =$ ____________ $I_T =$ ____________ $P_T =$ ____________

18. Did these readings change from the readings taken in the first circuit? __________ Why?

19. Does this verify the Ohm's law voltage-to-current relationship? ________ Explain.

20. Construct a new series circuit on the breadboard using the other resistors. Let $R_1 = 1.5\ \text{k}\Omega$, $R_2 = 2.2\ \text{k}\Omega$, and $R_3 = 3.3\ \text{k}\Omega$.
21. Calculate R_T and I_T.

$R_T =$ ________________, $I_T =$ ________________

22. Repeat steps 8 through 15, but list the values of the readings and calculations here.

$V_1 =$ __________ $I_1 =$ __________ $P_1 =$ __________

$V_2 =$ __________ $I_2 =$ __________ $P_2 =$ __________

$V_3 =$ __________ $I_3 =$ __________ $P_3 =$ __________

$V_T =$ __________ $I_T =$ __________ $P_T =$ __________

23. Does this verify the Ohm's law current-to-resistance relationship? ______ Explain.

24. Plot an I versus V graph for $R_T = 900\ \Omega$ and $R_T = 7000\ \Omega$ using the data you collected in the experiment. Draw the plots on the graph shown in Figure 4-14.

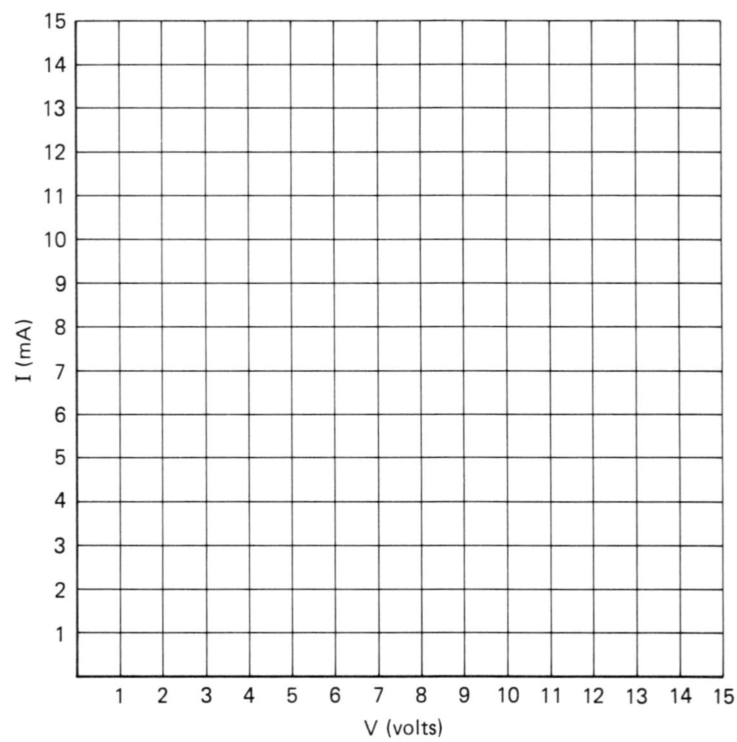

Figure 4–14 *I* versus *V* graph.

Fill-In-Questions:

1. The total resistance of a series circuit equals the ______ of all the resistors in the circuit.
2. The current in a circuit is directly proportional to the ______ across the circuit and inversely proportional to the ______ of the circuit.
3. The current in a series circuit is the ______ throughout the circuit.
4. The total voltage applied to a series circuit is equal to the ______ of the voltage drops around the circuit.
5. The amount of voltage dropped by a resistor is directly proportional to its __________ .
6. A larger-value resistor will drop more ______ in a series circuit than will a smaller-value resistor.
7. The total power dissipated by a series circuit is the ______ of all the individual powers consumed in the circuit.

EXPERIMENT 2. RESISTORS IN PARALLEL

Objective: To prove the parameters for a circuit with resistors in parallel and to prove Ohm's law.

Introduction: In this experiment you will verify the following formulas:

$$R_T = \frac{1}{\frac{1}{R_1} + \frac{1}{R_2} + \frac{1}{R_3}} \qquad I_T = \frac{V_T}{R_T}$$

$$V_T = V_1 = V_2 = V_3 \qquad I_T = I_1 + I_2 + I_3$$

You will also calculate the power dissipated by the resistors.

Materials Needed:

1 100-Ω ± 10% resistor at 0.5 W
1 330-Ω ± 10% resistor at 0.5 W
1 470-Ω ± 10% resistor at 0.5 W
1 1.5-kΩ ± 10% resistor at 0.5 W
1 2.2 kΩ ± 10% resistor at 0.5 W
1 3.3-kΩ ± 10% resistor at 0.5 W
1 Breadboard for contructing circuit and appropriate wires and leads

Equipment Required:

1 Variable power supply up to 30 V
1 Multimeter or DVM

Procedure:

1. In the space below, draw the circuit shown in Figure 4-5.

2. Measure the ohmic value of all the resistors to be sure that they are within tolerance.
3. Construct the circuit using the values shown but without a power source.
4. Calculate the total resistance of the circuit.

$$R_T = \frac{1}{\frac{1}{R_1} + \frac{1}{R_2} + \frac{1}{R_3}} = \underline{\qquad\qquad} \Omega$$

5. Place an ohmmeter across the points where the power source will be connected. List the reading here: ______________ Ω. Is the reading the same as the calculated value? _____ Why?

6. Remove the ohmmeter from the circuit and calculate the following formulas.

$I_T = \frac{R_T}{R_T} = \underline{\qquad\qquad}$ $V_T = V_1 = V_2 = V_3$

$V_1 = \underline{\qquad\qquad}$

$V_2 = \underline{\qquad\qquad}$

$V_3 = \underline{\qquad\qquad}$ $P_T = V_T I_T = \underline{\qquad\qquad}$

$I_1 = \frac{V_1}{R_T} = \underline{\qquad\qquad}$ $P_1 = V_1 I_1 = \underline{\qquad\qquad}$

$I_2 = \frac{V_2}{R_2} = \underline{\qquad\qquad}$ $P_2 = V_2 I_2 = \underline{\qquad\qquad}$

$I_3 = \frac{V_3}{R_3} = \underline{\qquad\qquad}$ $P_3 = V_3 I_3 = \underline{\qquad\qquad}$

$I_T = I_1 + I_2 + I_3 = \underline{\qquad\qquad}$ $P_T = P_1 + P_2 + P_3 = \underline{\qquad\qquad}$

7. Connect the power supply to the points on the circuit indicated by + and −.
8. Place a voltmeter across the power supply output terminals. Turn on the power supply and adjust the output voltage for +9 V.
9. Measure the voltage across each resistor.

$V_1 =$ ____________ $V_2 =$ ____________ $V_3 =$ ____________

10. Turn off the power supply. Open the circuit at the bottom of R_1 and connect an ammeter. Turn on the power supply and read the current: $I_1 =$ ____________ .
11. Turn off the power supply, remove the meter, and reconnect the original circuit.
12. Open the circuit at the bottom of R_2 and connect an ammeter. Turn on the power supply and read the current: $I_2 =$ ____________ .
13. Turn off the power supply, remove the meter, and reconnect the original circuit.
14. Open the circuit at the bottom of R_3 and connect an ammeter. Turn on the power supply and read the current: $I_3 =$ ____________ .
15. Turn off the power supply, remove the meter, and reconnect the original circuit.
16. Open the circuit to the + terminal of the power supply and connect an ammeter. Turn on the power supply and read the current: $I_T =$ ____________ .
17. Turn off the power supply and connect the original circuit.
18. Place the voltmeter across the power supply output terminals. Turn on the power supply and adjust the output voltage for +12 V.
19. Repeat steps 9 through 16, but list the values of the readings and calculations here.

$V_1 =$ ____________ $I_1 =$ ____________ $P_1 =$ ____________

$V_2 =$ ____________ $I_2 =$ ____________ $P_2 =$ ____________

$V_3 =$ ____________ $I_3 =$ ____________ $P_3 =$ ____________

$V_T =$ ____________ $I_T =$ ____________ $P_T =$ ____________

Notice that the same effects are observed when the applied voltage is increased to a series circuit or a parallel circuit.

20. Construct a new parallel circuit on the breadboard using the other resistors. Let $R_1 = 1.5\ \text{k}\Omega$, $R_2 = 2.2\ \text{k}\Omega$, and $R_3 = 3.3\ \text{k}\Omega$.

21. Calculate the following values:

$R_T =$ ________ $I_1 =$ ________

$I_2 =$ ________

$I_3 =$ ________

$I_T =$ ________

22. Repeat steps 5 through 16, but list the values of the readings and calculations here.

$V_1 =$ ________ $I_1 =$ ________ $P_1 =$ ________

$V_2 =$ ________ $I_2 =$ ________ $P_2 =$ ________

$V_3 =$ ________ $I_3 =$ ________ $P_3 =$ ________

$V_T =$ ________ $I_T =$ ________ $P_T =$ ________

Notice that if the total resistance of a parallel circuit increases, the current decreases.

Fill-In Questions:

1. The total resistance of a parallel circuit will be ________ than the smallest-value resistor.

2. The total current in a parallel circuit is equal to the ________ of all the branch currents.

3. The voltage of a parallel circuit is the ________ across each branch.

4. The total power dissipated by a parallel circuit is

the ________ of all the individual powers consumed in the circuit.

5. If the voltage to a circuit (whether it be series or parallel)

increases, the current in the circuit will ________ .

6. If the total resistance of a circuit decreases, the current in the

circuit will ________ .

SECTION 4-5 TROUBLESHOOTING APPLICATION: TESTING A BLOWN FUSE

Sometimes when electronic apparatus is turned on, nothing happens. The device will not respond and appears completely "dead." In most cases, the fuse is blown, creating an open circuit. One possible cause for this condition is that the power applied to the device experienced a voltage surge of sufficient amount to cause the fuse to operate. This is one reason for the use of the fuse: to protect the circuits and components of the device. Another cause for this condition is that some component in the device decreased in resistance, or shorted out.

This troubleshooting application involves blowing two fuses so that each condition mentioned above can be demonstrated. You will need three specific components: two fuses rated at $\frac{1}{4}$ A (0.250 mA) and two 50-Ω resistors rated at 10 W. Also needed is a SPST switch and a variable power supply with a voltage at least up to +15 V. A fuse holder would be appropriate to use, but clip leads can be used to connect the fuse in the circuit.

With the power supply off, connect the circuit shown in Figure 4-15. A voltmeter will be placed across the fuse later. The other resistor

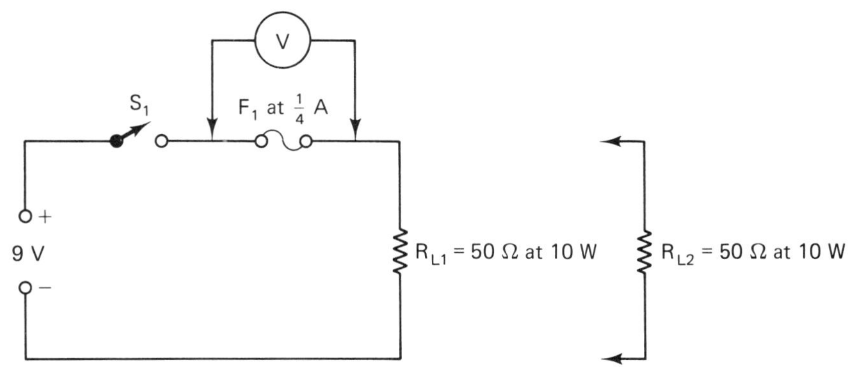

Figure 4–15 Testing a blown fuse.

R_{L_2} will be connected momentarily later. Make sure that the power supply output voltage control is at zero.

Turn on the power supply and slowly increase the voltage output control until +9 V is reached. Place the voltmeter across the fuse. This voltage should read 0 volts because the fuse is acting as an ideal conductor and there is no potential difference across it. With the voltmeter still across the fuse, advance the voltage control on the power supply to maximum. With this voltage increase, the fuse should have blown and the voltage across it will be reading the maximum voltage output of the power supply.

Return the output voltage control on the power supply to zero. Replace the blown fuse with the other good fuse. Again adjust the voltage to the circuit for +9 V. Place the voltmeter across the fuse and it should read 0 volts.

Now take the second 50-Ω resistor and place it across the 50-Ω resistor in the circuit (leads touching leads). This will cause the circuit resistance to decrease, the circuit current to increase, and the fuse should blow. The voltmeter should also be reading the power supply output voltage. As you can see, a good fuse (short circuit) will measure 0 volts across it and a blown fuse (open circuit) should measure the applied voltage.

Sometimes it is difficult and dangerous to make voltage readings across a fuse. The device can then be turned off and the fuse removed to be tested by an ohmmeter. A good fuse will measure zero resistance and a blown fuse will measure infinite resistance.

SECTION 4-6 INSTANT REVIEW

- Ohm's law states: The current in a circuit is directly proportional to the voltage across the circuit and inversely proportional to resistance of the circuit. $I = V/R$.
- The total resistance of a series circuit is the sum of its individual resistances: $R_T = R_1 + R_2 + R_3 + \cdots$.
- The total current in a series circuit is the same throughout the circuit: $I_T = I_1 = I_2 = I_3 = \cdots$.
- The sum of the voltage drops around a series circuit is equal to the applied (total) voltage: $V_T = V_1 + V_2 + V_3 - \cdots$.
- Kirshhoff's voltage law states: The sum of the voltages around a closed-loop circuit equals zero: $+V_T + (-V_1) + (-V_2) + (-V_3) + \cdots = 0$.
- The total resistance of a parallel circuit is always less than the smallest branch resistance.
- The formula for total resistance for more than two resistors in parallel is

$$R_T = \frac{1}{\frac{1}{R_1} + \frac{1}{R_2} + \frac{1}{R_3}} \cdots$$

- The formula for total resistance for equal-value resistors in parallel is $R_T = R/N_R$.
- The formula for total resistance when two resistors are in parallel is:

$$R_T = \frac{R_1 \times R_2}{R_1 + R_2}$$

- The total current in a parallel circuit is the sum of its individual branch currents: $I_T = I_1 + I_2 + I_3 + \cdots$.
- Kirchhoff's current law states: The sum of the currents entering and leaving a node or point in a conductor equals zero: $+I_T + (-I_1) + (-I_2) + (-I_3) + \cdots = 0$. A node is a specific reference point on a circuit.
- The total voltage is the same across each branch of a parallel circuit: $V_T = V_1 = V_2 = V_3 = \cdots$
- Work is the process of converting energy from one form to another form.
- A joule is a unit of work and expressed in electrical terms as 1 coulomb of electrons moving through a potential difference of 1 volt.
- Power is the rate of doing work and electrical power is defined as voltage times current: $P = VI$.
- The watt is the unit for electrical power.
- Current flowing through a resistive device develops a voltage drop. The end where the current exists is more positive than the end where it enters, which is more negative.

- A decrease in resistance causes an increase in current, which produces what is called a loading effect.
- An overload occurs when too much current flows in a circuit.
- A variable power supply can have positive and negative voltage output that is adjustable. It may also have one or more fixed voltages.
- A fuse is a circuit protector that melts, causing an open circuit when the circuit current exceeds the fuse current rating.
- A slow-blow fuse is used in a circuit so that momentary voltage surges will not affect the circuit.
- A ground connection is used as a common or reference point in a circuit.

SECTION 4-7 SELF-CHECKING QUIZ

Circle the most correct answer for each question. Questions 1 through 4 refer to Figure 4-16.

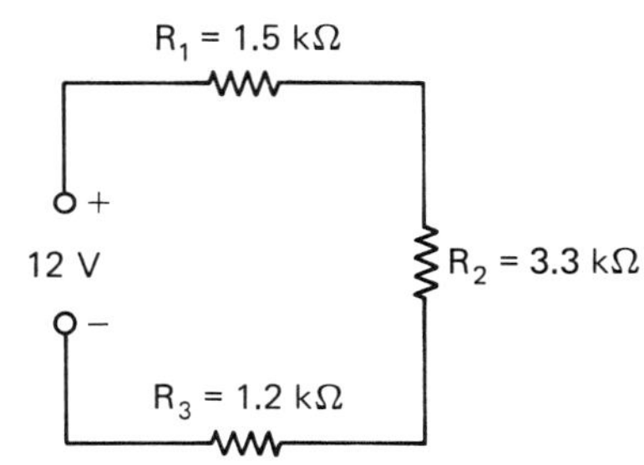

Figure 4–16

1. The total resistance of the circuit is:

a. 2 kΩ **b.** 555 Ω **c.** 12 kΩ **d.** none of the above

2. The total current of the circuit is:

a. 1 mA **b.** 2 mA **c.** 6 mA **d.** 1.2 A

3. The total power of the circuit is:

a. 1.2 W **b.** 6 mW **c.** 24 mW **d.** 2 mW

4. The voltage across R_2 is:

a. 12 V **b.** 6.6 V **c.** 3.3 V **d.** none of the above

Questions 5 through 8 refer to Figure 4-17.

Figure 4–17

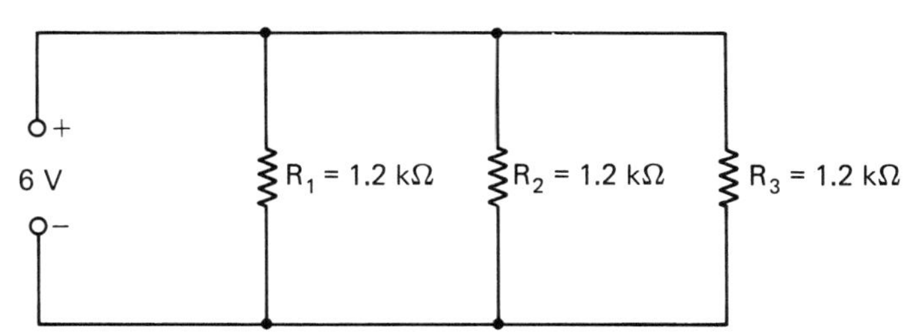

5. The total resistance of the circuit is:

a. 1.2 kΩ **b.** 800 Ω
c. 2.4 kΩ **d.** 400 Ω

6. The total current in the circuit is:

a. 15 mA **b.** 5 mA
c. 3 mA **d.** none of the above

7. The current through R_2 is:

a. 3 mA **b.** 1.2 mA
c. 6 mA **d.** 5 mA

8. The power dissipated by R_3 is:

a. 6 mW **b.** 30 mW
c. 90 mW **d.** 15 mW

9. The rate of doing work is called:

a. joule **b.** energy
c. power **d.** watt

10. The difference of potential that develops when current flows through a resistive device is called a:

a. voltage rise **b.** voltage drop
c. negative voltage **d.** voltage surge

11. The power dissipated by a registor is 61 mW, and the current flowing through it is 3 mA. The value of the resistor is approximately:

a. 6.8 kΩ **b.** 5.6 kΩ
c. 4.7 kΩ **d.** 3.3 kΩ

12. Referring to Figure 4-18, the value of R_3 is:

a. 1.2 kΩ **b.** 3.3 kΩ
c. 4.7 kΩ **d.** 1.5 kΩ

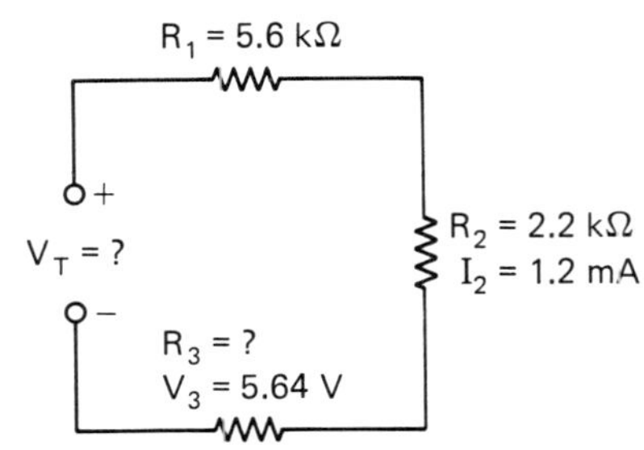

Figure 4–18

13. Referring to Figure 4-18, the value of V_T is:

a. 9 V **b.** 15 V
c. 12.5 V **d.** 6 V

14. When a device is placed in parallel with another device and the current increases, it is referred to as a:

a. voltage drop **b.** power loss
c. voltage surge **d.** loading effect

15. Refer to Figure 4-19. The total voltage is:

a. 6 V **b.** 9 V
c. 12 V **d.** 50 V

16. Refer to Figure 4-19. The total current is approximately:

a. 9 mA **b.** 12 mA
c. 15 mA **d.** 23 mA

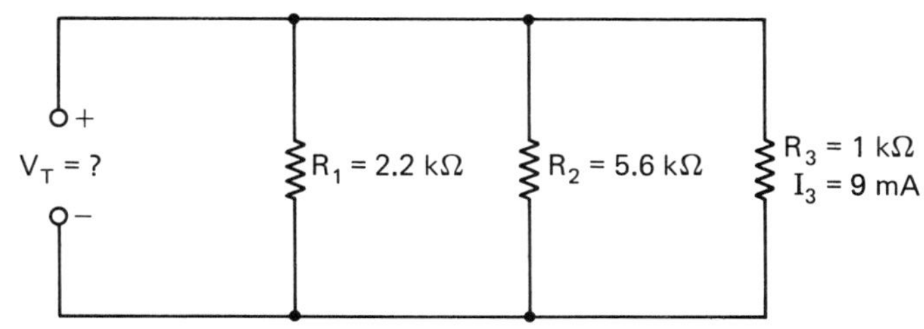

Figure 4–19

17. The power dissipated by a resistor is 36 mW and the voltage across it is 45 V. The value of the resistor is approximately:

a. 56 kΩ **b.** 68 kΩ
c. 4.7 kΩ **d.** 1.2 kΩ

18. Two resistors are in parallel; R_1 = 2.2 kΩ and R_2 = 3.3 kΩ. Their total resistance is:

a. 1.32 kΩ **b.** 5.5 kΩ
c. 2.75 kΩ **d.** 3.63 kΩ

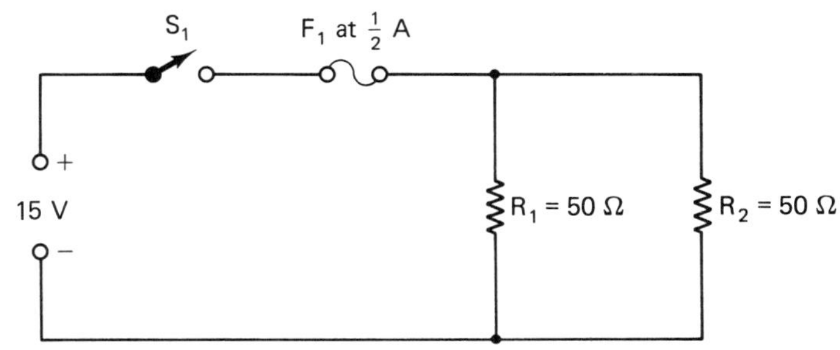

Figure 4–20

19. Refer to Figure 4-20. When switch S_1 is closed, the total current will equal:

a. 600 mA **b.** 150 mA
c. 0 mA **d.** none of the above

20. A 1.5-kΩ resistor dissipates 216 mW of power. The voltage across it and the current through it, respectively, are:

a. 9 V and 24 mA
b. 12 V and 18 mA
c. 12 V and 14.4 mA
d. 18 V and 12 mA

ANSWERS TO EXPERIMENTS AND SELF-CHECKING QUIZ

Experiment 1: **(1)** sum **(2)** voltage, resistance **(3)** same **(4)** sum **(5)** resistance **(6)** voltage **(7)** sum

Experiment 2: **(1)** less **(2)** sum **(3)** same **(4)** sum **(5)** increase **(6)** increase

Self-Checking Quiz: **(1)** d **(2)** b **(3)** c **(4)** b **(5)** d **(6)** a **(7)** d **(8)** b **(9)** c **(10)** b **(11)** a **(12)** c **(13)** b **(14)** d **(15)** b **(16)** c **(17)** a **(18)** a **(19)** c **(20)** d

Unit 5

Combination Resistor Circuits

INTRODUCTION

Complex resistive circuits may, at first, look very complicated; however, they can be simplified by following a few basic steps. By applying the basic Ohm's law and Kirchhoff's laws to small segments of the total circuit, all parameters of the circuit can be calculated.

UNIT OBJECTIVES

Upon completion of this unit, you will be able to:

1. Solve complex resistive circuits.
2. Calculate equivalent resistances.
3. Calculate all currents in a circuit.
4. Calculate all voltage drops in a circuit.
5. Calculate all power dissipated by a circuit.
6. Construct and test complex resistive circuits.
7. Show the difference between a balanced bridge circuit and an unbalanced bridge circuit.
8. Design positive and negative voltage dividers.
9. Recognize the loading effects of electronic test equipment.
10. Troubleshoot open and short circuits.

SECTION 5-1 FUNDAMETAL CONCEPTS

5-1a SERIES–PARALLEL CIRCUIT

The circuit shown in Figure 5-1a is a series–parallel combination circuit. Resistor R_1 is in series with the parallel combination of R_2 and R_3, which, in turn, are in series with R_4. To find the total current for the circuit, we must first calculate the total resistance. In this case we begin by finding the equivalent resistance (R_e) of R_2 and R_3 (see Figure 5-1b). Using the product-over-sum formula for two resistors in parallel, we have

$$R_e = \frac{R_2 \times R_3}{R_2 + R_3} = \frac{3.3\text{ k}\Omega \times 6.8\text{ k}\Omega}{3.3\text{ k}\Omega + 6.8\text{ k}\Omega} = \frac{3.3 \times 10^3 \times 6.8 \times 10^3}{3.3 \times 10^3 + 6.8 \times 10^3}$$

$$= \frac{22.44 \times 10^6}{10.1 \times 10^3} = 2.22 \times 10^3 = 2.22\text{ k}\Omega$$

We can now find the total resistance:

$$R_T = R_1 + R_e + \text{R}_4 = 4.7\text{ k}\Omega + 2.22\text{ k}\Omega + 2.2\text{ k}\Omega$$

$$= 4.7 \times 10^3 + 2.22 \times 10^3 = 9.12 \times 10^3 = 9.12\text{ k}\Omega$$

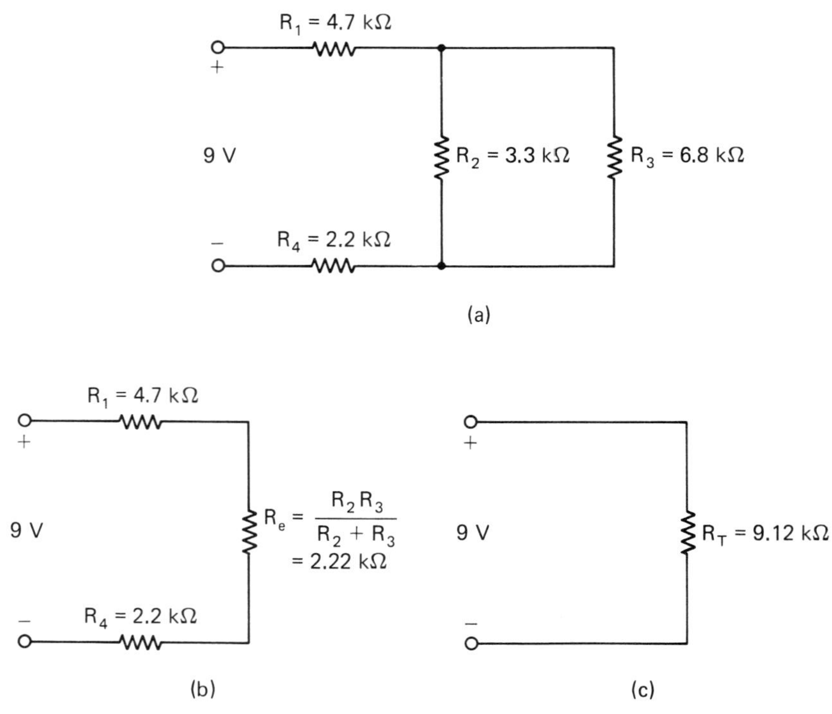

Figure 5–1 Series–parallel circuit: (a) normal circuit; (b) equivalent resistance of R_2 and R_3; (c) total resistance.

As far as the voltage source is concerned, it "sees" a single resistance of 9.12 kΩ, as shown in Figure 5-1c. The total current for the circuit is

$$I_T = \frac{V_T}{R_T} = \frac{9\text{ V}}{9.12\text{ k}\Omega} = \frac{9}{9.12 \times 10^3} = 0.9868 \times 10^{-3} = 0.987\text{ mA}$$

or nearly 1 mA.

The total current flows through R_1 and R_4; therefore, their voltage drops can be easily calculated as

$$V_1 = I_T R_1 = 0.987\text{ mA} \times 4.7\text{ k}\Omega = 0.987 \times 10^{-3} \times 4.7 \times 10^3 = 4.639\text{ V}$$

$$V_4 = I_T R_4 = 0.987\text{ mA} \times 2.2\text{ k}\Omega = 0.987 \times 10^{-3} \times 2.2 \times 10^3 = 2.171\text{ V}$$

Since R_2 and R_3 are in parallel, the voltage drop can be found by adding V_1 to V_4 and then subtracting this sum from V_T, thus:

$$V_2 \text{ or } V_3 = V_T - (V_1 + V_4) = 9\text{ V} - (4.639\text{ V} + 2.271\text{ V})$$
$$= 9\text{V} - 6.81 = 2.19\text{ V}$$

The current through R_2 and R_3 can now be found as

$$I_2 = \frac{V_2}{R_2} = \frac{2.19\text{ V}}{3.3\text{ k}\Omega} = \frac{2.19\text{ V}}{3.3 \times 10^3} = 0.664 \times 10^{-3} = 0.664\text{ mA}$$

$$I_3 = \frac{V_3}{R_3} = \frac{2.19\text{ V}}{6.8\text{ k}\Omega} = \frac{2.19\text{ V}}{6.8 \times 10^3} = 0.322 \times 10^{-3} = 0.322\text{ mA}$$

The sum of these two currents should approximately equal I_T. Any slight differences is because of the rounding off of the numbers.

The electron current flow in this circuit can be summarized as follows: The I_T of 0.987 mA is leaving the negative terminal of the voltage source and flows through R_4 toward the bottom junction of R_2 and R_3. At this point the I_T divides inversely proportionally to the resistance of the two branches, with 0.664 mA (I_2) flowing through R_2 and 0.332 mA (I_3) flowing through R_3. These two currents recombine at the top junction of R_2 and R_3 to form the I_T of 0.987 mA, which flows through R_1 to the positive terminal of the voltage source.

Now that the voltage across each resistor and the current through each resistor are known, the power dissipated by each resistor can be calculated, thus:

$$P_1 = V_1 I_1 = 4.639\text{ V} \times 0.987\text{ mA} = 4.639 \times 0.987 \times 10^{-3} = 4.579 \times 10^{-3} = 4.579\text{ mW}$$

$$P_2 = V_2 I_2 = 2.19\text{ V} \times 0.644\text{ mA} = 2.19 \times 0.644 \times 10^{-3} = 1.41 \times 10^{-3} = 1.41\text{ mW}$$

$$P_3 = V_3 I_3 = 2.19 \times 0.332\text{ mA} = 2.19 \times 0.332 \times 10^{-3} = 0.727 \times 10^{-3} = 0.727\text{ mW}$$

$$P_4 = V_4 I_4 = 2.171 \times 0.987\text{ mA} = 2.171 \times 0.987 \times 10^{-3} = 2.143 \times 10^{-3} = \underline{2.143\text{ mW}}$$

$$P_T = P_1 + P_2 + P_3 + P_4 = 8.854\text{ mW}$$

To check the total power, we could multiply V_T by I_T:

$$P_T = V_T \times I_T = 9\text{ V} \times 0.987\text{ mA} = 9 \times 0.987 \times 10^{-3} = 8.883 \times 10^{-3} = 8.883\text{ mW}$$

5-1b PARALLEL–SERIES CIRCUIT

A parallel–series combination circuit is shown in Figure 5-2a. In this circuit, resistors R_1 and R_2 are in parallel and are in series with R_3, which, in turn, is in series with the parallel combination of R_4 and R_5. To calculate the total resistance of this circuit, we must first find the equivalent resistances of $R_1 \parallel R_2$ and $R_4 \parallel R_5$ (the symbol $\parallel$ between the R's indicates that they are in parallel). We can use the product-over-sum method for $R_1 \parallel R_2$:

$$R_{e_1} = \frac{R_1 \times R_2}{R_1 + R_2} = \frac{2.2\text{ k}\Omega \times 4.7\text{ k}\Omega}{2.2\text{ k}\Omega + 4.7\text{ k}\Omega} = \frac{2.2 \times 10^3 \times 4.7 \times 10^3}{2.2 \times 10^3 + 4.7 \times 10^3} = \frac{10.34 \times 10^6}{6.9 \times 10^3}$$
$$= 1.498 \times 10^3 = 1.498\text{ k}\Omega \quad \text{or} \quad \text{approximately } 1.5\text{ k}\Omega$$

We can use the equal-value resistor method for finding $R_4 \parallel R_5$:

$$R_{e_2} = \frac{R}{N_R} = \frac{1\text{ k}\Omega}{2} = \frac{1 \times 10^3}{2} = 0.5 \times 10^3 = 0.5\text{ k}\Omega \quad (\text{or } 500\ \Omega)$$

Figure 5-2b shows the equivalent resistances. The total resistance can now be found with the series resistor method:

$$R_T = R_{e_1} + R_3 + R_{e_2} = 1.498\text{ k}\Omega + 10\text{ k}\Omega + 0.5\text{ k}\Omega$$
$$= 1.498 \times 10^3 + 10 \times 10^3 = 11.998 \times 10^3\ 11.998\text{ k}\Omega \quad \text{or} \quad \text{approximately } 12\text{ k}\Omega$$

The voltage source "sees" a total resistance of 12 kΩ, as shown in Figure 5-2c.

The total current is calculated from the formula

$$I_T = \frac{V_T}{R_T} = \frac{12\text{ V}}{12\text{ k}\Omega} = \frac{12}{12 \times 10^3} = 1 \times 10^{-3} = 1\text{ mA}$$

Since $R_4 = R_5$, the total current divides evenly through each resistor and can be found thus:

$$I_4 = \frac{I_T}{2} = \frac{1\text{ mA}}{2} = \frac{1 \times 10^3}{2} = 0.5 \times 10^{-3} = 0.5\text{ mA}$$

$$I_5 = \frac{I_T}{2} = \frac{1\text{ mA}}{2} = \frac{1 \times 10^3}{2} = 0.5 \times 10^{-3} = 0.5\text{ mA}$$

The voltage drops of R_4 and R_5 are the same and can be found by

$$V_4 = I_4R_4 = 0.5\text{ mA} \times 1\text{ k}\Omega = 0.5 \times 10^{-3} \times 1 \times 10^3 = 0.5\text{ V}$$
$$V_5 = I_5R_5 = 0.5\text{ mA} \times 1\text{ k}\Omega = 0.5 \times 10^{-3} \times 10^3 = 0.5\text{ V}$$

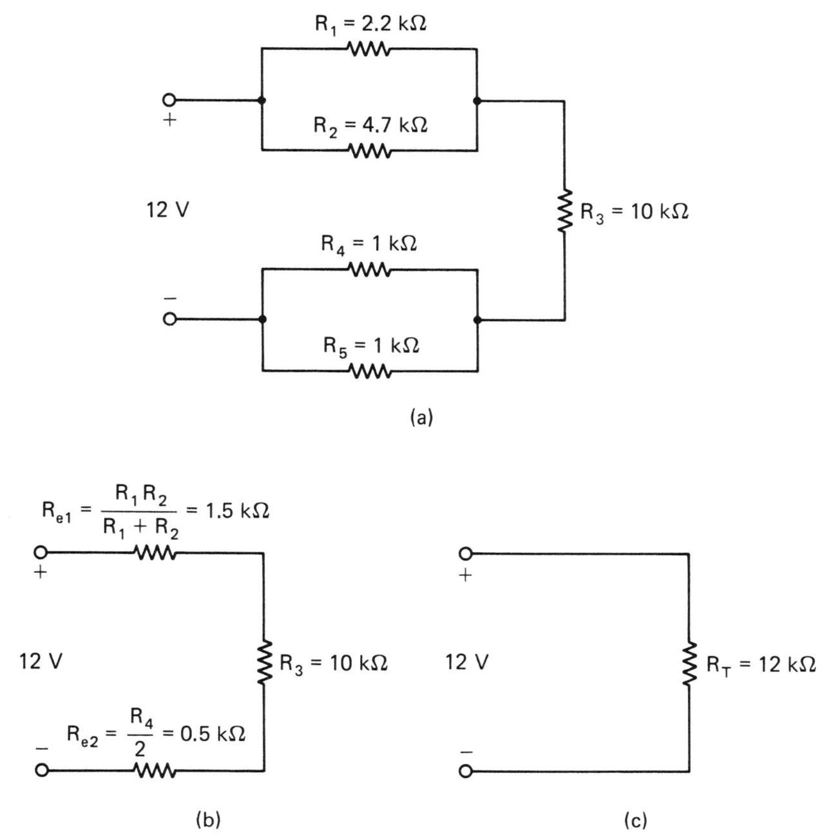

Figure 5–2 Parallel–series circuit: (a) normal circuit; (b) equivalent resistance of R_1, R_2 and R_4, R_5; (c) total resistance.

Since the total current flows through R_3 ($I_T = I_3$), its voltage drop can be found:

$$V_3 = I_3R_3 = 1 \text{ mA} \times 10 \text{ k}\Omega = 1 \times 10^{-3} \times 10^3 = 10 \text{ V}$$

The voltage drops across $R_1 \parallel R_2$ can now be calculated by the formulas

$$V_1 = V_T - (V_3 + V_4 \text{ or } V_5) = 12 \text{ V} - (10 \text{ V} + 0.5 \text{ V}) = 12 - 10.5 = 1.5 \text{ V}$$

$$V_2 = V_T - (V_3 + V_4 \text{ or } V_5) = 12\text{V} - (10 \text{ V} + 0.5 \text{ V}) = 12 - 10.5 = 1.5 \text{ V}$$

The currents flowing through R_1 and R_2 can now be calculated:

$$I_1 = \frac{V_1}{R_1} = \frac{1.5 \text{ V}}{2.2 \text{ k}\Omega} = \frac{1.5}{2.2 \times 10^3} = 0.682 \times 10^{-3} = 0.682 \text{ mA}$$

$$I_2 = \frac{V_2}{R_2} = \frac{1.5 \text{ V}}{4.7 \text{ k}\Omega} = \frac{1.5}{4.7 \times 10^3} = 0.319 \times 10^{-3} = 0.319 \text{ mA}$$

The sum of these currents should approximately equal I_T.

To summarize the electron current flow in this circuit: The I_T of 1 mA leaves the negative terminal of the voltage source and divides evenly at the left-side junction of R_4 and R_5 with 0.5 mA flowing through each branch. These two currents combine at the right-side junction of R_4 and R_5 to form the I_T of 1 mA, which flows through R_3. The current divides inversely proportional to the resistance at the right-side junction of R_1 and R_2 with 0.682 mA flowing through R_1 and 0.319 mA flowing through R_2. These currents recombine at the left-side junction of R_1 and R_2 to form the I_T of 1 mA, which flows toward the positive terminal of the voltage source.

The power dissipated by the individual resistors and the total circuit can be calculated as was done in the previous circuit:

$$
\begin{aligned}
P_1 &= V_1I_1 = 1.5\text{ V} \times 0.682\text{ mA} = 1.5 \times 0.682 \times 10^{-3} = 1.023 \times 10^{-3} = 1.023\text{ mW}\\
P_2 &= V_2I_2 = 1.5\text{V} \times 0.319\text{ mA} = 1.5 \times 0.319 \times 10^{-3} = 0.4785 \times 10^{-3} = 0.478\text{ mW}\\
P_3 &= V_3I_3 = 10\text{ V} \times 1\text{ mA} = 10 \times 1 \times 10^{-3} = 10 \times 10^{-3} = 10.000\text{ mW}\\
P_4 &= V_4I_4 = 0.5\text{ V} \times 0.5\text{ mA} = 0.5 \times 0.5 \times 10^{-3} = 0.25 \times 10^{-3} = 0.250\text{ mW}\\
P_5 &= V_5I_5 = 0.5\text{ V} \times 0.5\text{ mA} = 0.5 \times 0.5 \times 10^{-3} = 0.25 \times 10^{-3} = 0.250\text{ mW}\\
P_T &= P_1 + P_2 + P_3 + P_4 + P_5 = 12.00\text{ mW}
\end{aligned}
$$

To check the total power:

$$P_T = V_TI_T = 12\text{ V} \times 1\text{ mA} = 12\text{ V} \times 1 \times 10^{-3} = 12 \times 10^{-3} = 12\text{ mW}$$

Another parallel–series combination circuit is shown in Figure 5-3. We can keep the discussion and problem solving of this circuit simple, by realizing that the resistance will be in kilohms, the current in milliamperes, and the power in milliwatts. Now, let's begin to solve for all its parameters. We first find the R_{e1} of R_4 and R_5, as shown in Figure 5-3b, with the formula

$$R_{e1} = \frac{R_4 \times R_5}{R_4 + R_5} = \frac{5.6\text{ k}\Omega \times 3.3\text{ k}\Omega}{5.6\text{ k}\Omega + 3.3\text{ k}\Omega} = \frac{18.48\text{ k}\Omega}{8.9\text{ k}\Omega} = 2.076\text{ k}\Omega \approx 2\text{ k}\Omega$$

The R_{e2} of R_{e1} and R_3, as shown in Figure 5-3c, is found by the formula

$$R_{e2} = R_{e1} + R_3 = 2\text{ k}\Omega + 1\text{ k}\Omega = 3\text{ k}\Omega$$

Figure 5–3 More complex series–parallel circuit: (a) normal circuit; (b) equivalent resistance of R_4 and R_5; (c) equivalent resistance of R_1, R_2 and R_3, R_{e1}; (d) total resistance.

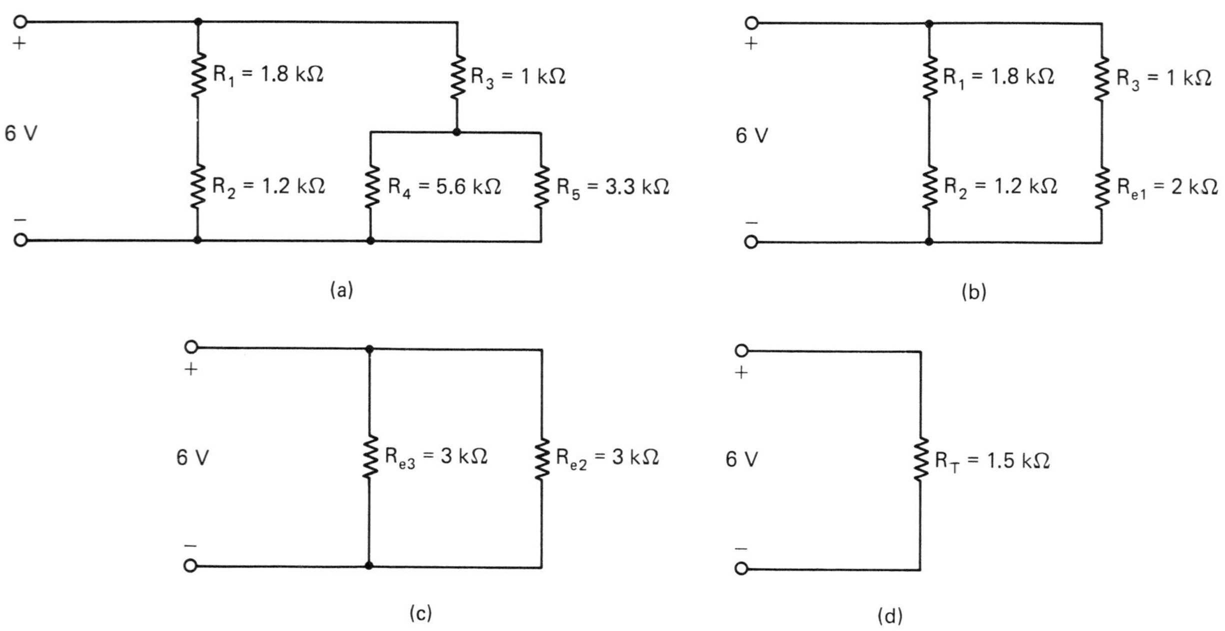

The total resistance R_T or R_{e3} and R_{e2}, shown in Figure 5-3d, is found by the formula

$$R_T = \frac{R_{e3}}{2} \text{ or } \frac{R_{e2}}{2} = \frac{3\ \text{k}\Omega}{2} = 1.5\ \text{k}\Omega$$

The I_T is found with the formula

$$I_T = \frac{V_T}{R_T} = \frac{6\ \text{V}}{1.5\ \text{K}\Omega} = 4\ \text{mA}$$

Since the resistance of the two major branches (see Figure 5-3b) are the same, each major branch has 2 mA of current; therefore, I_1 = 2 mA and I_2 = 2 mA. Then

$$V_1 = I_1R_1 = 2\ \text{mA} \times 1.8\ \text{k}\Omega = 3.6\ \text{V (the m and k cancel)}$$
$$V_2 = I_2R_2 = 2\ \text{mA} \times 1.2\ \text{k}\Omega = 2.4\ \text{V (the m and k cancel)}$$
$$V_3 = I_3R_3 = 2\ \text{mA} \times 1\ \text{k}\Omega = 2\ \text{V}$$
$$V_4 = V_5 = V_T - V_3 = 6\ \text{V} - 2\ \text{V} = 4\ \text{V}$$

Since R_4 and R_5 are in parallel, their voltage drops are the same. However, these two resistors are in series with R_3, which is across the total voltage V_T; therefore, the voltage drops V_4 and V_5 can be found thus:

$$V_4 \text{ or } V_5 = V_T - V_3 = 6\ \text{V} - 2\ \text{V} = 4\ \text{V}$$

Now the currents through R_4 and R_5 can be solved with the formulas

$$I_4 = \frac{V_4}{R_4} = \frac{4\ \text{V}}{5.6\ \text{k}\Omega} = 0.7143\ \text{mA} = 0.714\ \text{mA}$$
$$I_5 = \frac{V_5}{R_5} = \frac{4\ \text{V}}{3.3\ \text{k}\Omega} = 1.212\ \text{mA}$$

The sum of I_4 and I_5 should equal I_3.

$$I_3 = I_4 + I_5 = 0.71\ \text{mA} + 1.212\ \text{mA} = 1.926\ \text{mA}$$

The discrepancy here is because of rounding off the equivalent resistance of R_4 and R_5. The power dissipated by each resistor, and the total power, are:

$$\left.\begin{aligned} P_1 &= V_1I_1 = 3.6\ \text{V} \times 2\ \text{mA} = 7.2\ \text{mW} \\ P_2 &= V_2I_2 = 2.4\ \text{V} \times 2\ \text{mA} = 4.8\ \text{mW} \\ P_3 &= V_3I_3 = 2\ \text{V} \times 2\ \text{mA} = 4.0\ \text{mW} \\ P_4 &= V_4I_4 = 4\ \text{V} \times 0.715\ \text{mA} = 2.86\ \text{mW} \\ P_5 &= V_5I_5 = 4\ \text{V} \times 1.212\ \text{mA} = 4.848\ \text{mW} \end{aligned}\right\} + = 23.708\ \text{mW}$$
$$P_T = V_TI_T = 6\ \text{V} \times 4\ \text{mA} = 24\ \text{mW}$$

5-1c THE BRIDGE CIRCUIT

The bridge circuit is used in various applications of electronics and an understanding of its operation is useful for analyzing other circuits. A basic bridge circuit is nothing more than a simple parallel–series circuit, as shown in Figure 5-4a. The total resistance of each branch is 2 kΩ ($R_1 + R_2$ and $R_3 + R_4$); therefore, the current through each branch is 3 mA ($I = 6\text{ V}/2\text{ k}\Omega = 3\text{ mA}$). The voltage drop across each resistor is 3 V ($V = IR = 3\text{ mA} \times 1\text{ k}\Omega = 3\text{ V}$). Points A and B are at +3 V with reference to the negative terminal of the voltage source. If a voltmeter is placed across these points, it would indicate 0 V, because the difference in potential of +3 V and +3 V equals zero [+3 V − (+3 V) = 0 V].

Most often, the bridge circuit is drawn as shown in Figure 5-4b. This is still a balanced bridge value. The balanced condition exists because the ratio of R_1 to R_2 is equal to the ratio of R_3 to R_4:

$$\frac{R_1}{R_2} = \frac{R_3}{R_4} = \frac{100\ \Omega}{1000\ \Omega} = \frac{1000\ \Omega}{10{,}000\ \Omega}$$

When the numbers are reduced, we obtain

$$\frac{100\ \Omega}{1000\ \Omega} = \frac{1000\ \Omega}{10{,}000\ \Omega} = \frac{1\ \Omega}{10\ \Omega} = \frac{1\ \Omega}{10\ \Omega}$$

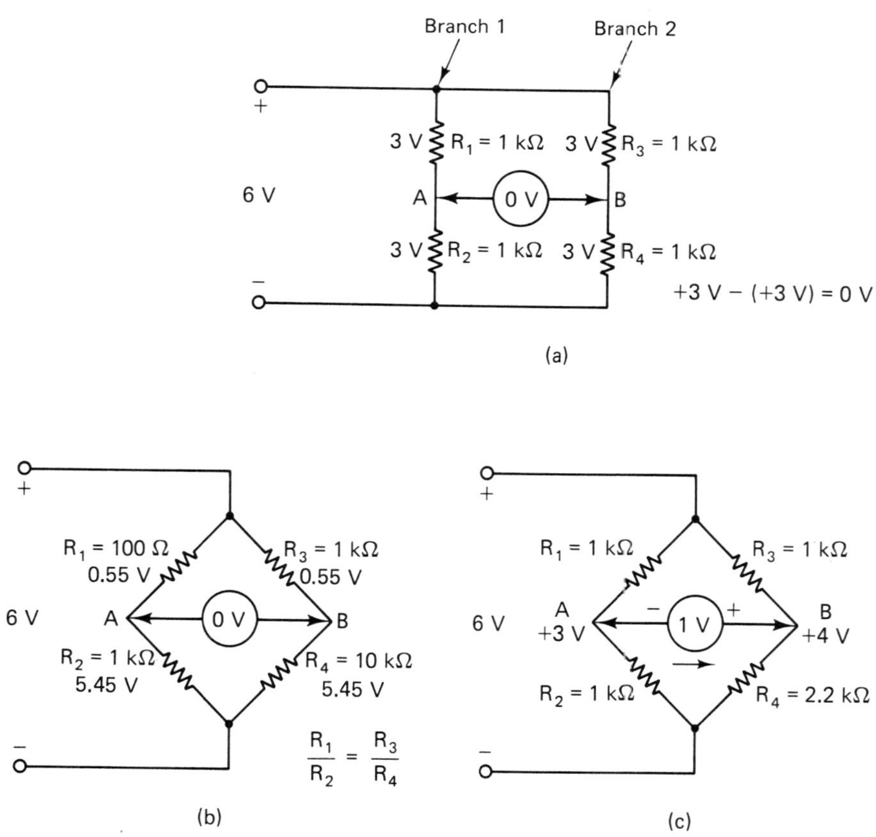

Figure 5–4 Bridge circuit: (a) balanced bridge with equal resistors; (b) balanced bridge with unequal resistors; (c) unbalanced bridge.

We can prove that the balanced condition exists by using Ohm's law. The current in the branch with R_1 and R_2 in series is the same through each resistor; therefore, let's find the current through R_2:

$$I_2 = \frac{V_T}{R_1 + R_2} = \frac{6\text{ V}}{100 + 1\text{ k}\Omega} = \frac{6\text{ V}}{1.1\text{ k}\Omega} = 5.45\text{ mA}$$

The branch with R_3 and R_4 in series is the same; thus

$$I_4 = \frac{V_T}{R_3 + R_4} = \frac{6\text{ V}}{1\text{ k}\Omega + 10\text{ k}\Omega} = \frac{6\text{V}}{11\text{ k}\Omega} = 0.545\text{ mA}$$

The voltage drops across V_2 and V_4 are found thus:

$$V_2 = I_2R_2 = 5.45\text{ mA} \times 1\text{ k}\Omega = 5.45\text{ V}$$
$$V_4 = I_4R_4 = 0.545\text{ mA} \times 10\text{ k}\Omega = 5.45\text{ V}$$

Since the voltages are equal, the voltage between points A and B is zero ($V_2 - V_4 = 0$, $5.45\text{ V} - 5.45\text{ V} = 0\text{ V}$).

The bridge is unbalanced when the ratio of the resistors is not equal, as shown in Figure 5-4c. The ratio of R_1/R_2 is not equal to the ratio R_3/R_4. Analyzing this circuit will enable you to better understand bridge circuits. First, let's find the current through each branch:

$$I_2 = \frac{V_T}{R_1 + R_2} = \frac{6\text{ V}}{1\text{ k}\Omega + 1\text{ k}\Omega} = \frac{6\text{ V}}{2\text{ k}\Omega} = 3\text{ mA}$$
$$I_4 = \frac{V_T}{R_3 + R_4} = \frac{6\text{ V}}{1\text{ k}\Omega + 2.2\text{ k}\Omega} = \frac{6\text{ V}}{3.3\text{ k}\Omega} = 1.818\text{ mA}$$

Therefore,

$$V_2 = I_2R_2 = 3\text{ mA} \times 1\text{ k}\Omega = 3\text{ V}$$
$$V_4 = I_4R_4 = 1.818\text{ mA} \times 2.2\text{ k}\Omega = 3.999\text{ V} \quad \text{or} \quad \approx 4.0\text{ V}$$

Using the negative terminal of the voltage source as a reference, point A is +3 V and point B is +4 V. If a high-impedance voltmeter is placed across these points, a small microampere current will flow from point A through the meter toward point B and will indicate a potential difference of 1 V.

5-1d VOLTAGE DIVIDERS

A basic series–parallel circuit application is the voltage divider, shown in Figure 5-5. The voltage divider is used at the output of a power supply to distribute various voltages to different circuits. An unloaded voltage divider is a series of resistors that drop voltage but do not have loads connected to them. When the loads are connected in parallel, the circuit parameters change. In other words, an unloaded voltage divider is of little use; therefore, we will consider how to design and use a voltage divider with loads.

Notice in Figure 5-5 the common ground connection. This ground could be a metallic chassis or a common bus wire. These ground sym-

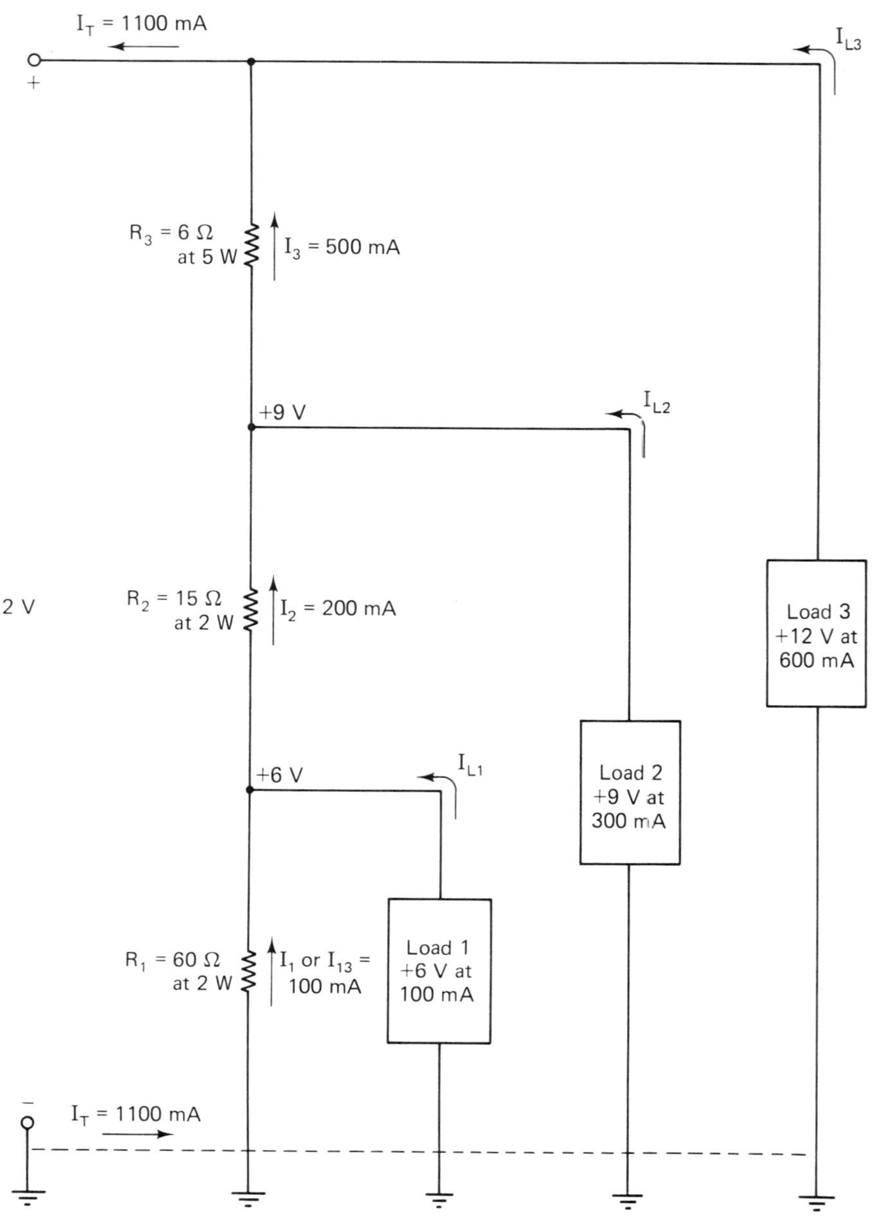

Figure 5–5 Voltage divider.

bols are used to simplify schematic drawings by eliminating the wire from the negative terminal of the voltage source. The dashed lines indicate the electron flow on this drawing. The circuit is supplied with a +12-V power source.

To construct a voltage divider, we have to know which voltages are required by the loads, and the current that each one will "draw" (*draw* is a term used to indicate how much current flows through a particular device)—in other words, the amount of current a device draws from the voltage source. The following procedures can be used to design the voltage divider in Figure 5-5.

Find the total current of all the loads:

$$I_{LT} = I_{L1} + I_{L2} + I_{L3} = 100 \text{ mA} + 300 \text{ mA} + 600 \text{ mA} = 1000 \text{ mA}$$

The current that flows through R_1 is called the *bleeder current* I_1 or I_B, and flows through all the other resistors in the divider. This resistor is often referred to as the bleeder resistor. Since the bleeder current does not flow through any load, its value is not critical and is usually set at 10% of the total load current:

$$I_B = 10\% \, I_{LT} = 0.1 \times 1000 \text{ mA} = 100 \text{ mA}$$

The top of R_1 must be +6 V in reference to ground to supply load 1; therefore, the value of R_1 is found by the formula

$$R_1 = \frac{V_{L1}}{I_B} = \frac{6 \text{ V}}{100 \text{ mA}} = \frac{6}{100 \times 10^{-3}} = 60 \ \Omega$$

The top of R_2 must be +9 V in reference to ground, but R_1 already drops 6 V; therefore,

$$V_2 = V_{L2} - V_{L1} = +9 \text{ V} - 6 \text{ V} = +3 \text{ V}$$

The currents that flow through R_2 are I_B and the current from load 1 is given as

$$I_2 = I_B + I_{L1} = 100 \text{ mA} + 100 \text{ mA} = 200 \text{ mA}$$

The resistance of R_2 can now be found by the formula

$$R_2 = \frac{V_2}{I_2} = \frac{3 \text{ V}}{200 \text{ mA}} = \frac{3 \text{ V}}{200 \times 10^{-3}} = 15 \ \Omega$$

or a composite formula can be given as

$$R_2 = \frac{V_{L2} - V_{L1}}{I_B + I_{L1}} = \frac{9 \text{ V} - 6 \text{ V}}{100 \text{ mA} + 100 \text{ mA}} = \frac{3 \text{ V}}{200 \text{ mA}} = 15 \ \Omega$$

The combined currents of I_B, I_{L1}, and I_{L2} flow through R_3 and the voltage dropped by this resistor is equal to the difference of the total voltage, less the voltage of load 2. The value of R_3 can be found by the formula

$$R_3 = \frac{V_T - V_{L2}}{I_2 + I_{L2}} = \frac{+12 \text{ V} - 9 \text{ V}}{200 \text{ mA} + 300 \text{ mA}} = \frac{+3 \text{ V}}{500 \text{ mA}} = \frac{+3 \text{ V}}{500 \times 10^{-3}} = 6 \ \Omega$$

The only remaining information needed is the power dissipated by each resistor, which tells us what size resistors to place in the voltage divider. These powers are:

$$P_1 = V_1 I_B = 6 \text{ V} \times 100 \text{ mA} = 600 \text{ mW}$$
$$P_2 = V_2 I_2 = 3 \text{ V} \times 200 \text{ mA} = 600 \text{ mW}$$
$$P_3 = V_3 I_3 = 3 \text{ V} \times 500 \text{ mA} = 1500 \text{ mW} \quad \text{or} \quad 1.5 \text{ W}$$

A positive and negative voltage with reference to ground can be developed by a voltage divider, as shown in Figure 5-6a. This circuit is

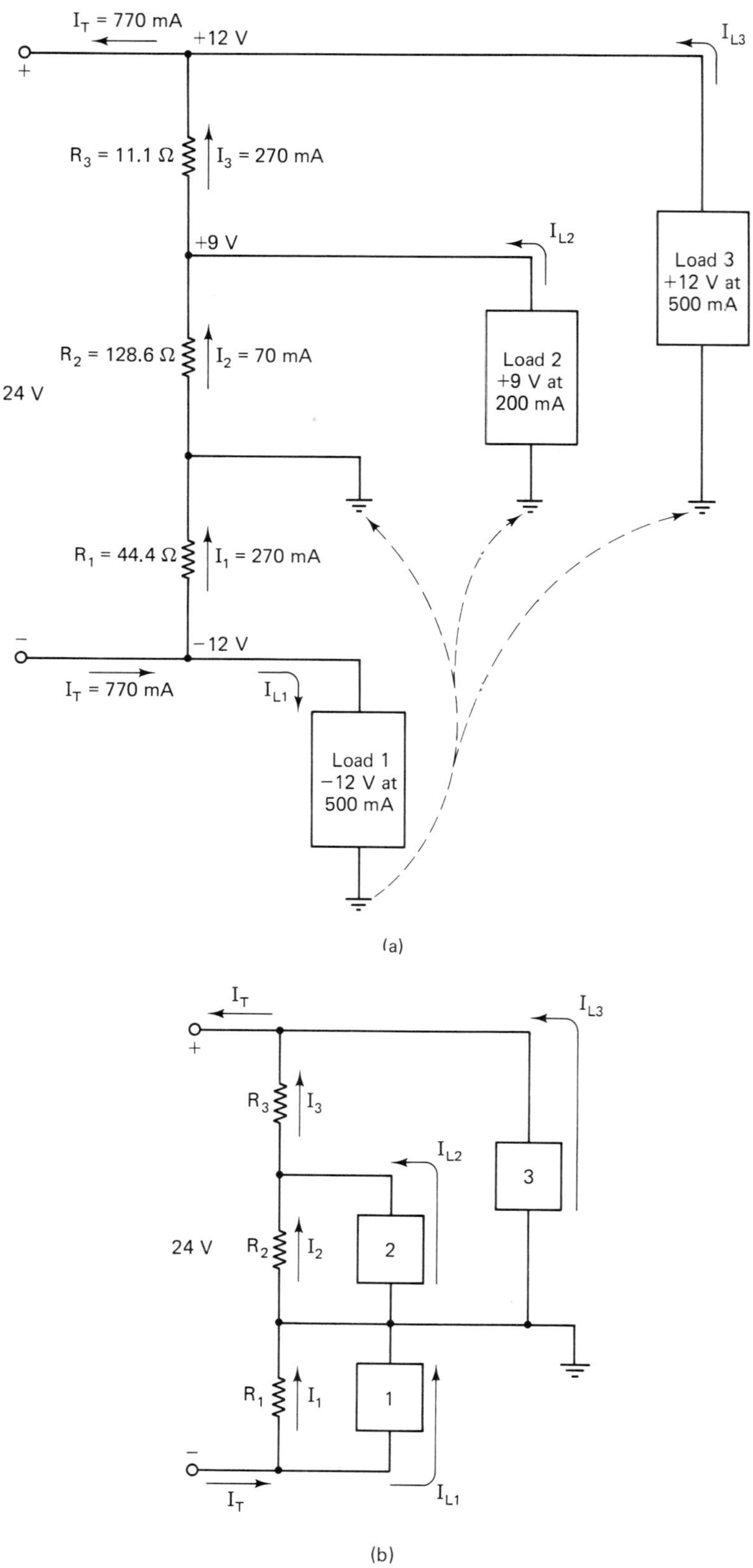

Figure 5–6 Positive and negative voltage divider: (a) normally drawn circuit; (b) rearranged circuit.

supplied with a 24-V power source and its required load voltages are + 12, +9, and −12 V. Notice that the ground reference is located at the top of R_1. The voltage at the bottom of R_1 is negative with respect to ground. The −12 V is developed across R_1. The current of load 1 flows from the negative terminal of the 24-V power source, through load 1 to ground. This current flowing through ground then travels up through R_2 and R_3 to the positive terminal of the 24-V power source. The +9 V is developed by R_2 and the additional 3 V developed by R_3 in series forms the +12 V needed.

Finding the resistances of this voltage divider is similar to the method used in the previous example, except that the total current first divides proportionately between R_1 and load 1 of the negative-voltage side of the divider before continuing to the positive-voltage sections of the divider. Figure 5-6b shows the electron paths through the circuit. As before, we first calculate the total load current for the positive voltage side:

$$I_{LT} = I_{L2} + I_{L3} = 200\text{ mA} + 500\text{ mA} = 700\text{ mA}$$

then the bleeder current of R_2:

$$I_2 = 10\%\, I_{LT} = 0.1 \times 700\text{ mA} = 70\text{ mA}$$

Therefore,

$$I_T = I_{LT} + I_2 = 700\text{ mA} + 70\text{ mA} + = 770\text{ mA}$$

At this point it is important to notice whether the required current of load 1 is at least 10% less than the total current. If it is not, then the total current should be set at least 10% greater than the current of load 1. In this case, the current of load 1 is less and we can proceed to find the value of the resistors.

To find the current through R_1:

$$I_1 = I_T - I_{L1} = 770\text{ mA} - 500\text{ mA} = 270\text{ mA}$$

Then

$$R_1 = \frac{V_{L1}}{I_1} = \frac{12\text{ V}}{270\text{ mA}} = 44.4\ \Omega$$

The remaining values are found thus:

$$R_2 = \frac{V_{L2}}{I_2} = \frac{9\text{ V}}{70\text{ mA}} = 128.6\ \Omega$$

$$I_3 = I_2 + I_{L2} = 70\text{ mA} + 200\text{ mA} = 270\text{ mA}$$

$$R_3 = \frac{V_{L3} - V_{L2}}{I_3} = \frac{3\text{ V}}{270\text{ mA}} = 11.1\ \Omega$$

The power dissipated by the resistors is

$$P_1 = V_1 I_1 = 12\text{ V} \times 270\text{ mA} = 3240\text{ mW} = 3.24\text{ W}$$

$$P_2 = V_2 I_2 = 9\text{ V} \times 70\text{ mA} = 630\text{ mW} = 0.63\text{ W}$$

$$P_3 = V_3 I_3 = 3\text{ V} \times 270\text{ mA} = 810\text{ mW} = 0.81\text{ W}$$

5-1e TEST EQUIPMENT LOADING EFFECT

Whenever test equipment is placed on a circuit for testing, a very small amount of current must flow through the testing device to give an indication. In an ideal situation, the input or probe resistance to the testing device is infinite. In many cases, the small amount of current drawn by the testing device does not appreciably affect the reading. However, in some cases, the testing device has a low input resistance and it draws sufficient current to upset the circuit being tested and an incorrect reading will occur. This can be illustrated by the voltmeter loading effect shown in Figure 5-7.

The total resistance of this circuit is 2 MΩ; therefore, the current flowing in it is 6 μA ($I_T = V_T/R_T = 12\text{ V}/2\text{ M}\Omega = 6\ \mu\text{A}$). The voltage from point A to ground should be +6 V ($V = IR = 6\ \mu\text{A} \times 1\text{ M}\Omega = 6\text{ V}$). When a voltmeter with an input resistance of 1 MΩ is placed at point A, it is in parallel with R_2, which results in an equivalent resistance (R_e) of 0.5 MΩ. The total resistance of the circuit is now

$$R_1 + \frac{R_2 \times R_{\text{meter}}}{R_2 + R_{\text{meter}}} = 1\text{ M}\Omega + 0.5\text{ M}\Omega = 1.5\text{ M}\Omega$$

Since the total resistance has decreased, the total current has increased:

$$I_T = \frac{V_T}{R_T} = \frac{12}{1.5\text{ M}\Omega} = 8\ \mu\text{A}$$

This I_T of 8 μA divides evenly between R_2 and R_{meter}, with 4 μA in each branch. The voltage across R_2 (and R_{meter}) or from point A to ground is now found to be

$$V_2 = I_2R_2 = 4\ \mu\text{A} \times 1\text{ M}\Omega = 4\text{ V}$$

The voltmeter is loading down the circuit and consequently an incorrect reading is indicated.

Generally, a testing device should have a much greater input resistance than that of the circuit being tested. A good rule of thumb should be that the testing device has an input resistance of 50 times greater and certainly no less than 10 times greater than the circuit being tested.

Figure 5–7 Voltmeter loading effect. Point A = +6 V before meter is attached and +4 V after meter is attached.

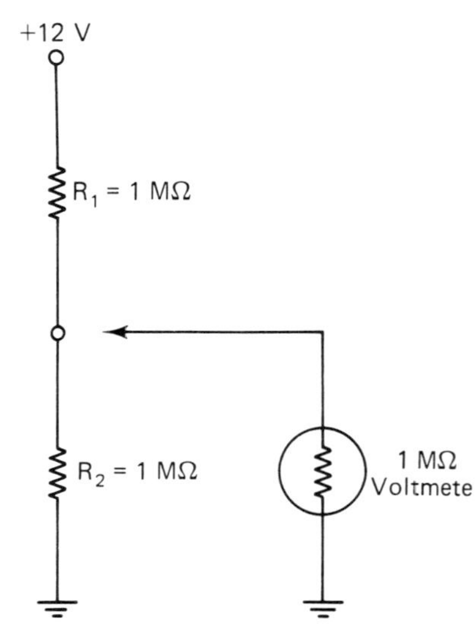

SECTION 5-2 DEFINITION EXERCISES

Write a brief description for each of the following terms.

1. Series–parallel circuit ______

2. Parallel–series circuit ______

3. Equivalent resistance ______

4. Total resistance ______

5. Balanced resistor bridge ______

6. Unbalanced resistor bridge ______

7. Voltage divider ______

8. Load ______

9. Draw (related to current) ______

10. Ground connection ______

11. Positive and negative voltage divider ______

12. Loading effect ______

SECTION 5-3 EXERCISES AND PROBLEMS

Complete this section before beginning the next section.

1. With the circuit shown in Figure 5-8, solve for the parameters required.

R_T = ______	V_3 = ______
I_T = ______	V_4 = ______
I_1 = ______	P_1 = ______
I_2 = ______	P_2 = ______
I_3 = ______	P_3 = ______
I_4 = ______	P_4 = ______
V_1 = ______	P_T = ______
V_2 = ______	

2. With the circuit shown in Figure 5-9, solve for the parameters required.

R_T = ______	V_3 = ______
I_T = ______	V_4 = ______
I_1 = ______	P_1 = ______
I_2 = ______	P_2 = ______
I_3 = ______	P_3 = ______
I_4 = ______	P_4 = ______
V_1 = ______	P_T = ______
V_2 = ______	

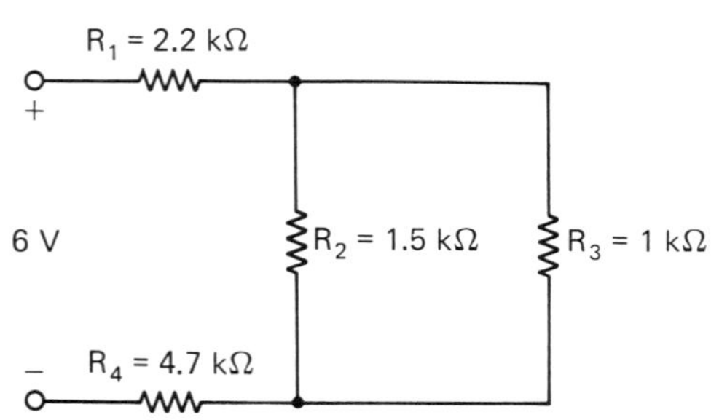

Figure 5–8

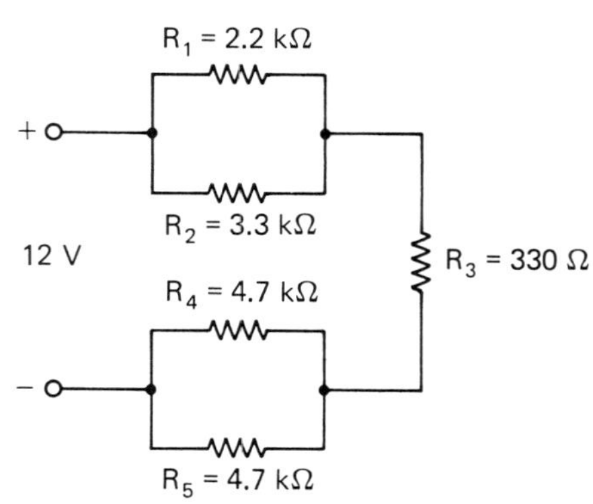

Figure 5–9

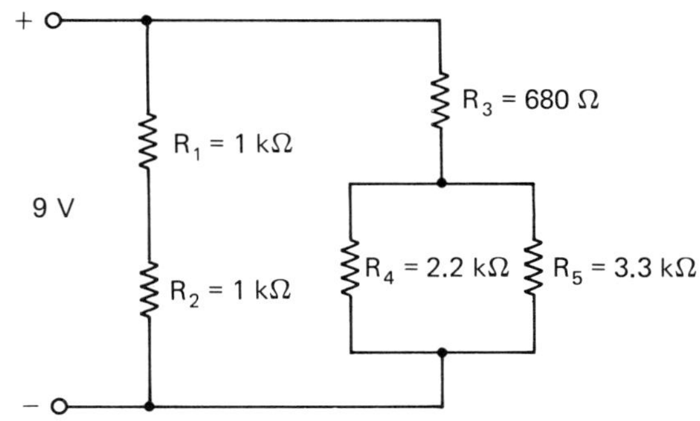

Figure 5–10

3. With the circuit shown in Figure 5-10, solve for the parameters required.

R_T = ________	I_5 = ________	P_1 = ________
I_T = ________	V_1 = ________	P_2 = ________
I_1 = ________	V_2 = ________	P_3 = ________
I_2 = ________	V_3 = ________	P_4 = ________
I_3 = ________	V_4 = ________	P_5 = ________
I_4 = ________	V_5 = ________	P_T = ________

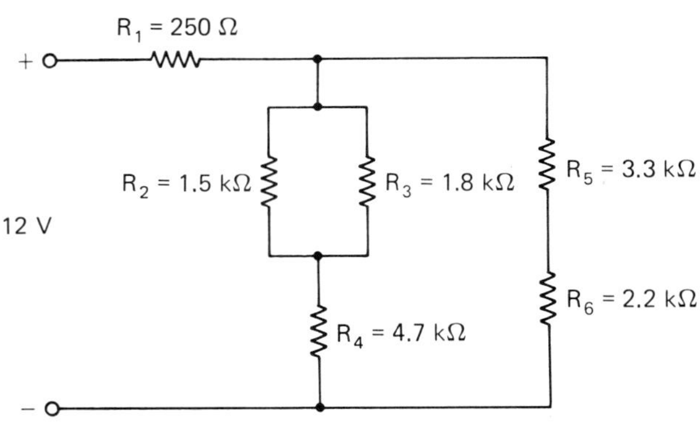

Figure 5–11

4. With the circuit shown in Figure 5-11, solve for the parameters required.

R_T = ________	I_6 = ________	P_1 = ________
I_T = ________	V_1 = ________	P_2 = ________
I_1 = ________	V_2 = ________	P_3 = ________
I_2 = ________	V_3 = ________	P_4 = ________
I_3 = ________	V_4 = ________	P_5 = ________
I_4 = ________	V_5 = ________	P_6 = ________
I_5 = ________	V_6 = ________	P_T = ________

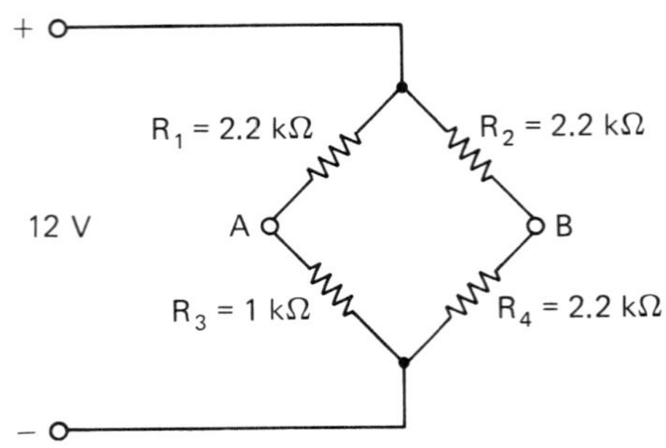

Figure 5-12

5. Answer the following questions pertaining to the circuit shown in Figure 5-12.

a. What is the approximate voltage between points A and B? ____________

b. Which point (A or B) is more positive? ____________

c. Is the circuit balanced or unbalanced? ____________

6. Design a positive voltage divider from the data given (refer to Figure 5-5). Input voltage = +9 V; output voltages = +9 V at 300 mA, +6 V at 200 mA, and +3 V at 100 mA.

7. Design a positive/negative voltage divider from the data given (refer to Figure 5-6). Input voltage = +30 V; output voltages = +15 V at 200 mA, +12 V at 50 mA, and −15 V at 100 mA.

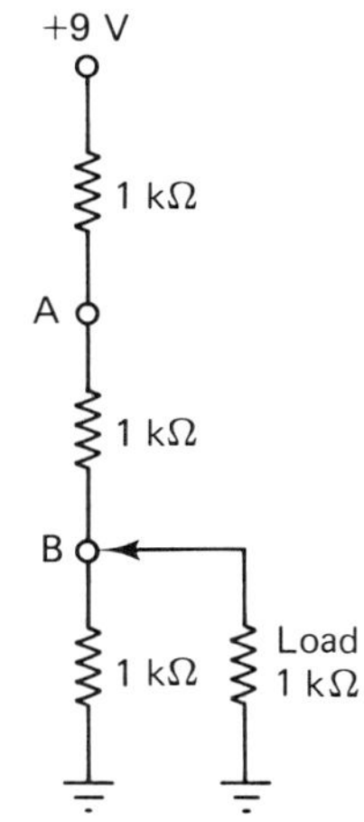

Figure 5-13

8. Answer the following questions pertaining to Figure 5-13.

a. What is the voltage from point A to ground? ____________

b. What is the voltage from point B to ground? ____________

After the load is placed at point B:

c. What is the voltage at point A to ground? ____________

d. What is the voltage from point B to ground? ____________

9. Match the description in column A with the proper term in column B.

Column A

____ **a.** A symbol used in a schematic diagram to indicate a common connection

____ **b.** Combining two or more resistances for calculating total resistance

____ **c.** The electrical result of current flowing through a resistance

____ **d.** A circuit that contains some components in series and others in parallel

____ **e.** A series circuit of resistors across a voltage source for providing voltages of other values

____ **f.** A condition where placing a meter on circuit actually changes the normal voltage reading

Column B

1. series–parallel circuit
2. equivalent resistance
3. voltage drop
4. ground
5. loading effect
6. voltage divider

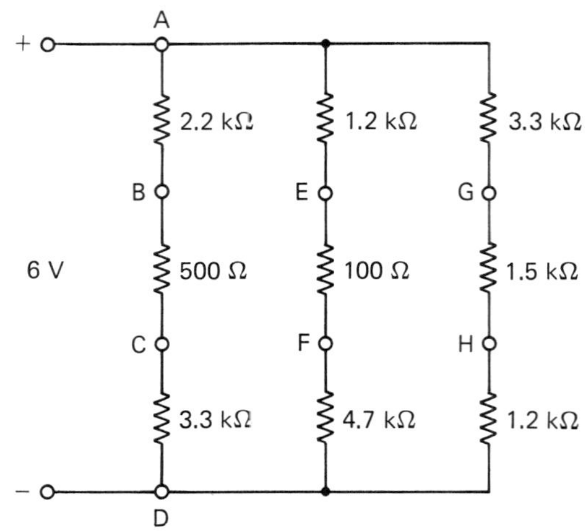

Figure 5–14

10. Calculate the voltage drops of all the resistors shown in Figure 5-14, then list the voltage between the indicated points. Indicate the most positive point of each pair of points.

Points	*Voltage*	*Most Positive Point*	*Points*	*Voltage*	*Most Positive Point*
AB	______	______	CE	______	______
AC	______	______	CF	______	______
AD	______	______	CG	______	______
AE	______	______	CH	______	______
AF	______	______	DE	______	______
AG	______	______	DF	______	______
AH	______	______	DG	______	______
BC	______	______	DH	______	______
BD	______	______	EF	______	______
BE	______	______	EG	______	______
BF	______	______	EH	______	______
BG	______	______	FG	______	______
BH	______	______	FH	______	______
CD	______	______	GH	______	______

SECTION 5-4 EXPERIMENTS

EXPERIMENT 1. SERIES–PARALLEL CIRCUIT

Objective: To show the voltage and current relationships in a complex series–parallel circuit.

Introduction: You will construct this circuit and record voltage and current measurements. These actual measurements will then be compared to the calculated values of the circuit.

Materials Needed:
1 2.2-kΩ ± 10% resistor at 0.5 W
1 3.3-kΩ ± 10% resistor at 0.5 W
1 4.7-kΩ ± 10% resistor at 0.5 W
1 6.8-kΩ ± 10% resistor at 0.5 W

Equipment Required:
1 Breadboard for constructing circuit and appropriate wires or leads
1 Variable power supply up to 30 V
1 Multimeter or DVM

Procedure:

1. In the space below, draw the schematic diagram of Figure 5-1a.

2. Calculate all of the following parameters and list them in their proper place.

R_T = ________	V_1 = ________	P_3 = ________
I_T = ________	V_2 = ________	P_4 = ________
I_1 = ________	V_3 = ________	P_T = ________
I_2 = ________	V_4 = ________	
I_3 = ________	P_1 = ________	
I_4 = ________	P_2 = ________	

3. Construct the circuit you have drawn, but before connecting the voltage source, measure the total resistance across the circuit terminals: R_T (measured) = ________. How does this value compare with the calculated value? ________

4. Connect the voltage to the circuit and perform the following measurements.

I_T = ________ I_3 = ________ V_2 = ________

I_1 = ________ I_4 = ________ V_3 = ________

I_2 = ________ V_1 = ________ V_4 = ________

5. Calculate the power dissipated by the resistors and for the total circuit from the measured values of voltage and current.

P_1 = ________ P_4 = ________

P_2 = ________ P_T = ________

P_3 = ________

Compared these values with the values calculated previously.

Fill-In Questions:

1. The measured circuit values are different from the calculated values mainly because of the resistors' value ________ .

2. Individual resistances are combined to solve for the total resistance of a circuit and are referred to as the ________ resistance.

EXPERIMENT 2. PARALLEL–SERIES CIRCUIT

Objective: To show the voltage and current relationship in a complex parallel–series circuit.

Introduction: The same procedures as those used in Experiment 1 are used in this experiment.

Materials Needed:
2 1-kΩ ± 10% resistor at 0.5 W
1 2.2-kΩ ± 10% resistor at 0.5 W
1 4.7-kΩ ± 10% resistor at 0.5 W
1 10-kΩ ± resistor at 0.5 W

Equipment Required:
1 Breadboard for constructing circuit and appropriate wires or leads
1 Variable power supply up to 30 V
1 Multimeter or DVM

Procedure:

1. In the space below, draw the schematic diagram of Figure 5-2a.

2. Calculate all of the following parameters and list them in their proper place.

R_T = ________	I_5 = ________	P_1 = ________
I_T = ________	V_1 = ________	P_2 = ________
I_1 = ________	V_2 = ________	P_3 = ________
I_2 = ________	V_3 = ________	P_4 = ________
I_3 = ________	V_4 = ________	P_5 = ________
I_4 = ________	V_5 = ________	P_T = ________

3. Construct the circuit you have drawn, but before connecting the voltage source, measure the total resistance across the circuit terminals: R_T (measured) = ________. How does this value compare with the calculated value? ________
4. Connect the voltage to the circuit and perform the following measurements.

I_T = ________	V_1 = ________
I_1 = ________	V_2 = ________
I_2 = ________	V_3 = ________
I_3 = ________	V_4 = ________
I_4 = ________	V_5 = ________
I_5 = ________	

5. Calculate the power dissipated by the resistors and for the total circuit from the measured values of voltage and current.

P_1 = ________	P_4 = ________
P_2 = ________	P_5 = ________
P_3 = ________	P_T = ________
P_4 = ________	

Fill-In Questions:

1. To find the equivalent resistance of two unequal-value resistors in parallel, you can use the ________ -over- ________ formula.

2. To find the equivalent resistance of equal-value resistors in parallel, you can ____________ the total number of resistors into the value of one ____________.

EXPERIMENT 3. BRIDGE CIRCUIT

Objective: To demonstrate the operation of the balanced and unbalanced bridge circuit.

Introduction: In this experiment you construct a bridge circuit and prove that the ratio of the resistors determines whether the circuit is balanced or unbalanced.

Materials Needed:

1 100-Ω ± 10% resistor at 0.5 W
4 1-kΩ ± 10% resistor at 0.5 W
1 2.2-kΩ ± 10% resistor at 0.5 W
1 10-kΩ ± 10% resistor at 0.5 W

Equipment Required:

1 Variable power supply up to 30 V
1 Multimeter or DVM

Procedure:

1. In the space below, draw the schematic diagram of Figure 5-4a.
2. Construct the circuit you have drawn.
3. Connect the voltage supply and place a voltmeter as indicated in the drawing. The meter reads ____________. Is the bridge balanced? ____________
4. Disconnect the voltage supply and change R_1 to 100 Ω and R_4 to 10 kΩ.
5. Reconnect the voltage and list the meter reading here: ____________. Is the bridge balanced? ____________
6. Disconnect the voltage supply and change R_1 to 1 kΩ and R_4 to 2.2 kΩ.
7. Reconnect the voltage and list the meter reading here: ____________. Is the bridge balanced? ____________

Fill-In Questions:

1. The ratio formula of resistors for a bridge circuit is ____________.

2. When the ratio of R_1 to R_2 equals the ratio of R_3 to R_4, the bridge circuit will be ____________ .

EXPERIMENT 4. VOLTAGE DIVIDER

Objective: To show the operation of a positive and negative voltage divider.

Introduction: When the circuit is constructed, you will be able to measure two separate positive voltages and a negative voltage.

Materials Needed:
2 100-Ω ± 10% resistor at 0.5 W
1 330-Ω ± 10% resistor at 0.5 W
3 1-kΩ ± 10% resistor at 0.5 W

Equipment Required:
1 Variable power supply up to 30 V
1 Multimeter or DVM

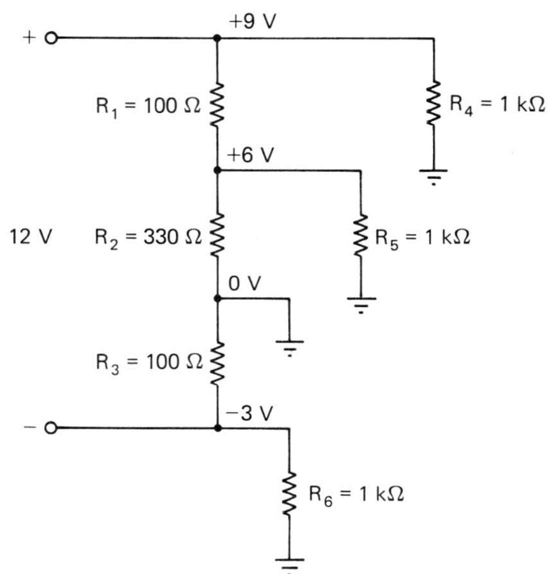

Figure 5-15

Procedure:
1. Construct the circuit shown in Figure 5-15.
2. Connect a +12-V power source to the circuit as shown.
3. Connect the black lead of the meter to the new ground point between R_2 and R_3.
4. Measure and list the voltage at the top of R_6: ____________ .
5. Measure and list the voltage at the top of R_2: ____________ .
6. Measure and list the voltage at the top of R_1: ____________ .

Fill-In Questions:

1. A voltage divider is used to ____________ various voltages to different circuits.

2. A positive and negative voltage divider can be produced by moving the ____________ to a selected point in the divider circuit.

SECTION 5-5 TROUBLESHOOTING APPLICATION: TESTING FOR OPEN AND SHORT CIRCUITS

The basis of electrical/electronic troubleshooting is making voltage measurements to determine proper circuit operation or finding faulty components. This application will enable you to solve basic circuit problems by understanding voltage indications that are not normal.

When an open circuit occurs, current stops flowing and voltage drops are nonexistent. The voltage measurements above the open will measure the source voltage. The voltage measurements below the open will measure zero volts.

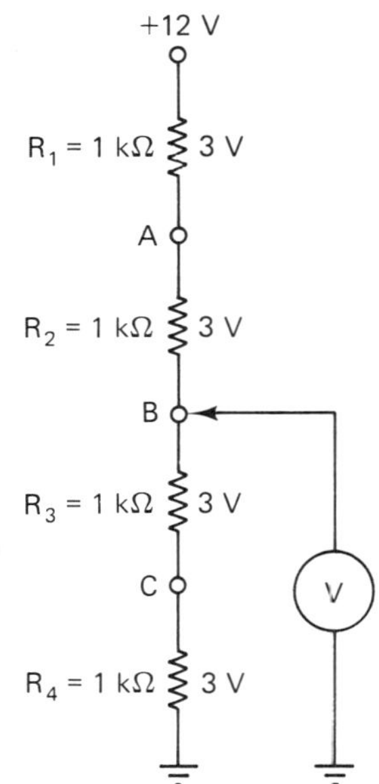

Figure 5-16

When a short circuit occurs, current increases in the circuit and voltage drops increase. The voltage measurements below a short will increase. The voltage measurements above a short will decrease.

Construct the circuit shown in Figure 5-16 and perform the conditions and voltage measurements given in the following table. List your voltage readings in the spaces provided.

Condition	*V = Point B to GND*	*Comment*
Normal	______	Voltage at B = $V_3 + V_4$
Point A open	______	Read point C to GND also
Point C open	______	Read point A to GND also
R_1 shorted	______	Place a wire across R_1 only
R_4 shorted	______	Place a wire across R_4 only

As a rule of thumb:

- A *low-voltage indication* suggests an open circuit in the current path toward the voltage source or a short circuit in the current path toward ground.
- A *high-voltage indication* suggests a short circuit in the current path toward the voltage source or an open circuit in the current path toward ground.

SECTION 5-6 INSTANT REVIEW

- Complex series–parallel and parallel–series circuits can be simplified by analyzing small segments of the total circuit.
- Resistors in series and/or parallel can be reduced to equivalent resistances.
- The total power consumed by a circuit is the sum of all the individual powers that are dissipated.
- A resistive bridge circuit is balanced when the ratio of R_1/R_2 is equal to the ratio R_3/R_4.
- A resistive bridge circuit is not balanced when the ratios are not equal.
- A voltage divider can have positive and negative voltages.
- The resistor values chosen for a voltage divider depends on the load currents.
- Placing a test instrument across a component in a circuit can cause a loading effect and give an erroneous reading.
- The loading effect results as an equivalent resistance of lower value that affects the voltage reading.
- An open circuit causes the current to stop.
- Voltage measurements above the open will be closer to or at the voltage source value.
- Voltage measurements below the open will be closer to or at ground potential.
- A short circuit causes the current to increase.
- Voltage measurements above the short will be lower in value or at ground potential.
- Voltage measurements below the short will be higher in value or toward the voltage source.

SECTION 5-7 SELF-CHECKING QUIZ

Circle the most correct answer for each question.

1. Analyzing complex resistive circuits is simplified by combining individual resistances and finding the:
   - **a.** loading effect
   - **b.** equivalent resistance
   - **c.** total power dissipated
   - **d.** none of the above.

2. The total power consumed in a complex resistive circuit can be found by:
   - **a.** taking the average of the power dissipated by the individual resistors
   - **b.** using the reciprocal formula for power
   - **c.** adding all of the powers dissipated by the individual resistors
   - **d.** none of the above

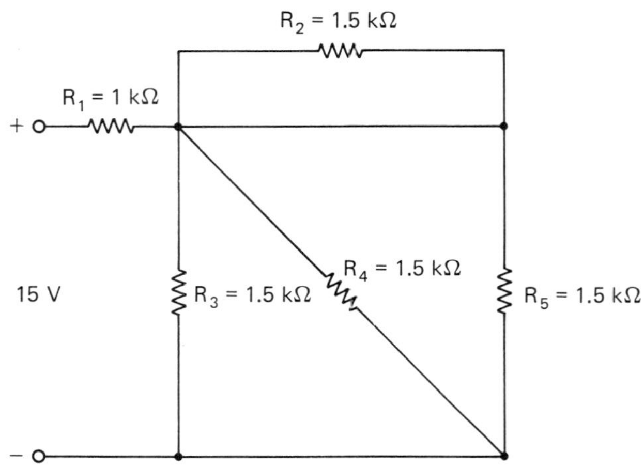

Figure 5–17

Questions 3 through 7 refer to Figure 5-17.

3. The I_T for the circuit is about:
   - **a.** 10 mA **b.** 11 mA
   - **c.** 1.5 mA **d.** 4 mA

4. The value of V_3 is about:
   - **a.** 1.5 V **b.** 10 V
   - **c.** 4 V **d.** 0 V
   - **e.** none of the above

5. The value of V_2 is:
   - **a.** 1.5 V **b.** 10 V
   - **c.** 4 V **d.** 0 V
   - **e.** none of the above

6. The value of I_4 is approximately:
   - **a.** 3.33 mA **b.** 2.7 mA
   - **c.** 4 mA **d.** 11.2 mA
   - **e.** none of the above

7. The value of P_5 is about:
   - **a.** 11 mW **b.** 5.5 mW
   - **c.** 12.4 mW **d.** 16.7 mW
   - **e.** none of the above

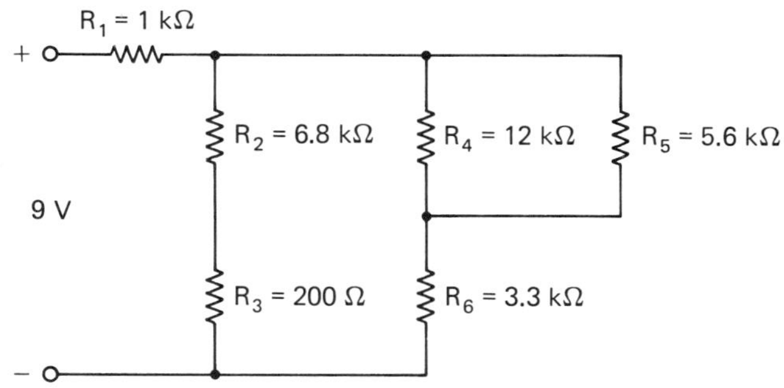

Figure 5-18

Questions 8 through 12 refer to Figure 5-18.

8. The R_T is approximately:

a. 4.5 kΩ **b.** 7 kΩ

c. 3.3 kΩ **d.** none of the above

9. The value of V_5 is about:

a. 3.5 V **b.** 2.8 V

c. 4.0 V **d.** 7 V

10. The value of I_6 is approximately:

a. 0.22 mA **b.** 0.90 mA

c. 0.6 mA **d.** 1.2 mA

11. The value of P_2 is about:

a. 7 mW **b.** 6.8 mW

c. 3.5 mW **d.** 2.2 mW

12. The total power for the circuit is about:

a. 12.4 mW **b.** 15. 8 mW

c. 18 mW **d.** 22 mW

13. A voltmeter is placed on a normal circuit, but the expected voltage reading is much less. A possible cause for this condition could be:

a. the equivalent resistance

b. the loading effect

c. power dissipation

d. none of the above

14. A balanced bridge circuit could be described as:

a. four resistors in a parallel–series combination circuit whose ratio of resistance allows the voltage between the resistors to equal zero volts

b. a circuit containing four resistors of equal value in a parallel–series combination

c. a circuit used to bridge the power levels between two circuits of different voltage levels.

d. none of the above

15. With reference to a specific point in a circuit using a positive voltage source, the voltage indication is lower than the normal reading. The possible cause could be:

a. an open circuit from that point to ground

b. a short circuit from that point to the voltage source

c. a short circuit from that point to ground

d. answers b and c are correct

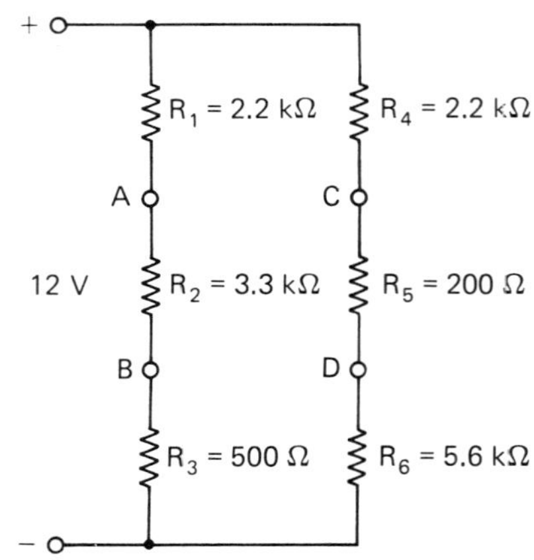

Figure 5-19

Questions 16 through 20 refer to Figure 5-19.

16. The voltage difference between points A and C is:

a. +8.7 V **b.** +7.6 V
c. +1.1 V **d.** +3.3 V

17. The points that are more positive with reference to ground are:

a. A more than B
b. D more than B
c. B more than C
d. C more than D
e. answers b and d are correct
f. answers a, b, and d are correct

18. The total current in the circuit is:

a. 2 mA **b.** 3.5 mA
c. 0.5 mA **d.** none of the above

19. If point A becomes open, the voltage from point B to ground will be:

a. 0 V **b.** +1 V
c. +7.6 V **d.** +12 V

20. If point D becomes open, the voltage from point C to ground will be:

a. 0 V **b.** +8.7 V
c. +12 V **d.** none of the above

ANSWERS TO EXPERIMENTS AND SELF-CHECKING QUIZ

Experiment 1: **(1)** tolerances **(2)** equivalent

Experiment 2: **(1)** product, sum **(2)** divide, resistor

Experiment 3: **(1)** $R_1/R_2 = R_3/R_4$ **(2)** balanced

Experiment 4: **(1)** distribute **(2)** ground

Self-Checking Quiz: **(1)** b **(2)** c **(3)** a **(4)** e **(5)** d **(6)** a **(7)** d **(8)** a **(9)** c **(10)** b **(11)** d **(12)** c **(13)** b **(14)** a **(15)** c **(16)** c **(17)** f **(18)** b **(19)** a **(20)** c

Unit 6

Magnetism and Electromagnetism

INTRODUCTION

The phenomenon of magnetism has been known for centuries. A natural magnet of the earth is found in a material called lodestone or magnetite. This natural form was first used as a compass for ships. Stronger magnets were later produced, but real advances in science and modern civilization were made when the interaction of magnetism and electromagnetism became apparent. Magnetic/electromagnetic devices include motors, generators, electronic meters, transformers, relays, microphones, loudspeakers, audio and video tape recorders, and many other industrial and consumer products.

UNIT OBJECTIVES

Upon completion of this unit, you will be able to:

1. Describe the nature of magnetism and electromagnetism.
2. Determine the north and south poles of a magnet and a solenoid.
3. Explain the flow direction of magnetic flux lines.
4. State the laws of magnetic/electromagnetic attraction and repulsion.
5. Describe the molecular structure of magnetic and nonmagnetic materials.
6. Compare magnetic conductivity of magnetic and nonmagnetic materials.
7. Describe magnetic induction.
8. Explain the magnetization curve or hysteresis loop.
9. Use the left-hand rule for finding the direction of magnetic flux around a wire and the north pole of a solenoid.
10. Describe the electromagnetic effect.
11. Demonstrate the operation of the relay.
12. Identify magnetic/electromagnetic applications.
13. Troubleshoot basic magnetic/electromagnetic devices.
14. Define the terms *magnetic flux, flux density, permeability, reluctance, retentivity, residual magnetism, ferromagnetic, diamagnetic,* and *paramagnetic.*

SECTION 6-1
FUNDAMETAL CONCEPTS

6-1a THE NATURE OF MAGNETISM

Magnetism is a property of certain materials, such as iron, nickel, and cobalt, which exerts an invisible force on other magnetic materials. The invisible magnetic field that surrounds a magnet is the force that influences other magnetic materials. An artificial magnet is created in a magnetic material when it encounters a magnetizing force. A permanent magnet results when the magnetizing force is removed and the material retains some if not all of its magnetism.

A permanent magnet comes in many shapes and sizes. It is made of iron, steel, or an alloy containing one of these materials. The easiest to analyze is the bar magnet, shown in Figure 6-1. An invisible force surrounds the magnet, as shown by the dashed lines. These invisible lines are referred to as *magnetic lines of force* or more correctly, as *flux lines.* The flux lines flow in a directional manner out of the designated north pole around the magnet and into the south pole. The same lines flow through the magnet toward the north pole and again exit the north pole in a continuous manner. Even though the figure is shown in two dimensions, remember that the flux lines are all around the magnet, in

Figure 6–1 Artificial magnet showing flux lines of force.

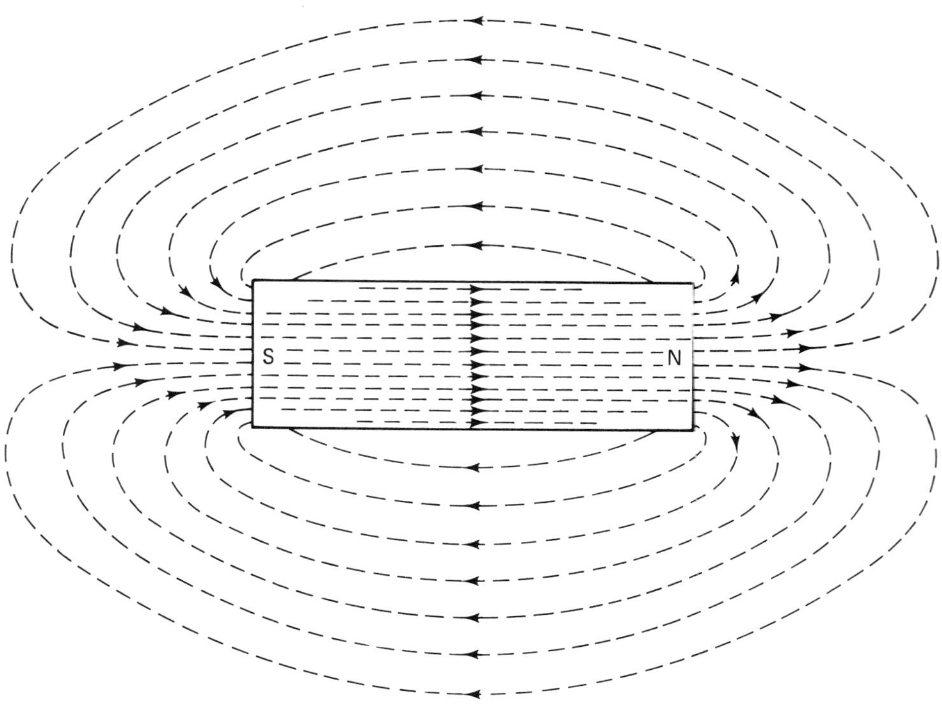

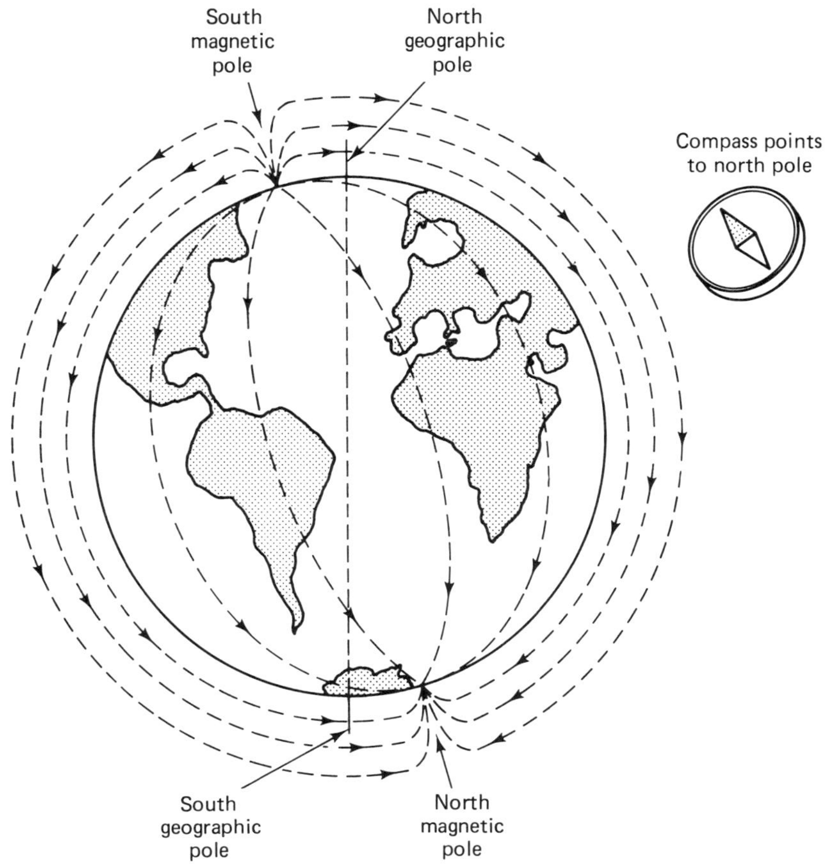

Figure 6–2 Magnetic phenomena of earth.

a three-dimensional manner. A single line is referred to as a *maxwell;* a group of 10 lines is called a *weber,* which helps to define the density of the lines. One weber filling a square meter of area is called a *tesla.* Therefore, the flux density of a magnet is expressed as the number of webers per square meter. A tesla is equal to 1 weber per square meter and *flux density* can be expressed by the equation

$$B = \frac{\phi}{A}$$

where B = flux density (teslas)
ϕ = total number of lines of flux (webers)
A = area (square meters)

If a string is tied around the center of the magnet and allowed to be suspended in midair, the magnet will align itself in a particular direction every time. This phenomenon gives evidence that the earth is also a form of magnet, but on a much greater scale. If a compass is held parallel with the earth's surface, its indicator will swing to the direction of the geographic north pole, as shown in Figure 6-2. Actually, the north pole of the compass is pointing to the south magnetic pole because of the laws of magnetic attraction and repulsion.

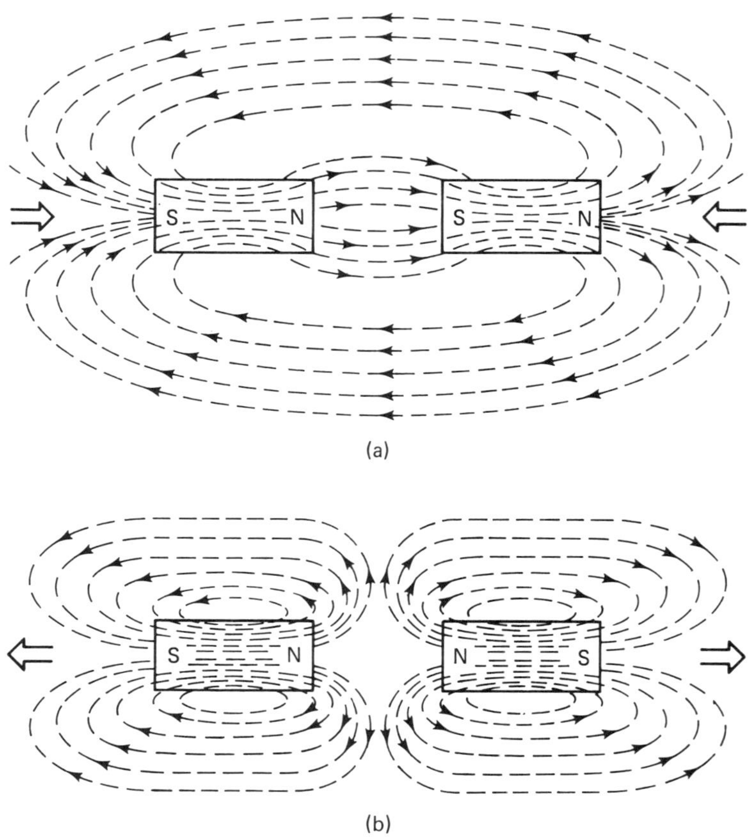

Figure 6–3 Magnetic attraction/repulsion: (a) unlike poles attract; (b) like poles repel.

6-1b LAWS OF MAGNETIC ATTRACTION AND REPULSION

The laws of magnetic attraction and repulsion state that unlike poles will attract each other and like poles will repel each other, as shown in Figure 6-3. Figure 6-3a shows how the flux fields of each magnet aid each other when a north pole and a south pole are brought together. The magnets have a tendency to move toward each other. In effect, a larger magnet is created with a more intense flux field. Figure 6-3b shows how the flux fields of each magnet oppose each other when two north poles are brought together. The magnets have a tendency to move away from each other. This also tends to reduce the total effect of the magnets. Similarly, if two south poles are brought together, they will repel each other and have a tendency to move away from each other.

6-1c MAGNETIC DOMAIN THEORY

A more modern theory of magnetism is based on the principle of the spin of the electron about its nucleus. This atomic movement creates a small magnet about it. As the atoms are grouped together to form molecules, the molecules also form magnets. In an unmagnetized material the molecules are in a random fashion, as shown in Figure 6-4a. When a magnetizing force is brought close to the material, the molecules will align themselves in a common north pole-to-south pole arrangement, as shown in Figure 6-4b.The molecules now aid each other and produce a larger magnet.

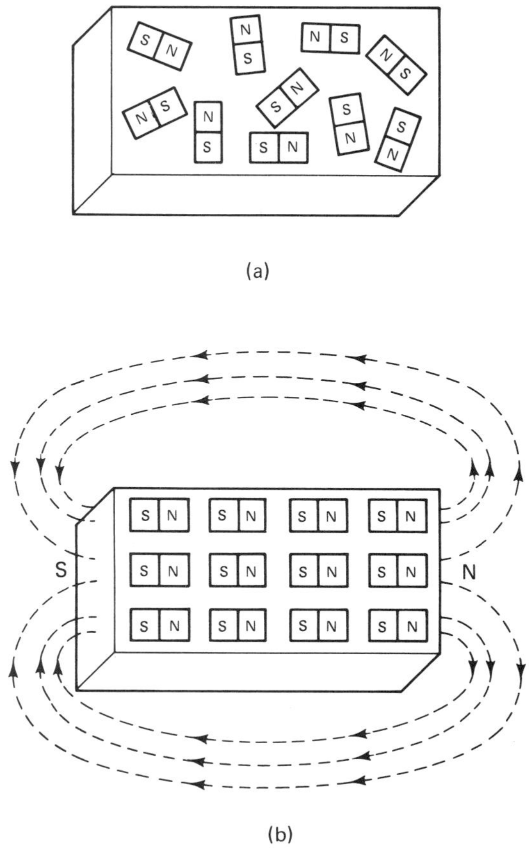

Figure 6–4 Magnetic domains: (a) unmagnetized material; (b) magnetized material.

The molecules of an iron bar can be aligned by bringing a magnet close to the bar. A stroking action with the two pieces will create a stronger magnet. Another method of magnetizing a bar of iron is with an electric current flowing in a coil of wire. This method is discussed in detail later in this unit. When the magnetizing force is removed, the molecules will remain this way, creating a permanent magnet. The ability of the material to remain magnetized is referred to as *retentivity*. Retentivity is even more apparent when a bar magnet is broken in half and each half exhibits magnetic qualities, thereby producing two small magnets from the original piece.

6-1d MAGNETIC MATERIALS

Iron is the most magnetic material known. Other materials related to iron, such as nickel and cobalt, have good magnetic properties and are called *ferromagnetic materials*. Materials having little or no magnetic properties, such as aluminum, platinum, and chromium, are referred to as *paramagnetic materials*. Copper, zinc, glass, plastic, wood paper, and similar materials have no magnetic properties and are referred to as *diamagnetic materials*.

Permeability is the relative ability of a material to concentrate magnetic lines of force within itself. Ferromagnetic materials have a high permeability, whereas paramagnetic and diamagnetic materials have little or no permeability. *Reluctance* is the opposition of a material that resists the concentration of magnetic lines within itself. Reluc-

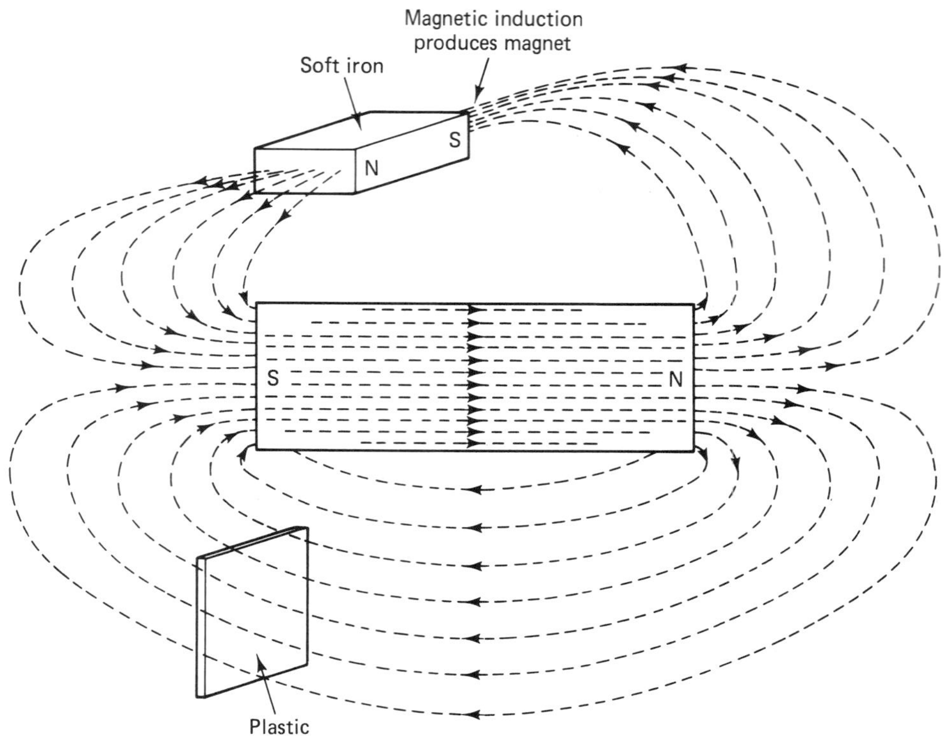

Figure 6–5 Magnetic flux conduction.

tance is inversely proportional to permeability. Aluminum has high reluctance and low permeability.

Figure 6-5 shows the relationship of materials with different reluctance and permeability. When a piece of soft iron is brought within the magnetic field of a magnet, the lines of force will actually bend toward it and concentrate through it. This action causes the soft iron to become a magnet also (note the north and south poles of the soft iron). When the soft iron is taken away from the field of influence, it will lose its magnetism because of low retentivity. Some ferromagnetic materials will have a higher retentivity, referred to as *residual magnetism.* An example of this is a screwdriver tip that has some magnetism. There is no effect on the lines of force of a magnet when a piece of plastic is brought within its field of influence. The lines of force pass right through the plastic.

Applications of magnetic principles, such as in audio, video, and computer data recording, require that magnetization occur in two directions—in other words, a reversing of the north and south poles. The actual magnetizing of a material lags the magnetizing force that is applied to it. Figure 6-6 shows a magnetization *B–H curve,* where B is the flux density or magnetization and H is the magnetizing force.

The magnetization curve is also referred to as a *hysteresis loop* because of the lagging influence between B and H. The numbers on the figure are used to aid in describing the magnetizing action. At point 0 the material is neutral and has no magnetism in either the $+B$ or $-B$ direction of flux density. The magnetizing force is also at 0. The magnetizing force H moves to the right in the $+H$ direction. The material becomes magnetized where it becomes saturated at point 1. The magnetizing force is decreased to 0, but the flux density $+\mathrm{B}$ does not decrease to 0 and remains at point 2, showing that the material has

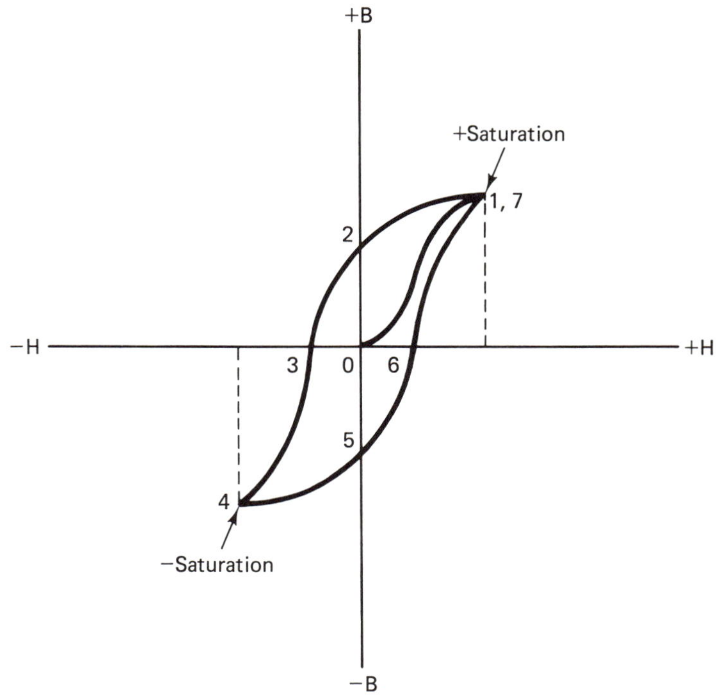

Figure 6–6 Magnetization *B–H* curve (hysteresis loop).

retentivity. The magnetizing force is then increased in the opposite direction toward $-H$, and at a specific place the magnetization of the material reverses direction, as indicated by point 3. As the force increases the material becomes more magnetized, until it reaches saturation at point 4. When the magnetizing force is again brought back to 0, the material remains magnetized in this direction, as shown by point 5. If the magnetizing force is again increased in the $+B$ direction, the flux density will shift toward the positive direction, as shown by point 6. As the magnetizing force increases, the flux density reaches saturation at point 7, which completes the hysteresis loop.

If the magnetizing force is placed into the $+H$ direction and then brought to zero, the material will remain magnetized in the $+B$ direction. Similarly, if the magnetizing force is placed in the $-H$ direction and then brought back to zero, the material will remain magnetized in the $-B$ direction. It is this principle that enables the use of many devices in the modern world, particularly magnetic tape recording.

6-1e ELECTROMAGNETIC INDUCTION

In the early nineteenth century another phenomenon was discovered which was directly related to magnetism. It was found that if a compass were brought close to a wire carrying an electrical current, the needle of the compass was affected. This led to the discovery that a magnetic field was produced about a conductor that was carrying electrical current. Figure 6-7 shows the magnetic flux resulting from electron flow in a conductor. When the flow of current is away from the viewer as indicated by the + (representing the feathers on an arrow going away from you) a magnetic flux develops in a counterclockwise direction. If the current is reversed and is coming toward the viewer, indicated by a • (represented as the point of an arrow coming toward you), the magnetic flux changed directions and now has a clockwise direction.

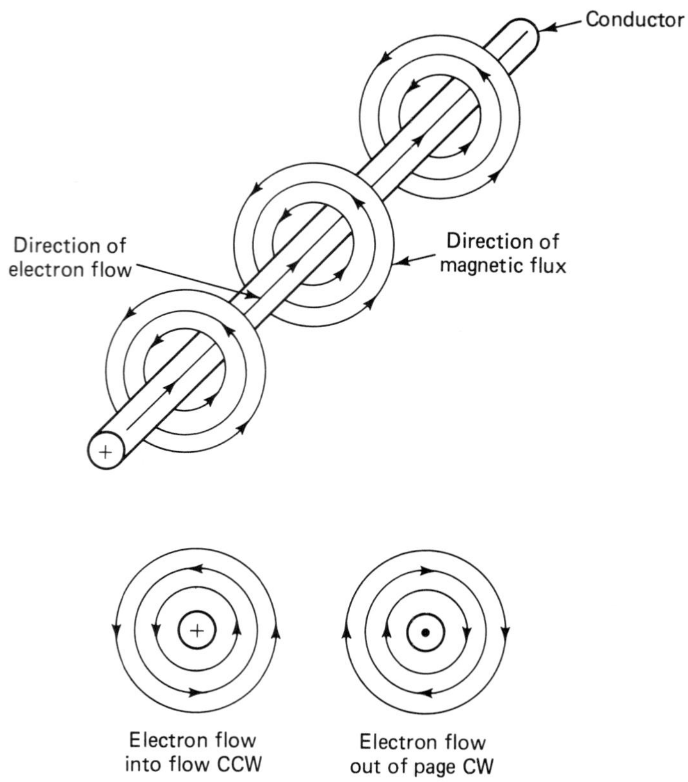

Figure 6–7 Magnetic flux resulting from electron flow in a conductor.

A left-hand rule for determining the direction of magnetic flux in a conductor with electron flow is shown in Figure 6-8. The thumb of the left hand points in the direction of electron flow. The other fingers of the hand are formed in a cupped manner which indicates the direction of the magnetic flux. This method of producing electromagnetism is called *induction.* As current flows through the wire, a magnetic flux is induced about the wire. In other words, electromagnetic induction is the magnetic field produced about a conductor that is carrying current.

Elecromagnetism also has the effects of attraction and repulsion, as shown in Figure 6-9. When two parallel conductors are carrying current in the same direction, as shown in Figure 6-9a, the magnetic fields aid each other and an attraction occurs between the conductors. The two conductors will tend to move toward each other. When the current

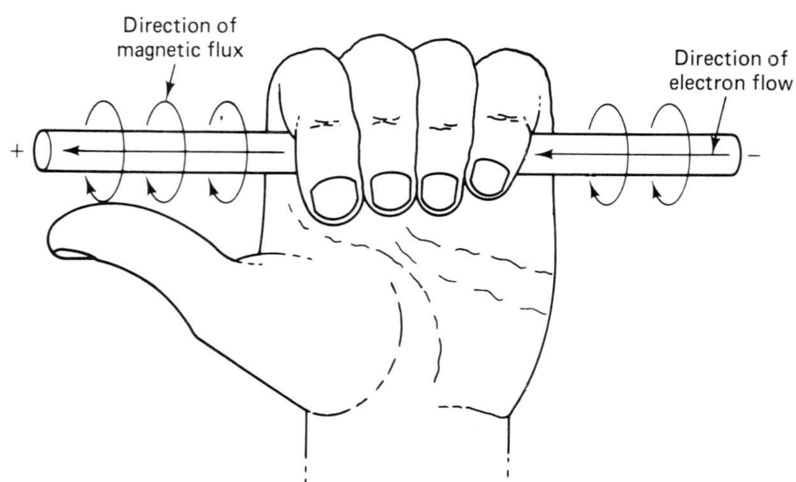

Figure 6–8 Left-hand rule for determining the direction of magnetic flux.

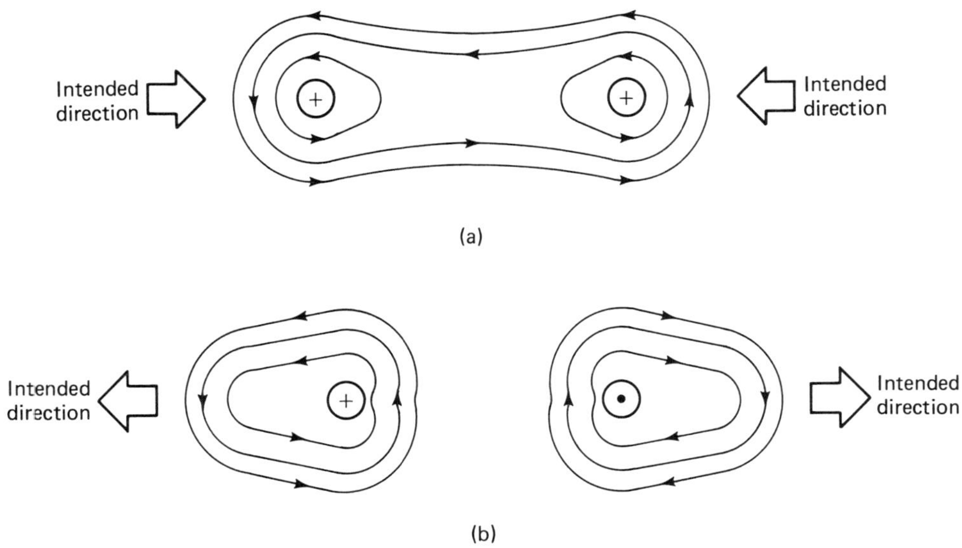

Figure 6–9 Electromagnetic effects of current-carrying conductors in parallel: (a) attraction; (b) repulsion.

in the conductors is moving in opposite directions, as shown in Figure 6-9b, the magnetic fields oppose each other and a repulsion occurs between the conductors. Now the two conductors will tend to move away from each other. This type of electromagnetic induction causes problems in some electronic circuits, and the wiring layout of a particular device is critical.

6-1f THE SOLENOID

It was found that when a wire was wound into a coil and connected to a voltage source, it displayed characteristics similar to those of a magnet. This electromagnetic device is caled a *solenoid.* The magnetic flux about the coils of wire aid each other and develop a magnet that has a north pole and a south pole. Figure 6-10 shows the left-hand rule for determining the north pole of a solenoid. When one end of the solenoid is connected to a positive terminal of a voltage source and other end to

Figure 6–10 Left-hand rule for determining the north pole of a solenoid.

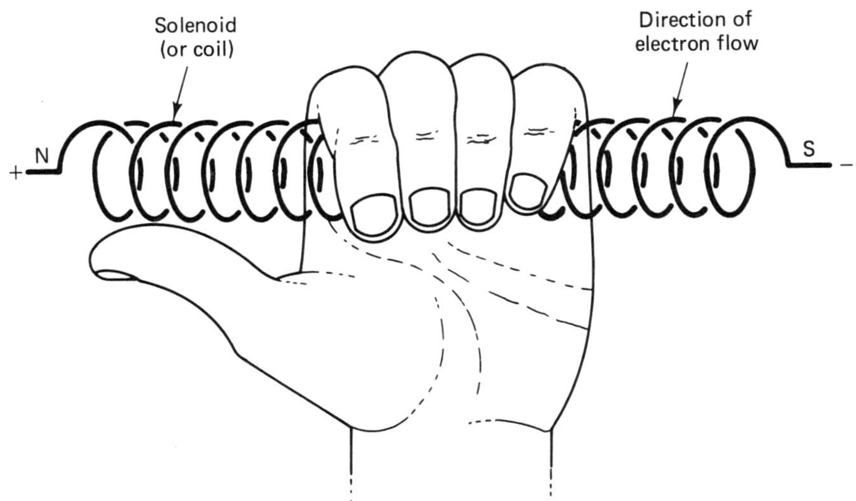

the negative terminal, current will flow in a counterclockwise direction, as shown in Figure 6-10 (from the little finger toward the thumb). The fingers of the hand represent the direction of electron flow and the extended thumb points to the north pole of the solenoid.

The solenoid makes a very good electromagnet, as shown in Figure 6-11a. The wire has many turns around a nonmagnetic form. The greater the number of turns of wire around the form, the greater the magnetic flux created, which in turn produces a stronger magnet. When a voltage source is applied to the solenoid it becomes an electromagnet and smaller pieces of magnetic materials will be attracted to it. When the voltage is removed the materials will fall from the electromagnet. The smaller magnetic pieces may retain some magnetism and be attracted to each other.

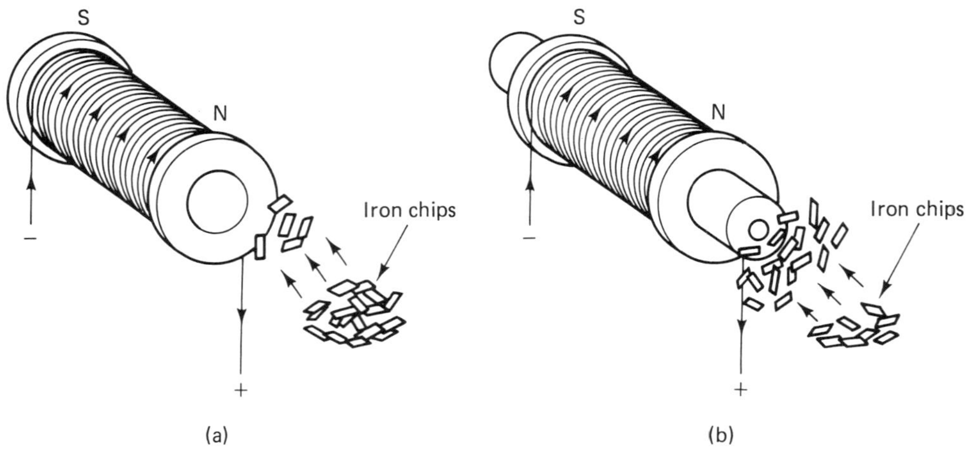

Figure 6–11 Electromagnet: (a) solenoid with air core; (b) solenoid with soft-iron core.

If a ferromagnetic material such as a soft-iron bar is placed inside the electromagnet, its magnetism becomes stronger, as shown in Figure 6-11b. The soft iron has a high permeability and concentrates the magnetic flux closer to the coil of wire. A greater force is now exerted on the other magnetic materials. A soft-iron bar is used because it has low retentivity and will become demagnetized when the voltage source is removed. Notice that with two solenoids of the same size and number of turns of wire, the one with a ferromagnetic core will have more magnetism and be a stronger magnet.

6-1g THE RELAY

One of the most used applications of the electromagnet is the relay, shown in Figure 6-12. The relay is used as a switching control device between circuits. It can be used to isolate a low-voltage control circuit from a high-voltage power circuit. There are many types of relays, which can have complex switching arrangements, but the simplest relay has one coil of wire, an armature, and three electrical contacts. The schematic symbol for a relay is shown in Figure 6-12a. It has one normally closed (NC) contact and one normally open (NO), contact with an armature contact that moves between them. The coil of wire has a soft-iron core and is mounted in the relay bracket. The movable armature

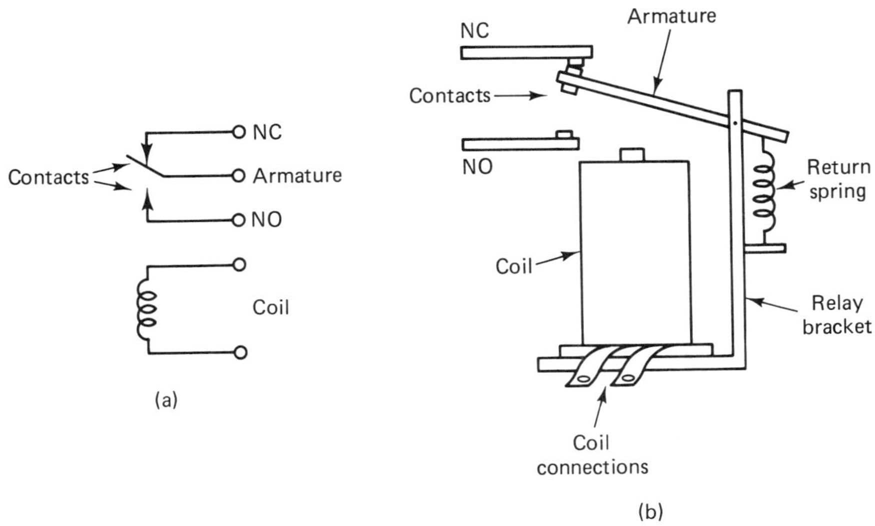

Figure 6–12 Relay:
(a) schematic symbol;
(b) physical representation.

is mounted above one end of the coil and has a return spring to keep it in the open position normally. The armature contact is touching the normally closed (NC) contact at this time. When voltage is applied to the coil connections, the armature will be attracted toward the end of the coil. The armature contact will be pulled from the NC contact and will now touch the normally open (NO) contact. In other words, the circuit has switched from the NC contact to the NO contact. When the voltage is removed from the coil connections, the armature will be pulled back to its normal position by the return spring. Now, the circuit has switched from the NO contact to the NC contact. Relays are found in use in automobiles, washing machines, dryers, electronic equipment, and manufacturing equipment. You will gain more experience with relays in the experiment section of this unit.

6-1h MAGNETIC/ELECTROMAGNETIC APPLICATIONS

A few practical applications using the principles of magnetism and electromagnetism are shown in Figure 6-13. A solenoid with a plunger, shown in Figure 6-13a, is used to electrically control the physical movement of an object. When voltage is applied to the coil connections the plunger will move into the core of the coil and any object connected to the plunger will be pulled toward the solenoid. When the voltage is removed from the coil connections, the object and plunger will be pulled back to the normal position by the return spring. Usually, this entire apparatus is referred to as a solenoid. You can find solenoids in automobiles, washing machines, and computers, and in factories where the movement and manufacturing of products requires electrical control.

A simple magnetic switch is shown in Figure 6-13b. The contacts are very thin strips of ferromagnetic material mounted in a tiny vacuum glass container. In this case the contacts are normally open. When the magnet is brought close to the container, the magnetic flux induces a small amount of magnetism in the contacts and they attract each other, thereby closing the switch. When the magnet is removed from the vicinity of the glass container, the contacts will open because they have little or no residual magnetism. These types of switches can be normally closed or have a combination of various contacts in a single

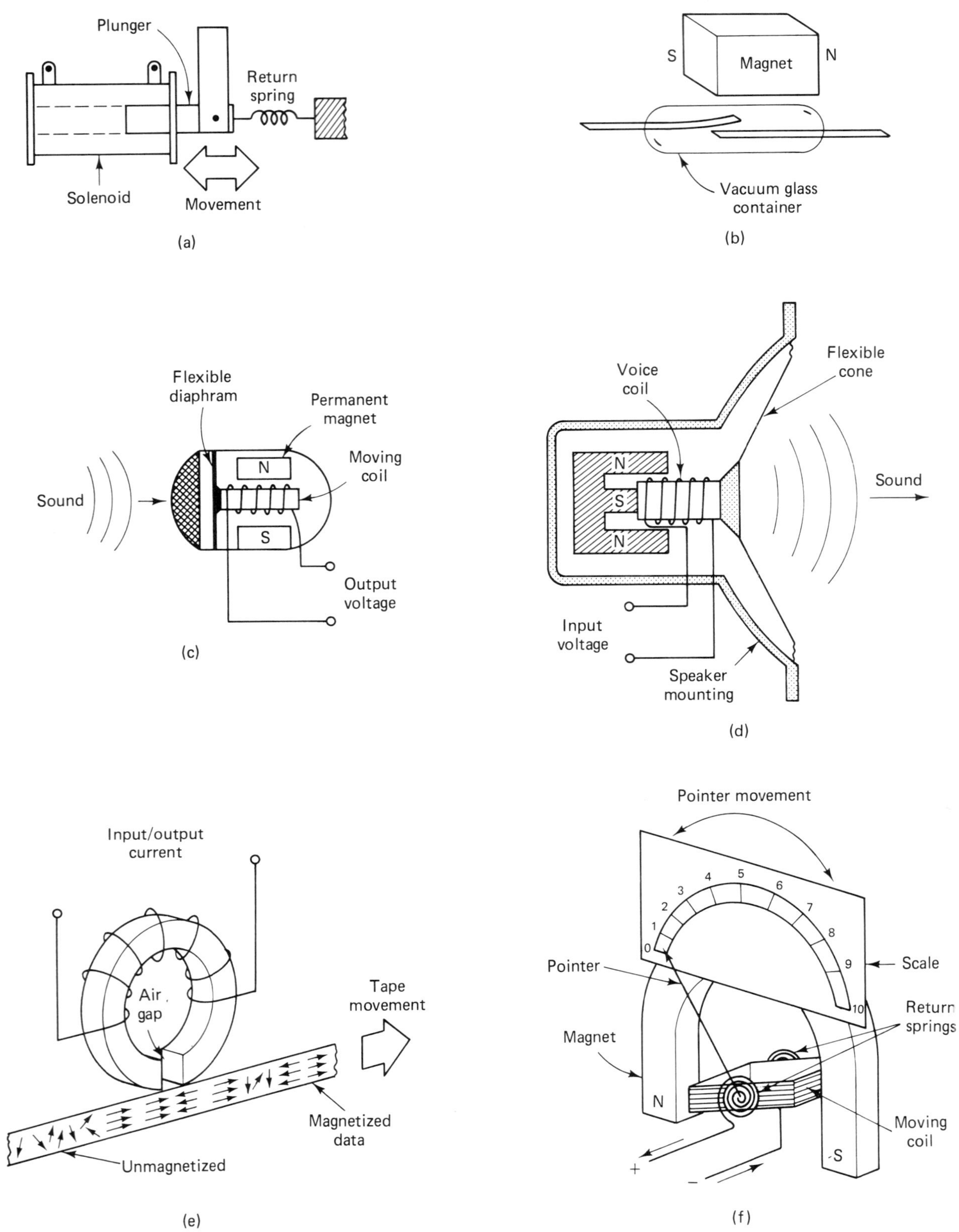

Figure 6–13 Magnetic/electromagnetic applications: (a) solenoid; (b) reed switch; (c) microphone; (d) loudspeaker; (e) recording tape; (f) meter movement.

container. Reed relays are similar to this but are vacuum-glass-enclosed contacts mounted inside small solenoids which operate with the application of voltage.

The well-known dynamic microphone is shown in Figure 6-13c. A flexible diaphragm is mounted to a coil of wire that is positioned between the poles of a magnet. The wire coil has external connections which supply the output voltage. When sound strikes the diaphragm it moves the coil back and forth between the magnet's poles, which induces a voltage in the wire of the coil. This output voltage is then amplified and used in other circuits. The principle of magnetically induced voltage is covered in detail in Unit 7.

Another well-known device, the loudspeaker, is shown in Figure 6-13d. It is similar to the microphone in that a coil of wire is connected to a flexible cone (or diaphragm) and also moves within the poles of a magnet. When an audio signal voltage is placed on the input voltage connections of the loudspeaker, an electromagnetic field is induced about the coil that aids and repels the permanent magnet field. This action, in turn, causes the cone of the loudspeaker to move in and out, thereby producing sound.

The principle of magnetic tape recording is shown in Figure 6-13e. The tape is plastic material that has a fine coating of a ferromagnetic material. The molecules of the coating are in a random direction with respect to magnetic polarization. A toroidal (round) piece of ferromagnetic material has a coil of wire wrapped around it. Voltage placed on the inputs to the wire create a current which induces a magnetic flux in the toroidal. Most of the flux is contained within the toroidal, but a small air gap in the structure is positioned close to the tape. At this point the magnetic flux can expand outward into the tape and change its molecular polarization. As the tape moves past the air gap, which is often referred to as the *recording head*, the molecules are magnetically aligned on the tape, depending on the type of signal voltage being presented at the input. The information that is recorded on the tape can represent music, speech, video, or other types of data. The recorded data on the tape can now be played back using the same recording head. As the tape moves by the air gap, a small amount of flux is induced into the toroidal by the data on the tape. This in turn induces a small voltage in the coil of wire and a small signal voltage will appear at the output of the wire.

The magnetic/electromagnetic principle was first used in meters, as shown in Figure 6-13f. A coil of wire is made in the shape of a small rectangle. Attached to the coil is a pointer that moves across a calibrated scale in the form of a dial. The coil also has return springs. The two wires from the coil go to other circuits and eventually come out of the meter as the test leads. The coil assembly is positioned between the poles of a permanent magnet so that it can turn freely. When the meter is measuring, a device or circuit current flows in the coil and creates a magnetic flux. This flux opposes the magnetic field of the magnet and the coil assembly rotates. The distance that the coil rotates is directly proportional to the amount of current flowing in the coil of wire. The pointer now indicates the measurement on the dial face. When the test leads are removed from the circuit, the return springs return the pointer to its original position.

SECTION 6-2 DEFINITION EXERCISES

Write a brief description for each of the following terms.

1. Lodestone ______________________________

____

2. Magnetite ______________________________

____

3. Magnetism ______________________________

____

4. Magnetic field ______________________________

____

5. Artificial magnet ______________________________

____

6. Permanent magnet ______________________________

____

7. Magnetic lines of force ______________________________

____

8. Flux lines ______________________________

____

9. Flux density ______________________________

____

10. Magnetic attraction ______________________________

____

11. Magnetic repulsion ______________________________

____

12. Magnetic domain theory ______________________________

____

13. Retentivity ____________________

____

14. Residual magnetism ____________________

____

15. Ferromagnetic ____________________

____

16. Paramagnetic ____________________

____

17. Diamagnetic ____________________

____

18. Permeability ____________________

____

19. Reluctance ____________________

____

20. Magnetization curve ____________________

____

21. Hysteresis loop ____________________

____

22. Electromagnetism ____________________

____

23. Electromagnetic induction ____________________

____

24. Solenoid ____________________

____

25. Relay ____________________

____

SECTION 6-3 EXERCISE AND PROBLEMS

Complete this section before beginning the next section.

1. Draw a bar magnet and label the north and south poles. Show the direction of magnetic lines of force (refer to Figure 6-1).

2. Explain the law of magnetic attraction and repulsion.

3. Using the formula $B = \phi/A$, solve for the flux density of the following magnets. Remember that A (area) = width × height; for example, a magnet that is 2.5 cm × 2 cm has an area of 5 cm^2.

   a. ϕ = 50 μWb (microwebers)
   size = 2 cm × 2 cm

   B = ____________ T (teslas)

   b. ϕ = 100 μWb
   size = 1.5 cm × 2.5 cm

   B = ____________ T

4. Fill in the blanks in the following statements.

   a. A ____________ magnet retains its magnetism indefinitely.

   b. The two magnetic poles of a magnet are called the ____________ pole and the ____________ pole.

   c. Magnetic lines of force flow from the ____________ pole to the ____________ pole outside the structure of a magnet.

   d. If a magnet is suspended by a thin string, it will align itself with the earth. Its north pole will point to the ____________ ____________ pole of the earth.

   e. Like poles ____________ each other.

   f. Unlike poles ____________ each other.

   g. When the molecules of a ferromagnetic material are not aligned, the material is ____________ magnetized.

h. ____________ is the relative ability of a material to concentrate magnetic lines of force within itself.

i. The opposition of a material to become magnetized is called ____________ .

j. Magnetism remaining in a material after the magnetizing force has been removed is called ____________ magnetism.

5. Draw a hysteresis loop and label all the parts (refer to Figure 6-6).

6. List beside each of the following materials its magnetic qualities. Let F = ferromagnetic, P = paramagnetic, and D = diamagnetic.

____ **a.** copper	____ **b.** cobalt	____ **c.** aluminum
____ **d.** iron	____ **e.** platinum	____ **f.** chromium
____ **g.** zinc	____ **h.** nickel	____ **i.** glass

7. Explain the use of the left-hand rule for determining the direction of magnetic flux in a current-carrying conductor.

8. Explain the use of the left-hand rule for determining the north pole of a current-carrying solenoid.

9. Fill in the blanks in the following statements.

a. A current-carrying conductor will have an ____________ magnetic field about it.

b. Two wires in parallel with the current going in the same direction will tend to ____________ each other.

c. Two wires in parallel with the current going in opposite directions will tend to ____________ each other.

d. Two solenoids are identical in size and have the same number of windings. If one has an iron core, it will produce the ____________ magnetism.

e. It is ____________ (true or false) that electromagnetic devices and magnets interact.

10. List at least six magnetic/electromagnetic applications.

a. **b.**

c. **d.**

e. **f.**

SECTION 6-4 EXPERIMENTS

EXPERIMENT 1. RELAY

Objective: To provide practical knowledge and skills in the use of relays.

Introduction: In this experiment you will see how a circuit with lower voltage can control a circuit using a higher voltage.

Materials Needed:

1 SPST switch
1 6-V relay (Radio Shack 275-004 or equivalent)
1 12-V lamp with socket
1 Breadboard for constructing circuit and appropriate wires or leads

Equipment Required:

1 6-V power supply
1 12-V power supply

Procedure:

1. Construct the circuit shown in Figure 6-14, making sure that S_1 is open. Use one armature contact and the NO contact to wire the lamp circuit. At this time S_1 is open, the relay is not energized, the contacts are not closed, and the lamp is off. The voltage connected to the switch is ____________ V. The voltage connected to the lamp is ____________ V.

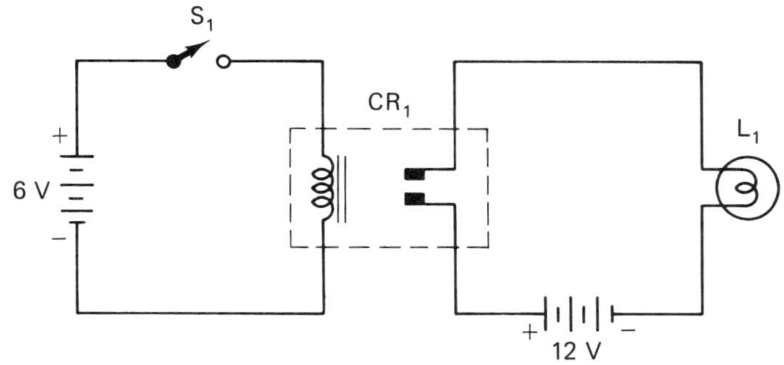

Figure 6-14

2. Close switch S_1. You may hear a slight click as the contacts of the relay close. What is the condition of the lamp at this time? ____________
3. Open switch S_1. What is the condition of the lamp at this time? ____________

Fill-In Questions:

1. A relay can be used to control a ____________ voltage with a low ____________ .

2. A relay can be used to ____________ two circuits with different voltages.

3. A relay is an ____________ device.

4. Relays are used in ____________, ____________, ____________, and manufacturing equipment.

SECTION 6-5 TROUBLESHOOTING APPLICATION: TESTING ELECTROMAGNETIC DEVICES

It is possible for us to make simple ohmmeter measurements on some electromagnetic devices to determine if they should function properly. The solenoid or relay can be checked as to whether there is continuity in its coil of wire. There should be several thousand ohms of resistance with a coil that has many turns. The ohmmeter can be connected to the contacts of the relay while it is energized to check for low resistance. Very often, with much use the contacts on relays can become dirty and pitted, which results in a poor connection between circuits.

Inoperative microphones, loudspeakers, and recording heads can be checked for continuity in their wire leads. Microphones and recording heads may have several thousand ohms of resistance, whereas loudspeakers will usually be 16 Ω or less. If a loudspeaker coil measures less than 4 Ω, it could have an internal short circuit. With a little thought and analysis you will be able to test the continuity of many electrical/electronic devices.

SECTION 6-6 INSTANT REVIEW

- Magnetism is a natural phenomenon whereby some materials are physically attracted to each other.
- Natural magnets are called lodestone and magnetite.
- Artificial magnets can be produced by stroking magnetic material with a magnet or with an electromagnetic field.
- A magnet has a north pole and a south pole.
- The magnetic field about the magnet consists of lines of force flowing from the north pole to the south pole.
- Magnetic lines of force are also called flux lines, and the flux density is the number of lines in 1 square meter.
- When magnets are brought close together, like poles repel each other and unlike poles attract each other.
- The magnetic domain theory basically states that the molecules in a magnetic material have a north and a south pole about the axis of their spin. When the molecules are in a random manner, the material is not magnetized. When the molecules are properly aligned, the material will become magnetized.
- The ease with which a material will concentrate the lines of magnetic force is called permeability.
- The opposition of a material to become magnetized is called reluctance.
- The amount of magnetism remaining in a material after the magnetizing force has been removed is referred to as retentivity.
- Residual magnetism is the same as retentivity.
- Permanent magnets have a high degree of residual magnetism.
- Ferromagnetic materials such as iron, nickel, and cobalt have a high degree of permeability and produce good magnets.
- Paramagnetic material such as aluminum, platinum, and chromium have very little permeability and cannot produce magnets of any quality.
- Diamagnetic materials such as copper, glass, and plastic have practically no permeability and cannot produce magnets.
- When electrical current passes through a conductor a magnetic field builds up about the conductor.
- Electromagnetism is the production of magnetism with electricity.
- Creating a magnetic field with electricity is called induction.
- Two parallel wires with current flowing in the same direction will be attracted to each other.
- Two parallel wires with current flowing in opposite directions will repel each other.
- A coil of wire with magnetic qualities is called a solenoid.
- A solenoid with a soft-iron core has stronger magnetic qualities than does a solenoid with an air core.
- A relay is an electromagnetic device, similar to a switch, with a movable armature that closes and/or opens electrical contacts.

SECTION 6-7 SELF-CHECKING QUIZ

Circle the most correct answer for each question.

1. Materials that produce excellent magnets are called:
   a. diamagnetic
   b. paramagnetic
   c. ferromagnetic
   d. magnetite

2. The external lines of force about a magnet flow from:
   a. north pole to south pole
   b. south pole to north pole
   c. in neither direction, but in a spiraling manner
   d. none of the above are correct

3. Like poles attract.
   a. True b. False

4. Plastic and glass will conduct magnetic lines of force.
   a. True b. False

5. The earth is similar to a gigantic magnet.
   a. True b. False

6. When a compass points toward the north it is really pointing toward the south magnetic pole.
   a. True b. False

7. The process of magnetizing a material by bringing it into the magnetic field of a permanent magnet or electromagnet is called:
   a. retentivity b. magnetic attraction
   c. hysteresis d. induction

8. A material that conducts magnetic flux very easily is:
   a. aluminum b. copper
   c. nickel d. plastic

9. The material iron has:
   a. high permeability and high reluctance
   b. high permeability and low reluctance
   c. low permeability and high reluctance
   d. low permeability and low reluctance

10. An example of residual magnetism is:
    a. a recording head
    b. an electromagnet
    c. a permanent magnet
    d. none of the above

11. The hysteresis loop is a:
    a. coil of wire used to show electromagnetism
    b. graph that indicates magnetization in two directions
    c. test device for sensing magnetic fields
    d. device for demagnetizing recording heads

12. Electromagnetism is:
    a. magnetism that produces electricity
    b. electricity that produces magnetism
    c. not very practical to use
    d. all of the above

13. A conductor that has current flowing away from you will produce a magnetic field in a:

a. clockwise direction

b. counterclockwise direction

c. direction the same as the current flow

d. none of the above

14. Two parallel conductors will tend to attract each other if the current in the conductors is flowing in:

a. the same direction

b. different directions

c. only one conductor

d. answers a and c are correct

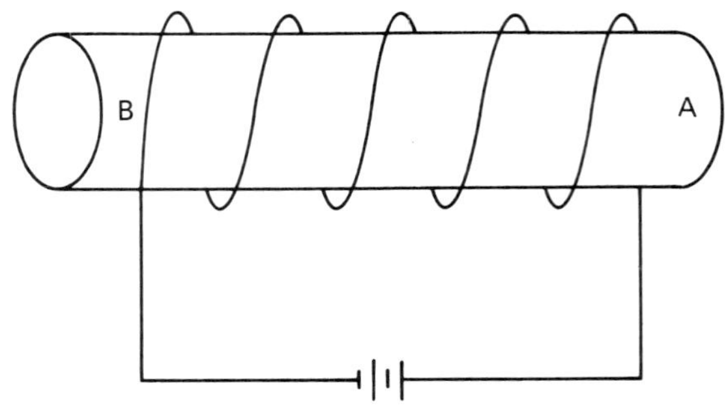

Figure 6–15

15. Referring to the solenoid in Figure 6-15, the north pole is the end marked:

a. A **b.** B

16. A solenoid that produces a stronger magnet is one with:

a. an air core **b.** a copper core

c. an iron core **d.** all of the above

17. If a circuit needed to be protected from magnetic lines of force, the best material to use as a shield would be:

a. copper **b.** aluminum

c. silver **d.** steel

18. If a relay is energized, the contacts that will be closed will be:

a. armature and NC

b. armature and NO

c. NO and NC

d. only the armature

19. The following item that is *not* an application of magnetism/electromagnetism is a:

a. a stereo sound system

b. a cassette tape recorder

c. a flashlight

d. a computer system

20. If a loudspeaker in a stereo sound system were distorting and your ohmmeter measurement across the input wires was 2.5 Ω, you could expect:

a. an open circuit of the voice coil

b. a partial short of the voice coil

c. a torn cone

d. a faulty permanent magnet

ANSWERS TO EXPERIMENT AND SELF-CHECKING QUIZ

Experiment 1: **(1)** high, low **(2)** isolate **(3)** electromagnetic **(4)** automobiles, washing machines, dryers, or electronic equipment

Self-Checking Quiz:

(1) c	**(2)** a	**(3)** b	**(4)** b	**(5)** a	**(6)** a	**(7)** d
(8) c	**(9)** b	**(10)** c	**(11)** b	**(12)** b	**(13)** a	**(14)** a
(15) a	**(16)** c	**(17)** d	**(18)** b	**(19)** c	**(20)** b	

Unit 7

Generators and Motors

INTRODUCTION

With the advent of robotics it is important for the electronics technician to understand the fundamentals of generators and motors. A technician must be able to determine if a problem in an electromechanical system is with the control circuits or in the mechanical parts of the system. This unit will provide the essential fundamentals to understanding generators and motors.

UNIT OBJECTIVES

Upon completion of this unit, you will be able to:

1. Explain how a current is induced in a conductor cutting lines of force.
2. Show how to use the left-hand generator rule.
3. Label the parts of an ac generator.
4. Describe how an ac sinusoidal wave is developed.
5. Label the parts of a dc generator.
6. Describe the voltage waveform of a dc generator.
7. Explain the motor motion of a current-carrying conductor in a magnetic field.
8. Demonstrate how to use the right-hand rule for motors.
9. Label the parts of a basic dc motor.
10. Analyze the armature rotation of a motor.
11. Identify the types of motor windings and their uses.
12. Explain how an induction motor operates.
13. Define the terms *field structure, armature, slip rings, brushes, commutator, torque, stator, rotor, lamination,* and *squirrel cage.*

SECTION 7-1 FUNDAMENTAL CONCEPTS

7-1a INDUCING A CURRENT IN A CONDUCTOR

In Unit 6 you learned how a conductor carrying a current would develop a magnetic field about it. Similarly, if a conductor is moved through a magnetic field, a current is induced in it as shown in Figure 7-1. When the conductor is moved upward through the flux lines as shown, electrons will flow toward the end of the conductor coming out of the page. The right end of the conductor will become positively charged. The voltage difference across the conductor is called an induced voltage or induced EMF. When the conductor stops in the field or is out of the field completely, there is no current induced. If the conductor is moved downward through the flux field, the induced current will flow in the opposite direction into the page.

The amount of EMF induced into the conductor is determined by four factors:

1. The strength of the magnetic flux field
2. The speed at which the conductor moves through the flux
3. The angle at which the conductor cuts through the flux
4. The length of the conductor in the flux field

For instance, the stronger the flux field, the greater the induced EMF will be. Also, the faster the conductor moves through the flux, the greater the induced EMF. The angle at which the conductor moves or cuts through the flux is very important. The maximum voltage is induced when the conductor moves at a right angle (or 90°) to the direction of the flux field. Less voltage is induced when the angle of cutting the flux field is less than 90°. If the conductor is moved in parallel to the direction of the flux field (from pole to pole), no EMF is induced. A longer conductor will also induce a greater EMF.

Figure 7–1 Inducing a current in a conductor.

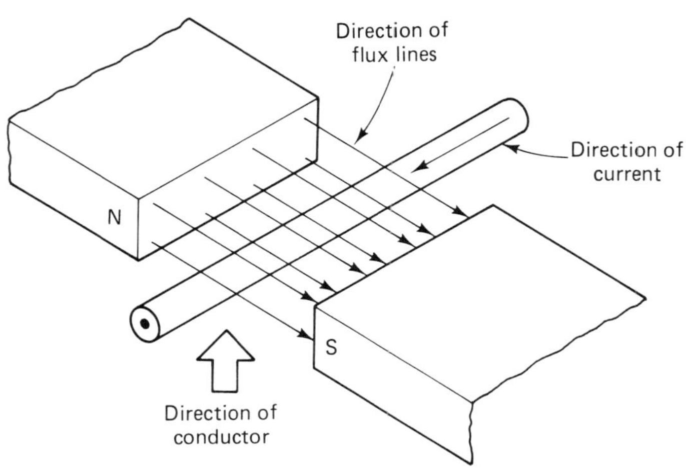

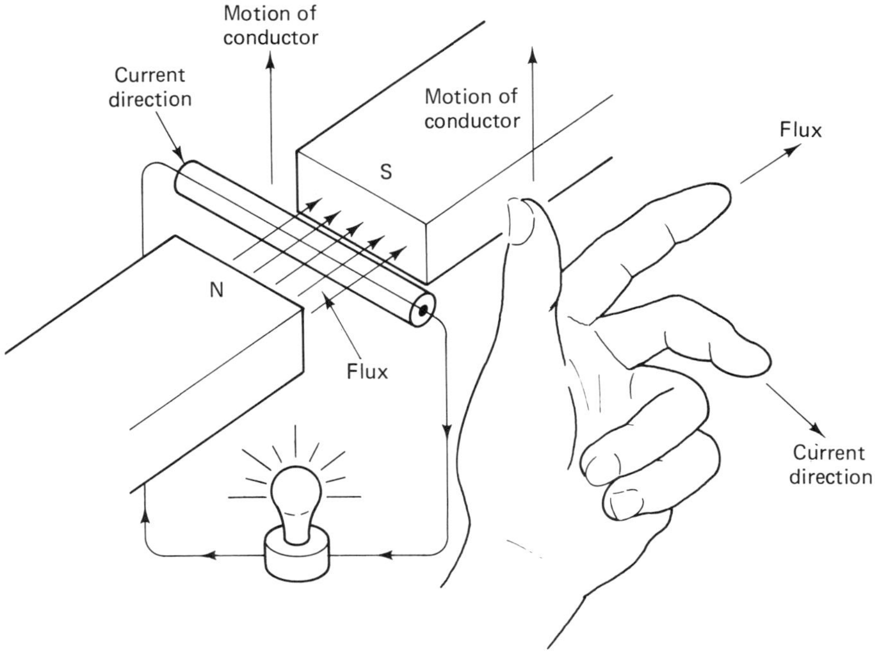

Figure 7–2 Left-hand generator rule.

Taking these four factors into account, a basic law still determines the amount of induced EMF. This law, called Faraday's law after its discoverer, Michael Faraday, states: The induced voltage of a conductor in a magnetic field is directly proportional to the rate at which the conductor cuts the magnetic lines of force. In other words, the more flux lines cut per second, the higher the induced EMF that will result.

The polarity of the induced EMF can be determined by another of the left-hand rules, called the left-hand rule for generator, which is shown in Figure 7-2. The thumb, index finger, and middle finger are positioned at right angles to each other, as shown. The index finger is pointed in the direction of the flux field. The thumb points to the direction the conductor is moving. The result is that the middle finger points to the direction of the induced current.

7-1b THE AC GENERATOR

Virtually all of the electrical power used in the world today is produced by electromagnetic induction. At present it is the most efficient and economical method of producing large amounts of electrical power. The basic device to accomplish this is the alternating-current (ac) generator, also called an alternator, shown in Figure 7-3a. This device converts mechanical energy into electrical energy by using electromagnetic induction. There are two magnetic pole pieces, called the field structure, which develop the magnetic field. Mechanical energy is required to establish relative motion between the magnetic field and a conductor. Either the magnetic field or the conductor can be movable. For this description a conductor named the armature rotates within the magnetic field. Slip rings are connected to the armature to prevent it from twisting and breaking while providing an electrical connection to the external circuit. Carbon brushes slide along the slip rings to complete the electrical circuit to the lamp. The schematic symbol for a generator is shown in Figure 7-3b. As the armature is rotated in the magnetic field, an induced current flows through the armature, to the slip rings,

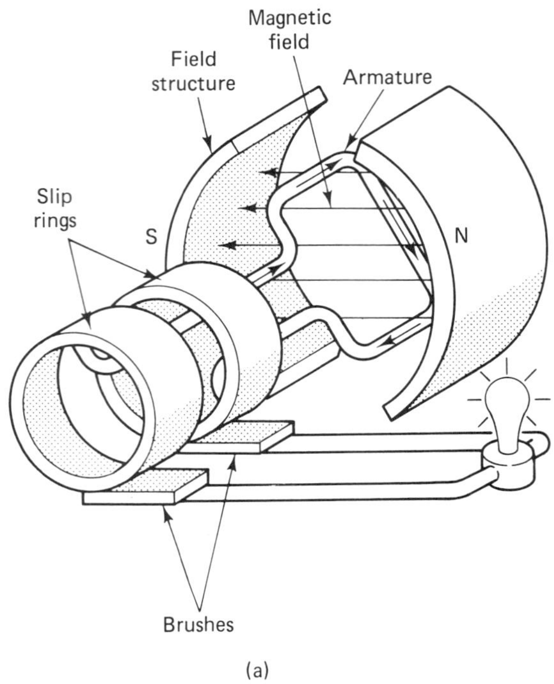

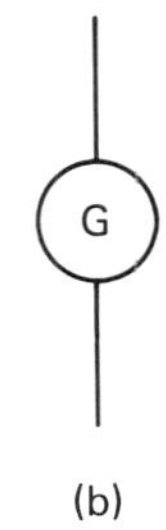

Figure 7–3 Ac generator (alternator): (a) pictorial representation; (b) schematic symbol.

through the brushes to the lamp. As long as the armature is rotated, the lamp will glow.

7-1c GENERATING ONE AC CYCLE

The current in the ac generator flows in two directions. Figure 7-4 will help to explain how one cycle of ac current or voltage is generated. The armature is at 0° and is not cutting any flux lines. There is no induced current at this time. The armature is then rotated in a counterclockwise direction and begins to cut lines of flux. A current is induced in the conductor with a direction of going into the A side and coming out of the B side. An EMF is developed at the brushes going in a positive direction. At 90° the armature is cutting the flux lines at the maximum angle; therefore, the maximum current and voltage are developed. The armature continues to rotate, but the cutting angle becomes less and the induced current decreases. The EMF at the brushes decreases. At 180° the armature is not cutting any flux lines and no current is induced. The armature continues to rotate in a counterclockwise direction, but now the induced current has changed directions. The current is flowing out of the A side and into the B side. A negative-going EMF is developed at the brushes, and at 270° the maximum angle of cutting occurs and the EMF is at maximum. As the armature continues to rotate the angle of cutting decreases, less current is induced, and the EMF decreases to zero at 360°. One cycle is now complete and the process begins over.

During one cycle the output voltage from the alternator has increased to +30 V, decreased back to 0 V, and then increased to −30 V. Figure 7-4b shows this relationship based on a circular rotation from 0° through 360° and the resulting output voltage waveform with reference to time. This type of voltage waveform is called a *sinusoidal waveform,* usually abbreviated *sine wave.* A sine wave has two alternations. The positive alternation occurs from 0 to 180° and its maximum ampli-

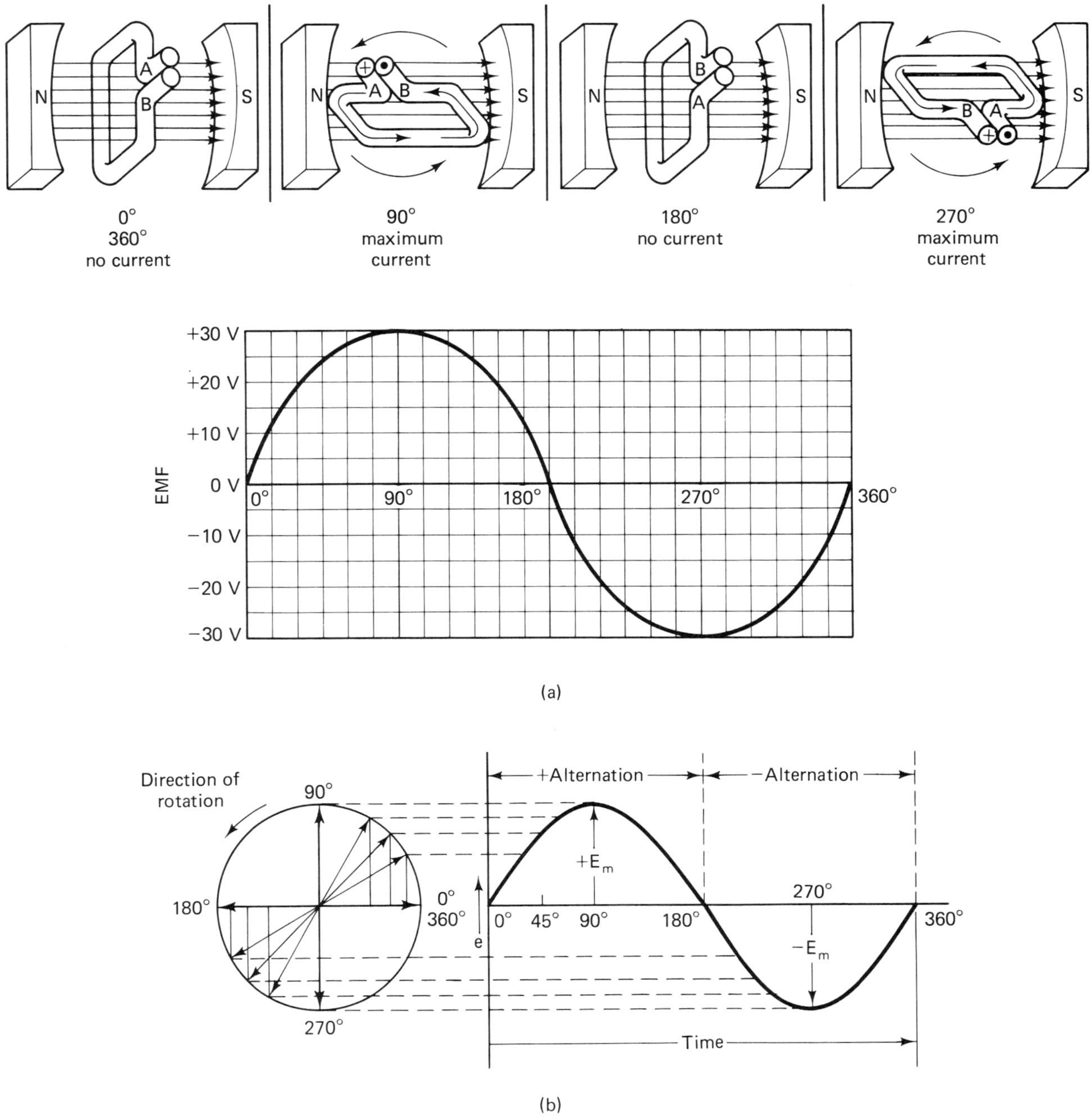

Figure 7–4 Generating one ac cycle: (a) cutting lines of force versus output voltage; (b) one revolution versus output voltage and time.

tude of voltage is indicated as $+E_m$. The negative alternation occurs from 180 to 360° and its maximum amplitude of voltage is indicated as $-E_m$. The current in a circuit using ac voltage will flow in one direction for the first alternation and in the opposite direction for the second alternation.

7-1d THE DC GENERATOR

The dc generator is similar to the ac generator except that the slip rings are replaced with a split contact known as a commutator. Figure 7-5a shows a dc generator with the brushes contacting each segment of the

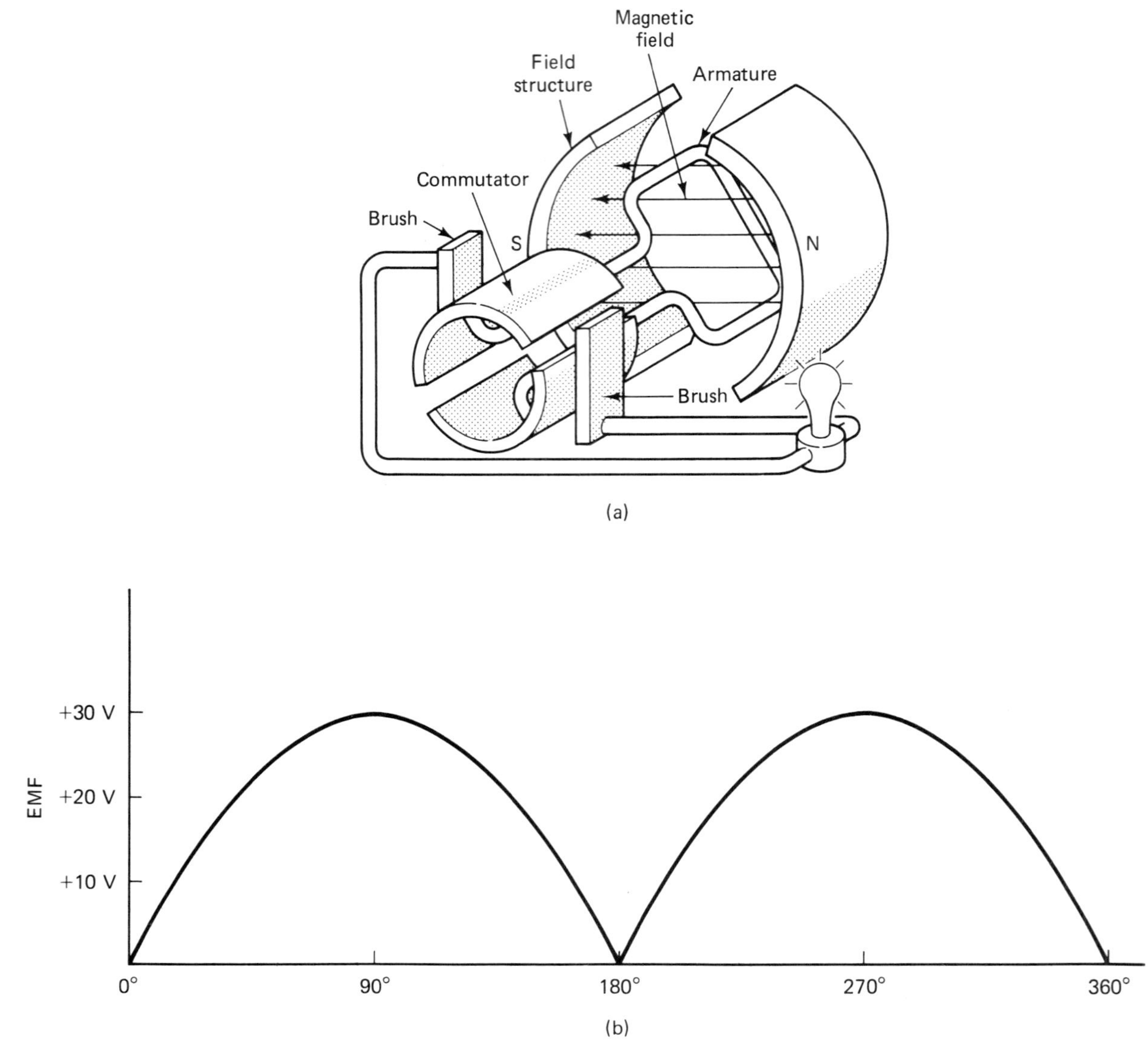

Figure 7–5 Dc generator: (a) pictorial showing commutator segments; (b) output voltage increasing in one direction.

commutator. Each end of the armature is connected to one segment of the commutator. As the armature is rotated a current is induced which flows to the commutator segments. A positive EMF is produced at the brushes for the first alternation. At the end of 180° the current in the armature reverses direction. However, the commutator segments are now contacting the opposite brushes; therefore, the output current continues to flow in the same direction. The resulting EMF at the brushes is also positive for the second alternation. The output voltage for a dc generator is called pulsating dc voltage, shown in Figure 7-5b. It begins at zero volts at 0°, increases to positive maximum voltage at 90°, decreases to zero volts at 180°, increases to maximum positive voltage at 270°, and then decreases to zero volts at 360° to complete the cycle. Remember that alternating current flows in one direction for 180° and then in the opposite direction for the next 180°, whereas direct current flows in the same direction for the 360° of the cycle.

7-1e MULTIPLE-POLE GENERATORS

In a practical generator the armature consists of many turns of varnish-coated single-strand wire that is wound on a steel or iron armature core. This increase in armature coils will produce a greater output voltage. The armature is placed into a circular case or yoke which contains the magnetic pole pieces. Figure 7-6a shows a simplified endview of a generator. The brushes are contacting slip rings or commutator segments attached to the mounting shaft. This generator has four pole pieces, which means that it will produce more output voltage. The resulting output of the four-pole ac generator would be a complex voltage waveform, but theoretically, what occurs is two sine waves 90° apart, as shown in Figure 7-6b. When one output is at maximum the other

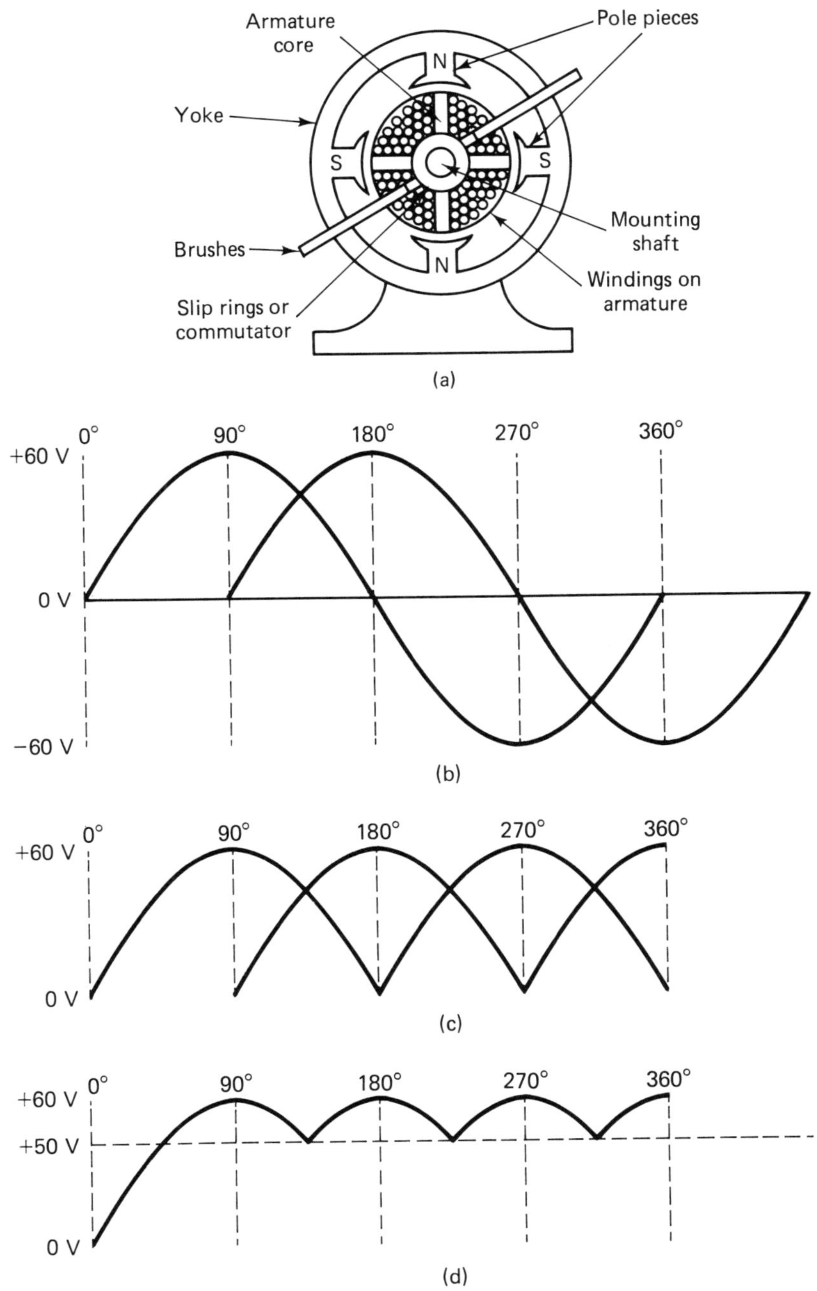

Figure 7–6 Four-pole generator: (a) end view of generator; (b) ac theoretical output; (c) dc theoretical output; (d) resulting practical output.

output is at zero, and vice versa. These voltages are said to be 90° out of phase.

For a four-pole dc generator the same action occurs, but the pulsating dc is out of phase by 90°, a shown in Figure 7-6c. The actual resulting output voltage of the dc generator is shown in Figure 7-6d. Before the voltage of the first alternation decreases back to zero, the voltage of the alternation shifted 90° pushes it back up to maximum, and vice versa. The resulting output voltage is a more stable dc voltage with a small varying dc or ripple voltage on the top. In other words, there is a constant +50 V dc with a varying dc voltage going back and forth from +50 V dc to +60 V dc.

7-1f MOTOR ACTION OF A CONDUCTOR

Electrical motors convert electrical energy into mechanical energy. There are many types of motors in existence, some of which are highly complex. To be an expert in motors and motor repair would be a lifetime occupation in itself. However, electronics technicians should know something about the theory of motors so that they can perform the simple ordinary maintenance problems associated with motors.

Generator action results when a conductor is moved through a magnetic field, which induces a current in the conductor. Motor action is the opposite of this. When current is sent through a conductor that is in a magnetic field, the conductor will tend to move in one of two directions. Figure 7-7a shows an example of motor action. The magnetic field is flowing from right to left and the current in the conductor is coming out of the page toward you. The lines of flux below the conductor are distorted, but exert a force on the conductor causing it to move upward. If the current in the conductor is reversed, the conductor will move downward. An easy aid for remembering this action is the right-hand rule for motors, shown in Figure 7-7b. Three fingers of the right hand are positioned at right angles to each other. The index finger points in the direction of the flux, the middle finger points in the direction of the current flow in the conductor, and the resulting movement of the conductor is indicated by the direction of the thumb.

Figure 7–7 Motor action: (a) motor motion of a current-carrying conductor; (b) right-hand rule for motors.

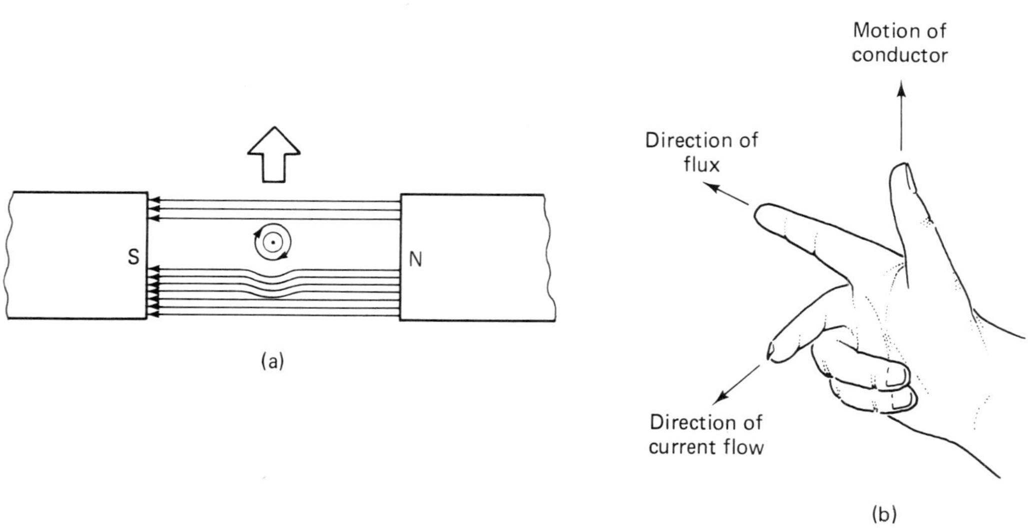

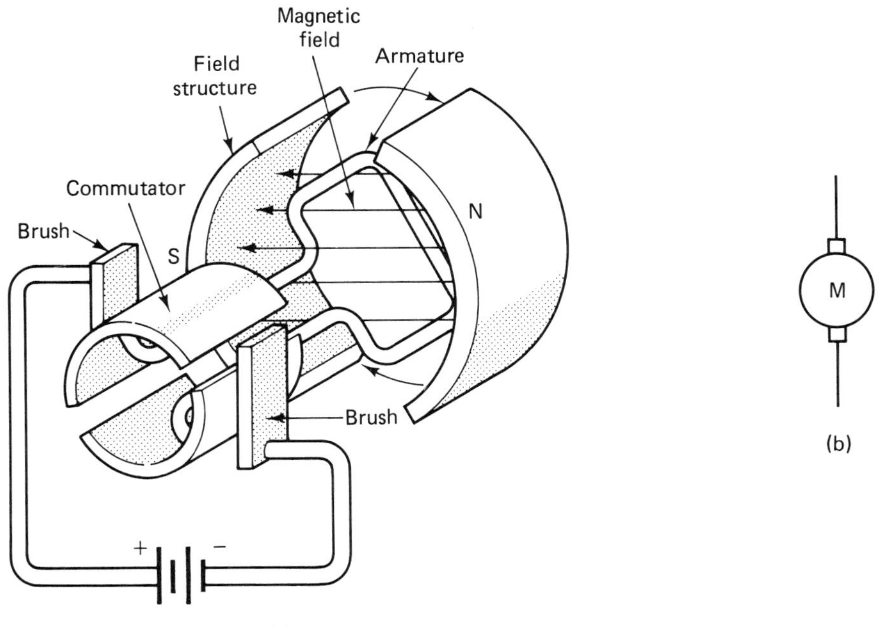

Figure 7–8 Basic dc motor: (a) pictorial representation; (b) schematic symbol.

7-1g THE BASIC DC MOTOR

A basic dc motor is very similar to the dc generator shown in Figure 7-8a. The only difference between the basic generator and the basic motor is that the load or lamp is replaced with a battery. Now, current is being sent into the armature instead of being induced in it. The resulting effect is that the armature begins to turn within the magnetic field. The battery supplies the electrical energy, which the motor converts into mechanical energy by the rotating of the armature. The schematic symbol for the motor is shown in Figure 7-8b.

Figure 7-9 illustrates another type of armature using a soft-iron

Figure 7–9 Rotation of an armature: (a) north pole and south pole develops; (b) repulsion; (c) attraction; (d) 180° later, commutator segments reverse current developing some polarity.

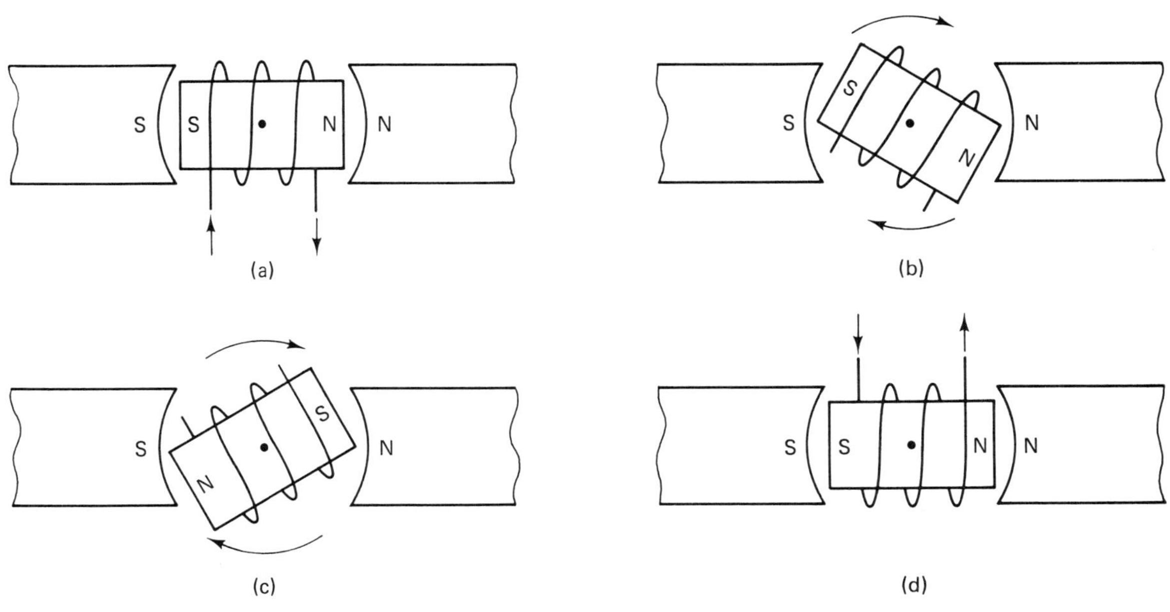

core upon which many turns of wire are placed. Initially, current is sent into the armature and a magnetic field develops (Figure 7-9a). The south poles and north poles repel each other and the armature begins to rotate (Figure 7-9b). As the armature rotates past 90° the south–north pole pairs begin to attract each other (Figure 7-9c). When the armature is at 180° the commutator segments are reversed and the polarity of the armature is also reversed (Figure 7-9d). However, since it is moving in a specific direction when this occurs it will continue in the same direction with a repeat of the previous action.

7-1h PRACTICAL MOTORS

The primary purpose of a motor is to develop sufficient force to cause the armature to rotate. The force that produces rotation or twisting is called *torque*. Therefore, motors are built to develope the torque needed to operate machinery and other devices, such as fans, drills, washing machines, dryers, cassette players, and videotape recorders.

The armature must have many turns of wire mounted on a soft-iron core, and the field poles must contain a large amount of magnetism. Figure 7-10 shows a practical dc motor structure. Enamel-insulated wire is used for the windings on the armature and the core is made up of plates or layers of soft iron in a process called *lamination.* Lamination reduces eddy current losses, which are discussed in the next section. The windings are connected to the commutator segments, all of which are mounted on the motor shaft (Figure 7-10a). The magnetism of the field poles is enhanced by having coils of wire wound around laminated pole pieces and connected to the input voltage to the motor. These coils of wire, referred to as *field coils,* are placed into the motor case (Figure 7-10b). Sometimes the armature is called the *rotor* and field coils the *stator.*

The manner in which the field coils and the armature windings are connected electrically determines the practical use of a motor. Figure 7-

Figure 7–10 Practical dc motor: (a) armature (rotor) assembly; (b) field (stator) assembly.

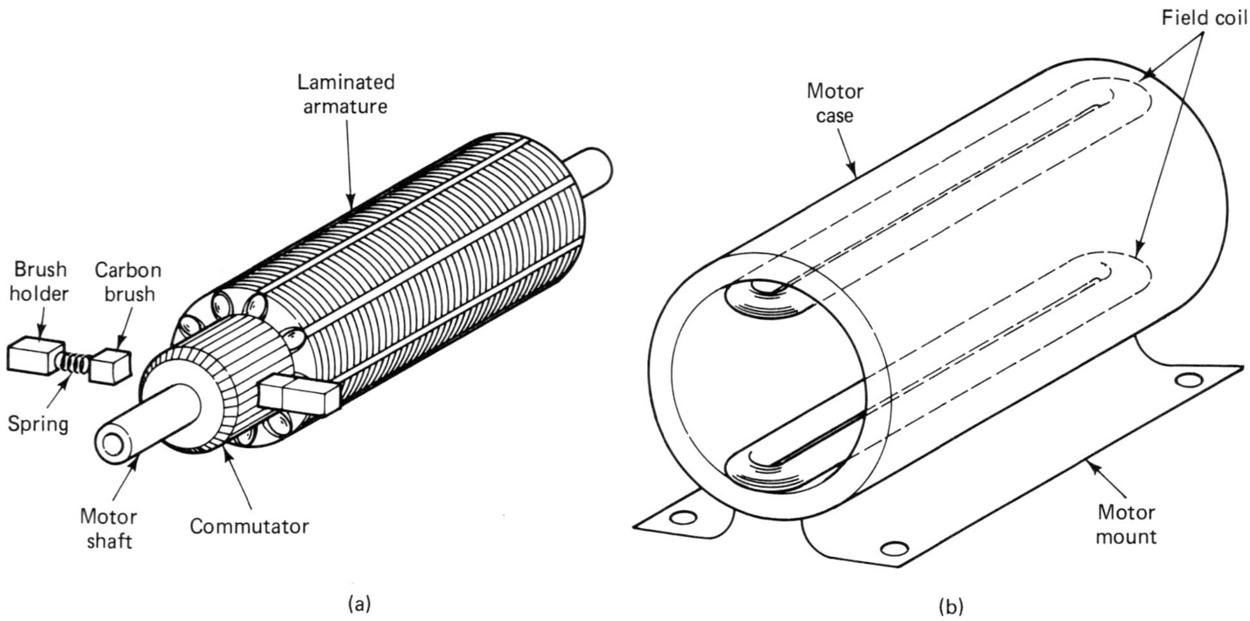

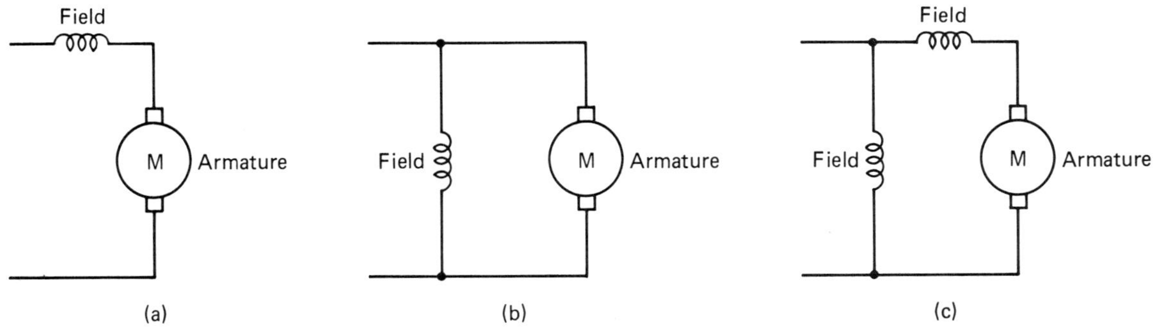

Figure 7–11 Types of motor windings: (a) series; (b) parallel; (c) compound.

11 shows the three basic types motor windings. If the field coil and the armature are wired in *series* (Figure 7-11a), the motor will have a high starting torque but very poor speed regulation. In fact, if the load on the motor varies, so will the speed of the motor. A typical example of this is a small hand drill. If the drill bit encounters more physical resistance, the speed of the motor tends to decrease. On the other hand, if this type of motor does not have a load, it will continue to increase its speed until it literally destroys itself.

A motor with the field windings and the armature wired in parallel, commonly called *shunt wound* (Figure 7-11b), has excellent speed regulation but very low torque. The fact that the current through the field windings does not flow through the armature results in better speed regulation. This type of motor is used where the load in minimal but the speed is critical.

A *compound motor* combines the characteristics of the series and shunt motors by using series and shunt field windings (Figure 7-11c). Some armature current flows in the series field windings, which results in a high-torque feature. The current in the shunt windings is separate from the armature current, which results in good speed regulation characteristics.

The *universal motor* is essentially a dc motor that is designed to operate at approximately the same torque and speed on either dc or single-phase ac. The universal motor is used in hand drills, vacuum cleaners, sewing machines, food fixers, and many other consumer products found around the house.

The induction motor is basically an ac motor. Figure 7-12a shows how the field windings develop the north and south poles in the stator. The ac normally used for this type of motor is three-phase and its application to the windings causes a rotating field about the stator. This rotating field induces current in the rotor, which in turn develops polarities and is virtually pulled around by the rotating stator field. The rotor is not constructed in the same manner as a normal armature, as shown in Figure 7-12b.

The rotor consists of circular laminated steel with large-gauge copper wire or bars. The copper bars form shorted loops, and as the magnetic flux cuts across the rotor, the induced current of the bars create a reverse flux that causes the rotor to turn. This type of rotor is referred to as a *squirrel cage,* because it resembles a device used for pet squirrels. The induction principle is used in many types of motors and can be found in some dc motors.

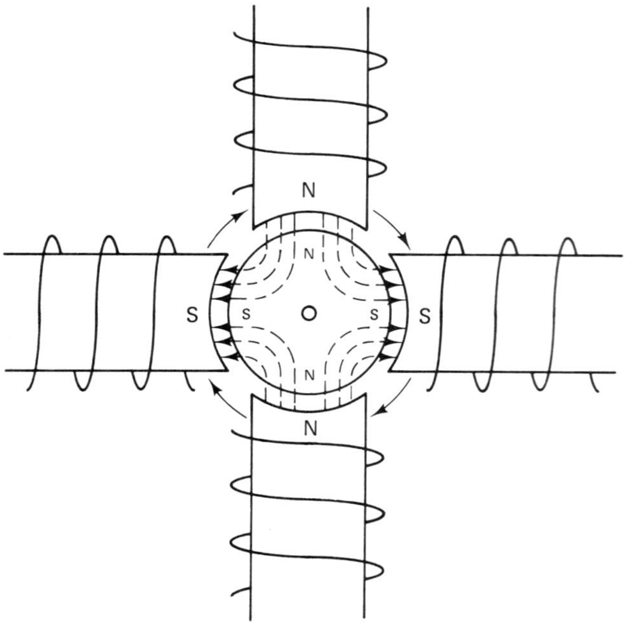

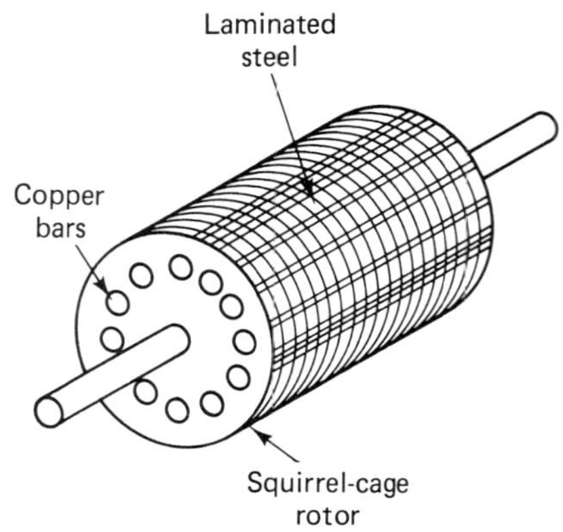

Figure 7–12 Induction motor: (a) magnetic fields; (b) rotor.

7-1i ELECTRICAL MACHINERY RATINGS AND LOSSES

Electrical machinery such as generators and motors have listed ratings so that they are used properly for a particular job or application. One of the most important ratings is horsepower. *Horsepower* (hp) is a unit of mechanical power, or the capacity of a mechanism to do work. It is equivalent to raising 550 pounds 1 foot off the ground in 1 second. Expressed as a formula: hp = 550 lb/1 ft/s.

One horsepower is equal to 746 W of electrical power; therefore, the output power (P_o) of a generator or motor can be determined. For example, a 2-hp motor will develop 1492 W of electrical power, as found by the formula

$$P_o = (\text{hp} \times 746)\ \text{watts}$$
$$= 2 \times 746 = 1492\ \text{W}$$

All devices have power losses. These losses in electrical/mechanical devices take several forms:

1. *Copper losses:* the resistance of the wire, which limits current flow.
2. *Counter EMF losses:* a reverse voltage produced when magnetic induction occurs, which creates a reverse polarity that opposes the normal and desired polarity.
3. *Hysteresis losses:* the opposition of a material to resist a change in the direction of the magnetizing force.
4. *I^2R or heat losses:* the power dissipated as a result of current flowing through a conductor, which is given up as heat.

5. *Eddy current losses:* the magnetic flux paths that flow through solid ferromagnetic materials produce heat, which results in the form of loss. The loss is reduced by the use of lamination. The thin plates used in lamination are coated with an insulating varnish to prevent eddy currents from flowing from one to another.
6. *Bearing or bushing friction losses:* the friction of the motor shaft in its bearings or bushings.

All of these losses determine the efficiency of a device. Efficiency is described as the output power (P_o) divided by the input power (P_i) or by the formula

$$\text{efficiency (percent)} = \frac{\text{output power}}{\text{input power} \times 100}$$
$$= \frac{P_o}{P_i} \times 100$$

The input power can be calculated from the Ohm's law power formula:

$$P_i = V_i \times I_i$$

For example:

$$P_i = 110 \text{ V} \times 15 \text{ A} = 1650 \text{ W}$$

The output power can be calculated from the horsepower-to-watts conversion formula. For example:

$$P_o = 2 \text{ hp} \times 746 \text{ W} = 1492 \text{ W}$$

Now the efficiency of the device can be calculated from the formula

$$\text{Efficiency} = \frac{P_o}{P_i} \times 100$$
$$= \frac{1492}{1650 \text{ W} \times 100}$$
$$= 0.9042 \times 100$$
$$= 90.42\%$$

The device has a very good efficiency of 90.42%. This efficiency may vary under various load conditions, and the actual efficiency be less than the calculated value.

An electromechanical device usually has a small metal plate attached to it that gives specifications and ratings. For example, a motor could have the following ratings:

1. *Hp:* the horsepower rating of the motor
2. *Volt:* the voltage required to run the motor
3. *Freq:* the frequency of the voltage if the motor operates on ac

4. *Amp:* the amperage the motor will draw when running
5. *Phase:* if the motor operates on ac, whether it is single phase or three phase
6. *RPM:* revolutions per minute, the normal operating speed of the motor
7. *AMB:* ambient or working temperature of the motor when it is running
8. *DC:* duty cycle, the time the motor is off (not running), as opposed to its on (running) time, since some motors do not operate continuously
9. *FN:* frame number, which designates the precise type of frame the motor is mounted on, since there are many sizes even for the same horsepower

SECTION 7-2 DEFINITION EXERCISES

Write a brief description for each of the following terms.

1. Induced current ______________________________

____

2. Generator ______________________________

____

3. Alternator ______________________________

____

4. Field structure ______________________________

____

5. Armature ______________________________

____

6. Slip rings ______________________________

____

7. Brushes ______________________________

____

8. Commutator ______________________________

____

9. Cycle ______________________________

____

10. Sinusoidal waveform ______________________________

____

11. Sine wave ______________________________

____

12. Alternation ______________________________

____

13. Alternating current (ac) ______________________________

____

14. Pulsating dc ______________________________

____

15. Pole piece ______________________________

____

16. Multiple-pole generator ______________________________

____

17. Motor ______________________________

____

18. Torque ______________________________

____

19. Lamination ______________________________

____

20. Field coil ______________________________

____

21. Rotor ______________________________

____

22. Stator ______________________________

____

23. Shunt wound ______________________________

____

24. Universal motor ______________________________

____

25. Induction motor ______________________________

____

26. Squirrel cage ______________________________

____

27. Horsepower ______________________________

____

28. Copper losses ______________________________

____

29. Counter EMF losses ______________________________

____

30. Hysteresis losses ______________________________

____

31. I^2R losses ______________________________

____

32. Eddy current losses ______________________________

____

33. Friction losses ______________________________

____

34. Device efficiency ______________________________

____

SECTION 7-3 EXERCISES AND PROBLEMS

Complete this section before beginning the next section.

1. List the four factors that determine the amount of current induced in a conductor that is moving through a magnetic field.

   a.

   b.

   c.

   d.

2. Describe the left-hand rule for generator action.

3. List the parts used in a basic ac generator.

   a.

   b.

   c.

   d.

4. Draw the schematic symbol for a generator.

5. Draw one cycle of a sine wave and label the parts (refer to Figure 7-4c).

6. List the parts used in a basic dc generator.

   a.

   b.

   c.

   d.

7. Draw a pulsating dc voltage waveform and label the parts (refer to Figure 7-5b).

8. Draw two sine waves with a 90° phase difference (refer to Figure 7-6b).

9. Describe the right-hand rule for motor action.

10. Explain the difference between a basic dc generator and a basic dc motor.

11. Draw the schematic symbol for a motor.

12. List the advantages of the three types of motor windings.

a. Series wound:

b. Shunt or parallel wound:

c. Compound wound:

13. Describe the basic principle of the induction-type motor.

14. Solve the outpower power for the following motor ratings.

a. 2.5 hp, P_o = ____________ W

b. 4.0 hp, P_o = ____________ W

c. 0.25 hp, P_o = ____________ W

15. Solve for the motor efficiency with the data given.

a. P_i = 2200 W, P_o = 1800 W, efficiency = ____________

b. P_i = 1200 W, P_o = 900 W, efficiency = ____________

c. V_i = 110 V, I_i = 12 A, P_o = 1200 W, efficiency = __________

d. V_i = 110 V, I_i = 15 A, P_o = 1.75 efficiency = __________

16. List the specifications and ratings you might find on a motor plate.

a. **b.**

c. **d.**

e. **f.**

g. **h.**

17. Match the description in column A with the proper term in column B.

Column A	*Column B*
____ **a.** A device that converts electrical energy to mechanical energy	**1.** armature
____ **b.** Force in a rotational direction	**2.** pulsating dc
____ **c.** Unit of measure for mechanical force	**3.** motor
____ **d.** The rotating part of a dc motor	**4.** generator
____ **e.** Current that reverses periodically with positive and negative values	**5.** efficiency
____ **f.** A device that converts mechanical energy to electrical energy	**6.** commutator
____ **g.** The brushes contact this in an ac motor	**7.** torque
____ **h.** The moving part of a motor	**8.** rotor
____ **i.** The brushes contact this in a dc generator	**9.** stator
____ **j.** A voltage that increases and decreases periodically, but whose current continues in the same direction	**10.** horsepower
____ **k.** The stationary part of a motor	**11.** ac
____ **l.** Output power divided by input power	**12.** slip rings

SECTION 7-4 EXPERIMENTS

EXPERIMENT 1. BASIC DC MOTOR/GENERATOR

Objective: To demonstrate the principles of a dc motor and a dc generator.

Introduction: In this experiment you will apply electrical energy to a motor and observe the conversion to mechanical energy. You will load down the motor and observe the increase of input current. You will be able to observe generator action with a reverse application of the motor.

Materials Needed:
1 1.5- to 3-V dc motor (Radio Shack 273-231 or equivalent)
2 Clip leads

Equipment Required:
1 3-V power supply
1 Multimeter or DMM

Procedure:

1. Connect the circuit shown in Figure 7-13.

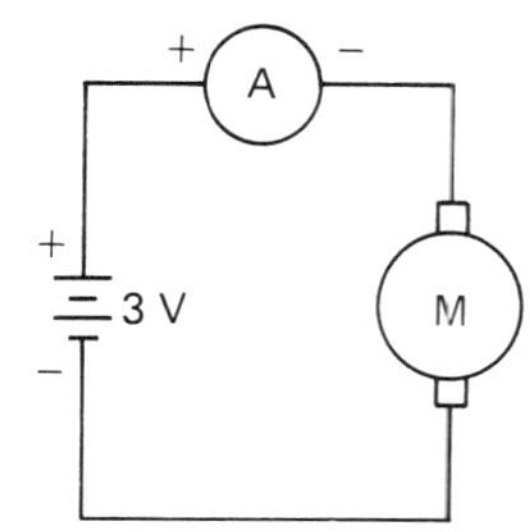

Figure 7–13

2. Observe the direction the motor is turning: clockwise or counterclockwise (circle one).
3. List the motor current here: ____________ mA (no load).
4. Grasp the motor shaft and slow the rotational movement down, but do not stop the motor.
5. List the motor current here: ____________ mA (loaded).
6. Reverse the power supply connections.
7. Observe the direction the motor is turning: clockwise or counterclockwise (circle one).

Figure 7–14

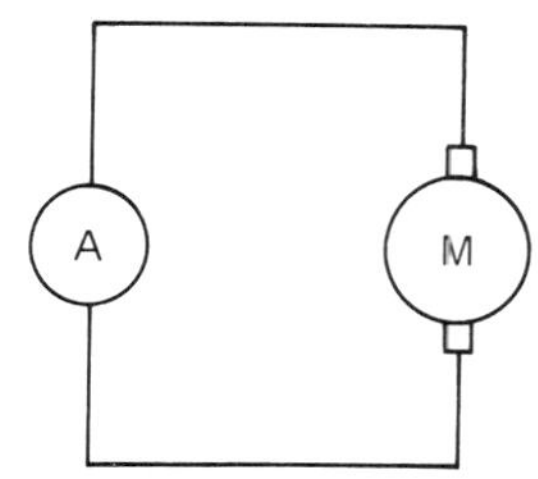

8. Connect the circuit as shown in Figure 7-14. The meter may be set to read current or voltage.
9. Spin the motor shaft with the fingers and notice the meter reading.

10. Spin the motor shaft in the reverse direction and notice the meter reading.

Fill-In Questions:

1. A motor converts ____________ energy to ____________ energy.

2. When the load on a motor increases, the current supplied to the motor ____________ .

3. Reversing the power supply connections to a motor will cause it to ____________ directions.

4. A generator converts ____________ energy to ____________ energy.

SECTION 7-5 TROUBLESHOOTING APPLICATION: BASIC MAINTENANCE ON MOTORS AND GENERATORS

In most instances the electronics technician does not become involved with the repair or rebuilding of motors and generators. However, there are some basic maintenance procedures that can be applied to motors and generators that may enable a system to start or at least keep it operating efficiently. Figure 7-15 illustrates the points on a motor/generator that can be checked by an electronics technician.

1. Make a visual inspection of all electrical connections and any broken mechanical parts, or even cracks in the case.

Figure 7–15

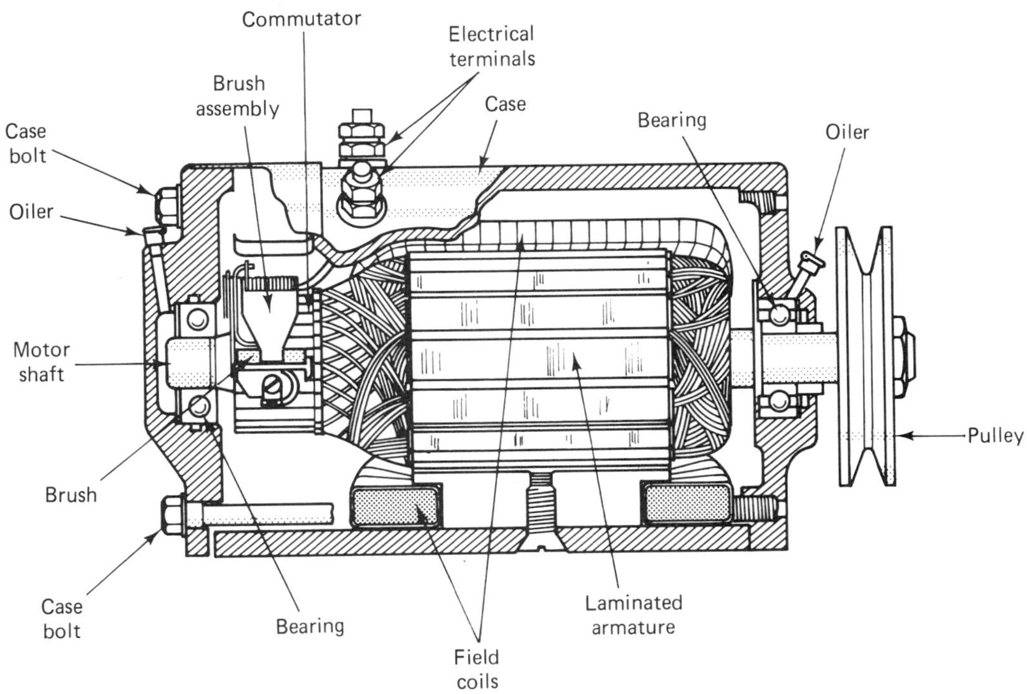

2. With some devices lubrication is required at the bearings. Care should be taken not to overlubricate these points so that oil gets into the electrical parts of the device.
3. Check for any shaft movement that may indicate worn bearings or bushings. Unless the bearings can be changed, the device may have to be replaced.
4. Check for worn brushes and that they are seating properly against the slip rings or commutator.
5. Clean off any oil or accumulation of dirt from the inside of the device.
6. The commutator may be dirty and can be cleaned by sanding it lightly with a very fine sandpaper. Never use emery paper or steel wool, since any metal particles could cause a short in the motor when it is running later.
7. The insulation between the segments on a commutator may wear more slowly than the copper segments, leaving the brushes to ride on insulated ridges. This may cause sparking. The insulation can be undercut below the level of the surface of the commutator segments with a very thin object such as a hacksaw blade.
8. Carbon tetrachloride can be used to clean the metallic parts of the device, but use it sparingly because it is highly toxic. *Do not touch the liquid or breathe the fumes.*
9. Careful observation very often reveals simple problems.

SECTION 7-6 INSTANT REVIEW

- Current is induced in a conductor when it is moved through a magnetic field.
- The amount of induced current depends on the (1) strength of the magnetic field, (2) speed the conductor is moving, (3) angle of cutting of the magnetic flux, and (4) length of the conductor.
- This induced effect is called generator action.
- There is a left-hand rule to describe generator action.
- A generator converts mechanical energy to electrical energy.
- A basic generator consists of field structure, armature, slip rings, or commutator and brushes.
- An ac generator or alternator uses slip rings.
- A dc generator uses a commutator.
- An ac voltage waveform, also called a sinusoidal waveform (sine wave) has two alternations, for a total of 360° or one cycle.
- The current of a sine wave flows one direction for 180°, and then in the other direction for the next 180°.
- Pulsating dc increases and decreases, but the current always flows in the same direction.
- Multiple-pole generators can produce multiple-voltage outputs that are out of phase.
- If a conductor sitting in a magnetic field has current passed through it, it will tend to move in one of two directions.
- This movement of the conductor is called motor action.
- There is a right-hand rule to describe motor action.

- An electrical motor converts electrical energy to mechanical energy.
- A basic motor has the same parts as a basic generator.
- A generator produces a voltage at its brushes, while a voltage is placed at the brushes of a motor.
- Torque is a rotational or twisting force.
- Lamination is used to reduce eddy currents.
- A series-wound motor has high torque.
- A shunt-wound motor has good speed regulation.
- A compound-wound motor has both high torque and good speed regulation.
- A rotor is the moving part of an electrical machine.
- A stator is the stationary part of an electrical machine.
- A squirrel cage is used in induction motors.
- The universal motor is used in hand drills, fans, vacuum cleaners, and other consumer products.
- One horsepower is defined as the amount of power capable of moving 550 pounds 1 foot off the ground in 1 second.
- One horsepower is equal to 746 watts of electrical power.
- The efficiency of a device is its output power divided by its input power.

SECTION 7-7 SELF-CHECKING QUIZ

Circle the most correct answer for each question.

1. If the speed of a conductor moving through a magnetic field at 90° is decreased, the induced current:

a. decreases **b.** increases

c. remains the same **d.** does not exist

2. A device that converts mechanical energy to electrical energy is the:

a. universal motor **b.** induction motor

c. alternator **d.** rotor

3. A left-hand rule can be used to explain motor action.

a. True **b.** False

4. An ac generator uses slip rings.

a. True **b.** False

5. A voltage that increases and decreases with a current that changes direction every 180° is called pulsating dc.

a. True **b.** False

6. If the voltage applied to a motor decreases, the speed of the motor will:

a. increase **b.** decrease

c. not change **d.** be undetermined

7. A particular motor is rated at 0.25 hp. Its output power in watts will be:

a. 746 W **b.** 1865 W

c. 186.5 W **d.** 373 W

8. A shunt-wound motor has high torque.

a. True **b.** False

9. One alternation lasts for 180°.
   a. True b. False

10. A sine wave has two alternations.
    a. True b. False

11. A particular motor has an input power of 1200 W and an output power of 900 W. Its efficiency is:
    a. 92% b. 87%
    c. 60% d. 75%

12. The operating speed of a motor is indicated as:
    a. AMB b. RPM
    c. FN d. Amp

13. A device that converts electrical energy to mechanical energy is the:
    a. induction motor b. generator
    c. stator d. none of the above

14. A loss incurred in a motor as a result of the changing of direction of magnetic fields is called:
    a. copper loss b. I^2R loss
    c. hysteresis loss d. eddy current loss

15. A particular motor is rated at 3.5 hp. The input voltage is 110 V with an input current of 25 A. The efficiency of this motor is approximately:
    a. 78% b. 95%
    c. 87% d. 70%

16. Reversing the voltage connections to a dc motor will cause it to:
    a. do nothing b. stop
    c. change directions d. none of the above

17. A right-hand rule can be used to explain motor action.
    a. True b. False

18. Force in a rotational direction is called:
    a. rotor b. horsepower
    c. torque d. hysteresis

19. A motor becomes inoperable and does not turn. One simple test an electronics technician could perform is:
    a. to replace the bearings on the shaft
    b. to rewind the armature
    c. to replace the field coils
    d. to make sure the brushes are seated properly

20. The best solvent to use in cleaning electrical devices is:
    a. water b. alcohol
    c. gasoline d. carbon tetrachloride

ANSWERS TO EXPERIMENT AND SELF-CHECKING QUIZ

Experiment 1: (1) electrical, mechanical (2) increases (3) change (4) mechanical, electrical

Self-Checking Quiz (1) a (2) c (3) b (4) a (5) b (6) b (7) c (8) b (9) a (10) a (11) d (12) b (13) a (14) c (15) b (16) c (17) a (18) c (19) d (20) d

Unit 8

Alternating-Current Analysis

INTRODUCTION

Modern societies exist partially, if not mainly, because of alternating current (ac) and its applications to everyday life. The principles of alternating current are used in the transmission of electrical power, radio and television broadcasting, manufacturing of goods, medical equipment, home entertainment systems, space exploration, and countless other uses to aid existence on this planet. An electronics technician must understand the various voltage waveforms and terminology to effectively troubleshoot and service electrical/electronic equipment and systems.

UNIT OBJECTIVES

Upon completion of this unit, you will be able to:

1. Define an ac voltage waveform and list its nomenclature.
2. Convert from peak-to-peak voltage to instantaneous, effective, and average voltage.
3. Explain frequency and calculate its value from a given period of time.
4. Define frequency ranges and their general use.
5. Explain complex voltage waveforms.
6. Show the relationship of harmonics to sound.
7. Define and solve sound levels in terms of decibels.
8. Label nonsinusoidal voltage waveforms.
9. Use a basic function generator.
10. Use a basic frequency counter.
11. Explain the basic operation of an oscilloscope.
12. Use an oscilloscope to measure various voltage waveforms.
13. Recognize basic voltage waveform distortions.
14. Define the terms *vector, phasor, right triangle, sin, frequency, pulse repetition rate, harmonics, square wave, sawtooth wave, triangle wave, leading edge, trailing edge, pulse width, amplitude,* and *cathode ray tube.*

SECTION 8-1 FUNDAMENTAL CONCEPTS

8-1a SINE-WAVE NOMENCLATURE

In Section 7-1c you were shown how a sine wave is generated by an alternator. The rotation of 360° was given to the relationship of time (refer to Figure 7-4b). Some parts of the sine wave were given at that point; however, a more complete nomenclature is shown in Figure 8-1. One *cycle,* also called the *period,* of a sine wave lasts for 360°. An *alternation* lasts for 180° and there are two alternations for each cycle. The level or volume of a voltage or current waveform that increases and decreases from a reference point, usually ground, is called *amplitude.*

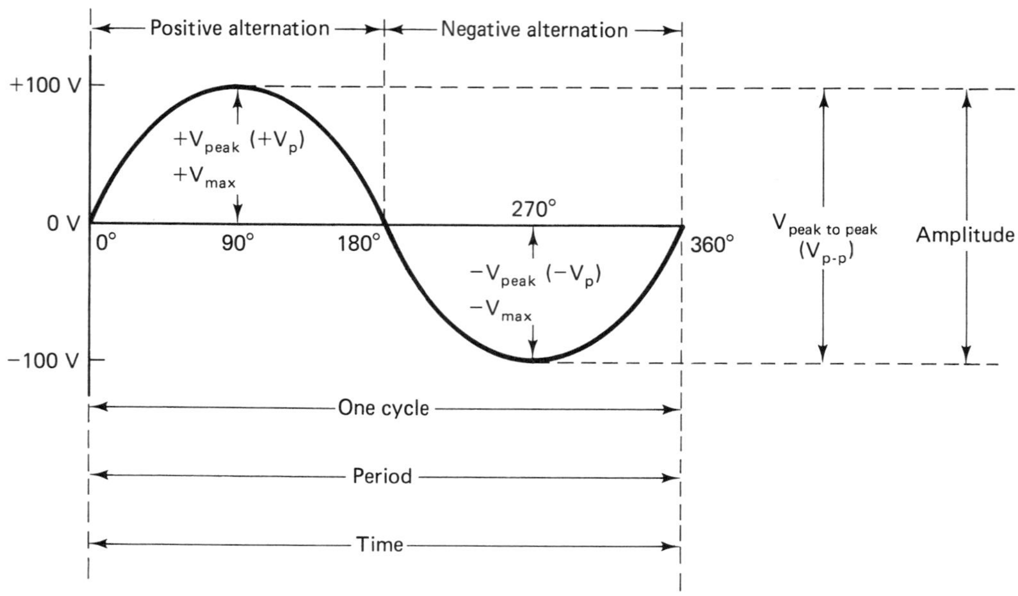

Figure 8-1 Sine-wave nomenclature.

The maximum amplitude of an alternation is referred to as the maximum voltage (V_{max}) or peak voltage (V_p). The total voltage amplitude of the waveform is called the peak-to-peak voltage ($V_{p\text{-}p}$). Notice that the peak voltage is equal to the peak-to-peak voltage divided by 2, or expressed as a formula, $V_p = V_{p\text{-}p}/2$.

8-1b VOLTAGE AND CURRENT CONVERSIONS

Very often in working with electrical/electronic circuits it is necessary to convert peak-to-peak voltages and currents to other equivalent voltages and currents. Since the calculations are the same for current, only voltage conversions will be discussed. There are three common types of voltages that you may need to calculate: instantaneous, effective, and average.

8-1b.1 Calculating Instantaneous Voltage

An *instantaneous voltage* is the voltage at any point of a sine wave. Since the amplitude of a sine wave is constantly changing, the voltage at any point will be present only for an instant. A *vector* is a representation of a force that has magnitude and direction. The length of the line indicates magnitude and the arrowhead shows direction. The armature of the alternator is a rotational force that can be represented with a vector, but a newer term assigned to a rotating vector is *phasor*. Figure 8-2 shows the concept of the phasor.

The phasor begins at 0° and rotates in a counterclockwise direction. The voltage at any instant that the armature is rotating can be found by using the principles of the right triangle from trigonometry.

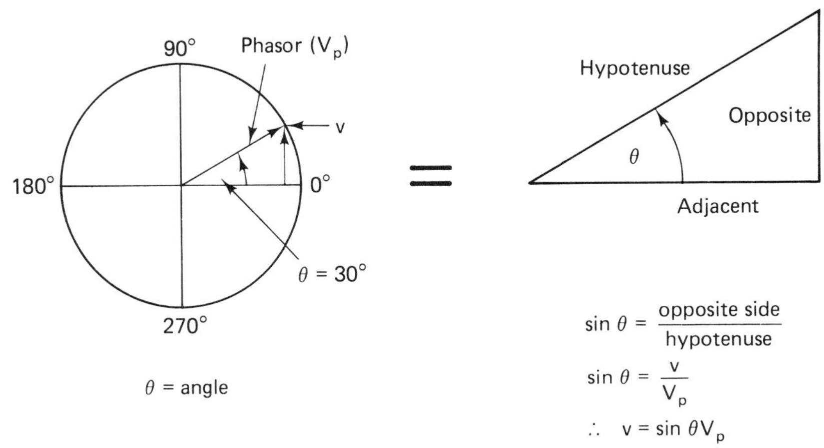

Figure 8-2 Relationship of right triangle to ac.

There are six basic functions of the right triangle, but we need only concern ourselves with one for finding the value of an instantaneous voltage. The *sine function* (abbreviated "sin") is a number that is related to an angle and depends on the ratio of the opposite side of a right triangle to its hypotenuse or ramp side. The sine of an angle is found by dividing the opposite side by the hypotenuse, or expressed as a formula,

$$\sin\theta = \frac{\text{opposite side}}{\text{hypotenuse}}$$

In Figure 8-2 you can see that the instantaneous voltage v represents the opposite side of a right triangle and the peak voltage V_p indicates the hypotenuse. From this information the sine of the angle can be found and is given by the formula

$$\sin\theta = \frac{v}{V_p}$$

Once the sine is found, the number can be looked up in a table of trigonometric functions or solved by a scientific hand-held calculator to find the angle.

If the formula is rearranged, we can find the value of an instantaneous voltage when the angle and peak voltage is known as given by the formula

$$v = \sin\theta \times V_p$$

For example: it is desired to know the instantaneous voltage of a sine wave with a peak-to-peak voltage of 200 V when it reaches 20° of the cycle. Remember, the peak voltage is one-half of the peak-to-peak voltage and therefore equals 100 V. Plugging these values into the formula will give us the result:

$$\begin{aligned} v &= \sin 20° \times V_p \\ &= 0.342020 \times 100 \text{ V} \\ &= 34.2020 \text{ V} \end{aligned}$$

Several angles and their equivalent sines are shown in Table 8-1. The voltage at 45° is found thus:

$$\begin{aligned} v &= \sin \theta \times V_p \\ &= 0.707107 \times 100 \text{ V} \\ &= 70.107 \text{ V} \end{aligned}$$

The sine of 90° is equal to 1; therefore, this is the peak voltage:

$$\begin{aligned} v &= \sin \theta \times V_p \\ &= 1.000000 \times 100 \text{ V} \\ &= 100 \text{ V} \end{aligned}$$

To find the instantaneous voltage between 90 and 270° you subtract the given angle from 180° and then solve the formula in the normal manner. For example, the instantaneous voltage for an angle of 150° is found as: 180° − 150° = 30°. Therefore,

$$\begin{aligned} v &= \sin 30° \times V_p \\ &= 0.500000 \times 100 \text{ V} \\ &= 50.0000 \text{ V} \end{aligned}$$

For an angle of 240°: 180° − 240° = −60°. Therefore,

$$\begin{aligned} v &= \sin -60° \times V_p \\ &= -0.866825 \times 100 \text{ V} \\ &= -86.6825 \text{ V} \end{aligned}$$

The negative sign indicates that the voltage is in the negative alternation. For an angle between 270 and 360° you subtract 360° from the

TABLE 8-1

Sample Angle–Sine Functions

Angle, θ (deg)	*Sin*
5	0.087155
10	0.173648
15	0.258819
20	0.342020
30	0.500000
45	0.707107
60	0.866825
80	0.984808
90	1.000000

given angle; for example, for an angle of 315°: 315° − 360° = −45°. Therefore,

$$\begin{aligned} v &= \sin -45^\circ \times V_p \\ &= 0.707107 \times 100\ \text{V} \\ &= -70.7107\ \text{V} \end{aligned}$$

8-1b.2 Calculating Effective Voltage

Effective voltage is the ac voltage required to dissipate the same amount of power in a circuit that would be dissipated by a dc voltage. Figure 8-3 illustrates this principle. Since an ac voltage is constantly changing, the power dissipated by the circuit is also varying and is

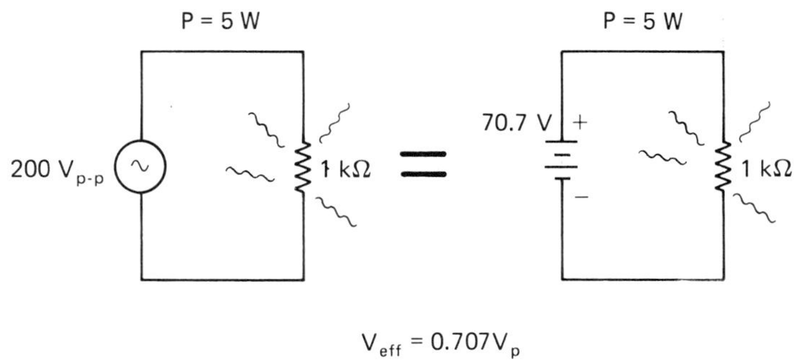

Figure 8–3 Comparison of effective voltage to dc voltage.

never constant for any moment. However, the overall heating value is compared to the dc voltage level that would produce the same power in the circuit. Mathematically, the value of this effective voltage is 70.7% of peak voltage, or expressed as a formula,

$$V_{\text{eff}} = 0.707 V_p$$

As an example, what is the effective voltage of a sine wave with 200 V peak to peak? Remembering that V_p is one-half of $V_{\text{p-p}}$, then, 200 V/2 = 100 V; therefore,

$$\begin{aligned} V_{\text{eff}} &= 0.707 V_p \\ &= 0.707 \times 100\ \text{V} \\ &= 70.7\ \text{V} \end{aligned}$$

The peak voltage can be found from the effective voltage by rearranging the formula:

$$\begin{aligned} V_p &= \frac{V_{\text{eff}}}{0.707} \\ &= \frac{70.7\ \text{V}}{0.707} \\ &= 100\ \text{V} \end{aligned}$$

or

$$V_p = V_{\text{eff}} \times \frac{1}{0.707}$$

and 1/0.707 = 1.414; therefore,

$$\begin{aligned} V_p &= 1.414\,V_{\text{eff}} \\ &= 1.414 \times 70.7\ \text{V} \\ &= 99.9698\ \text{V} \end{aligned}$$

The peak-to-peak voltage is twice the value of the peak voltage; therefore, the formula can be written as

$$\begin{aligned} V_{\text{p-p}} &= 2.828\,V_{\text{eff}} \\ &= 2.828 \times 70.7\ \text{V} \\ &= 199.9396\ \text{V} \end{aligned}$$

Effective voltage is also called *rms voltage,* after the mathematical method used to find the value 0.707. The letters "rms" stand for "root mean square." As a guide, remember that

$$V_{\text{eff}} = V_{\text{rms}}$$

8-1b.3 Calculating Average Voltage

Since an ac voltage is constantly changing, its average value can be found by summing instantaneous voltages in one alternation and then dividing this number by the total instantaneous voltages. Using integral calculus to solve this calculation would be the most accurate method, but in either case the average voltage is equal to approximately 63.7% of the peak voltage. Expressed as a formula,

$$V_{\text{av}} = 0.637\,V_p$$

As an example, what is the average voltage if the peak voltage is 50 V? Using the given formula yields

$$\begin{aligned} V_{\text{av}} &= 0.637\,V_p \\ &= 0.637 \times 50\ \text{V} \\ &= 31.85\ \text{V} \end{aligned}$$

By rearranging the formula, we can find the peak voltage:

$$\begin{aligned} V_p &= \frac{V_{\text{av}}}{0.637} \\ &= \frac{31.85\ \text{V}}{0.637} \\ &= 50\ \text{V} \end{aligned}$$

Notice that the average voltage was calculated for one alternation. In a complete cycle there is one positive alternation and one negative alternation; therefore, the average voltage would be zero. However, in the case of pulsating dc, as discussed in Section 7-1d, the average voltage value could be used since the current flows in one direction.

The various formulas for converting ac voltages can be summarized as follows:

$$V_{\text{p-p}} = 2V_p$$
$$V_p = \frac{V_{\text{p-p}}}{2}$$
$$V_p = \frac{V_{\text{eff}}}{0.707}$$
$$V_p = \frac{V_a}{0.637}$$
$$v = \sin\theta \times V_p$$
$$V_{\text{eff}} = 0.707V_p$$
$$V_{\text{av}} = 0.637V_p$$

8-1c FREQUENCY

Frequency is defined as the number of cycles of a waveform that occur in a specified time. Usually, the period of time is based on the second or fraction of a second. Figure 8-4 shows the concept of frequency.

When two cycles occur in 1 second (Figure 8-4a), the frequency of the waveform is said to be two cycles per second. Cycles per second has been replaced with a single term caled *hertz,* named after Heinrich Hertz, who initially performed experiments with frequency. A hertz is the unit of frequency equal to 1 cycle per second and is abbreviated Hz. The period or time duration of one cycle can be calculated by the formula $p = 1/f$, where f is the frequency in hertz. For example, the period of the frequency for 2 Hz is

$$p = \frac{1}{f}$$
$$= \frac{1}{2\text{ Hz}}$$
$$= 0.5\text{ s}$$

If the frequency doubles (Figure 8-4b), its period will be one-half of the original value and can be proven thus:

$$p = \frac{1}{f}$$
$$= \frac{1}{4}\text{ Hz}$$
$$= 0.25\text{ s}$$

If four cycles occur during 1 ms (Figure 8-4c), the period can be found by dividing 1 ms by 4, or $p = 0.25$ ms. Since the period is 0.25 ms, the frequency can be calculated using the formula

$$f = \frac{1}{p}$$
$$= \frac{1}{0.25\text{ ms}}$$
$$= \frac{1}{0.25 \times 10^{-3}\text{ s}}$$
$$= 4 \times 10^3\text{ Hz}$$
$$= 4\text{ kHz}$$

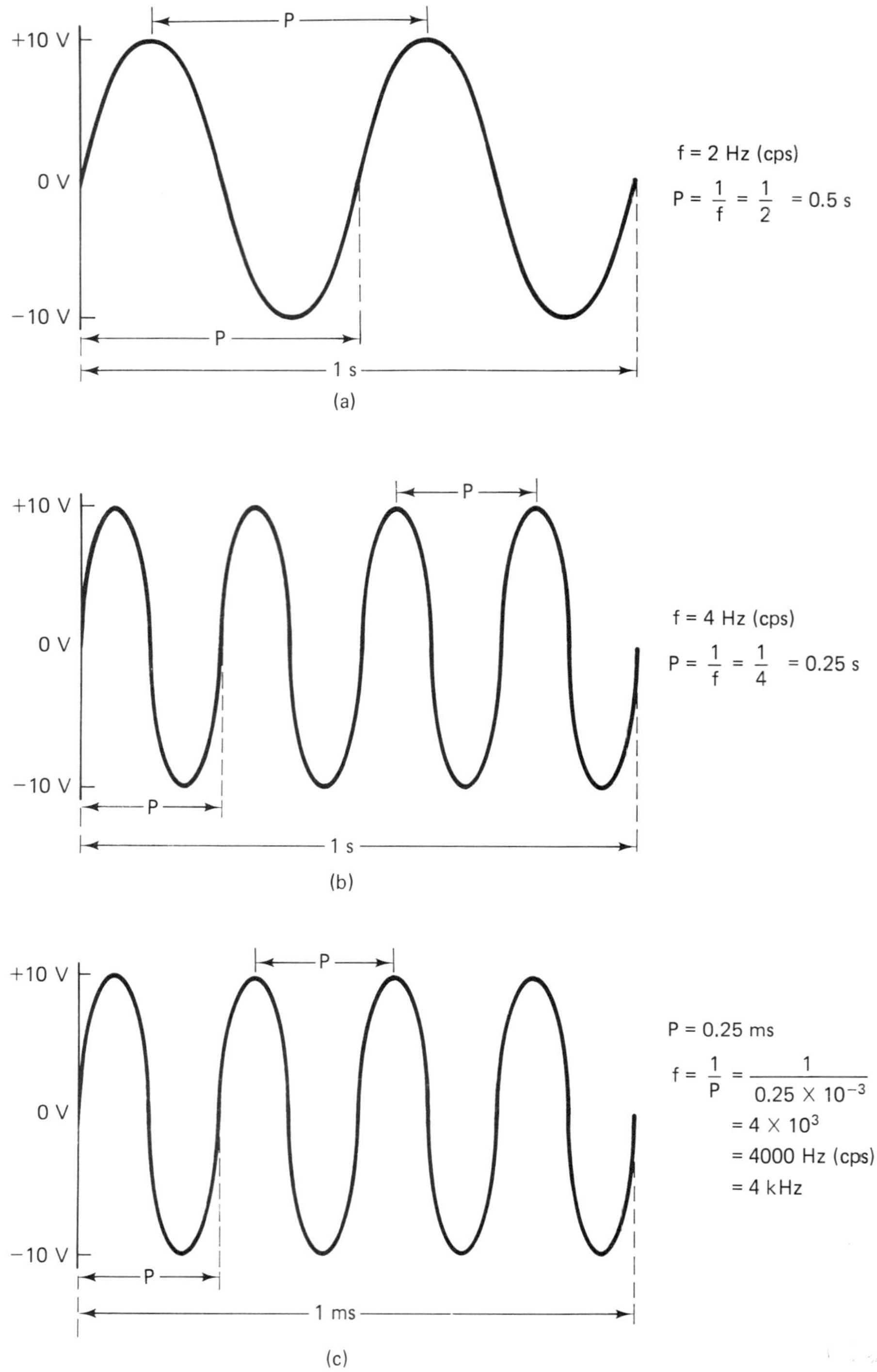

Figure 8–4 Frequency: (a) $f = 2$ Hz; (b) f = 4 Hz; (c) f = 4 kHz.

Applications using frequencies have been assigned or designated to various ranges. Table 8-2 shows a general designation of the frequency ranges. The lower frequencies below 100 Hz are used for commercial power transmission, the United States being 60 Hz and Europe, 50 Hz. The audio (meaning sound) range is from 20 Hz to 20 kHz. Radio and television broadcasting ranges extend from 30 kHz up to 300 GHz. Frequencies in this group are called radio frequencies (RF). Within this range are the medium frequency (MF), very high frequency (VHF), and

TABLE 8-2
General Designation of Frequency Ranges

Designation	*Frequency Range*
Power	Less than 100 Hz U.S. = 60 Hz Europe = 50 Hz
Audio	20 Hz to 20 kHz
Maritime radio	30 to 535 kHz
Broadcast AM radio (MF band)	536 to 1605 kHz
Amateur radio	1.605 to 30 MHz
Citizen's band radio	26.48 to 470 MHz
Broadcast FM radio (VHF band)	88 to 108 MHz
Aeronautical navigation	108 to 122 MHz
Standard television (VHF and UHF)	54 to 890 MHz
Government, radar, experimental, misc.	890 to 300 GHz
Infrared light rays	300 to 400,000 GHz
Visible light rays	400 to 750 THz
Ultraviolet light rays	750 to 3000 THz
X-rays	10^{18} to 10^{20} Hz
Gamma rays	10^{20} to 10^{21} Hz
Cosmic rays	10^{21} to 10^{22+} Hz

ultrahigh frequency (UHF) bands. Next are the light ranges, with infrared light rays (pertaining to heat) going from 300 GHz to 400 THz, visible light from 400 to 750 THz, and ultraviolet light from 750 to 3000 THz. Beyond this are the frequencies used for medical purposes, which are X-rays and gamma rays. The highest frequencies are in the cosmic range, which are minute particles from outer space and are capable of penetrating nearly every known material on earth. Some cosmic rays have been known to penetrate several feet of solid lead.

8-1d COMPLEX VOLTAGE WAVEFORMS

Many practical voltage waveforms are not pure sine waves but rather, complex waveforms. Figure 8-5 shows two examples of how a complex waveform is produced. Under certain conditions when a sine wave is produced, other frequencies may be created which are called harmonics. A *harmonic* is a sine wave that is an integral multiple frequency of the fundamental (original) frequency. Figure 8-5a shows a fundamental frequency; a frequency twice the fundamental frequency, which is called the *second harmonic;* and the waveform produced when two frequencies are summed or added together (which looks somewhat like a shark's fin). A similar situation is shown in Figure 8-5b, except that the *fourth harmonic* is added to the fundamental frequency and produces a waveform that begins to resemble what is called a *square wave.*

Many harmonics are combined in everyday sounds that give them their special characteristics, such as the differences in people's voices. These differences can best be understood through the tonal qualities of musical instruments, as shown in Figure 8-6. A pure musical "A" note, which is used to tune orchestras, is a perfect sine wave at 440 Hz. All the instruments of the orchestra can play the same note, but will definitely sound distinctly different because each instrument has characteristics of its own. The size and manner in which each instrument is constructed determines the harmonics that are generated and give the

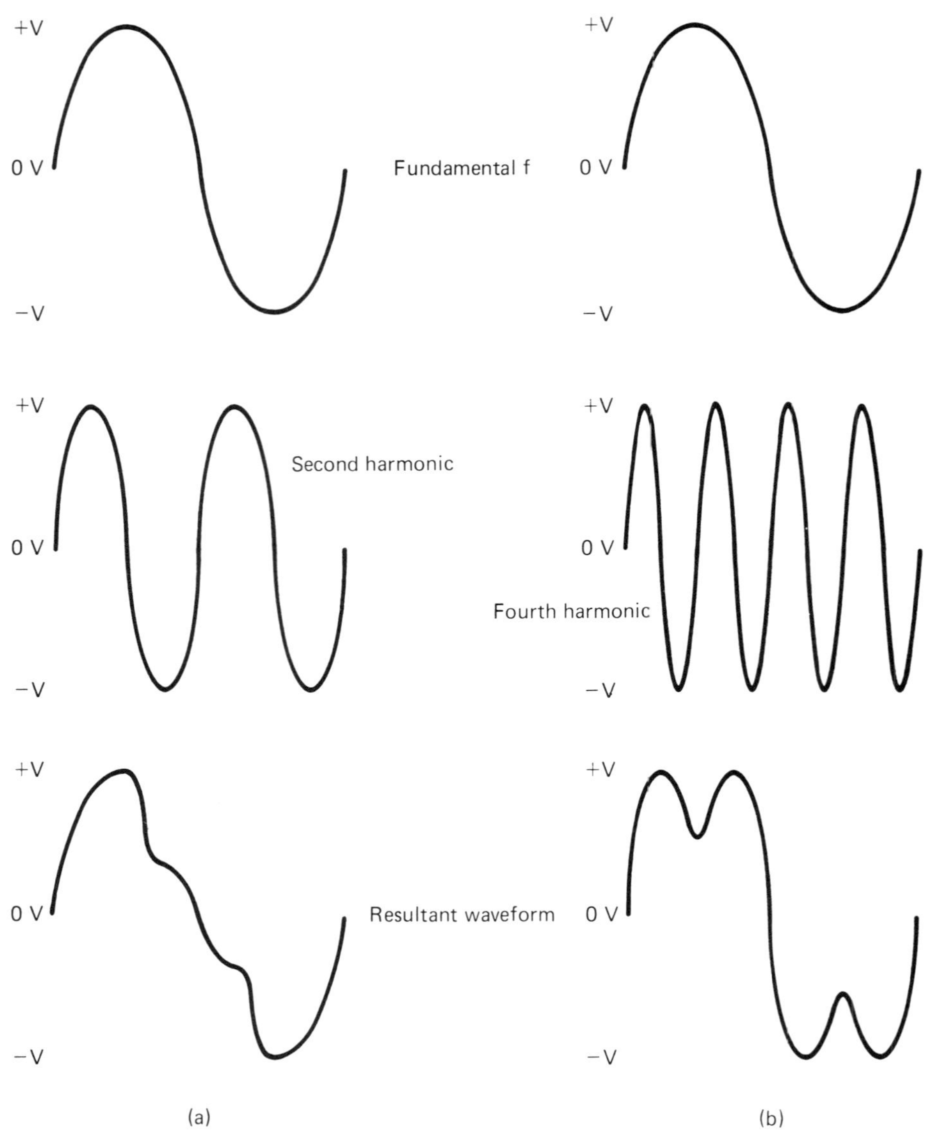

Figure 8–5 Complex voltage waveforms.

instrument its characteristic sound. The flute comes close to producing a pure sine wave. Notice the waveforms of the other instruments.

Certain sounds have a specific frequency range. A comparison of sounds is given in Table 8-3. Notice that a male voice can go lower in frequency than a female voice, and the female voice can go higher in frequency. A *bass* sound refers to low frequencies and a *treble* sound refers to high frequencies. Some stereo sound systems have both a bass control and a treble control, which allows you to set the low and high frequencies to suit your personal preference.

8-1e LOUDNESS AND THE DECIBEL

The loudness of a sound depends on the amplitude of its waveform. However, the human ear is slow to respond to changes in amplitude. For instance, a sound that is increased to twice its original power sound only slightly louder. To make the sound twice as loud, the power will have to be increased to 10 times the original value. Each time the power

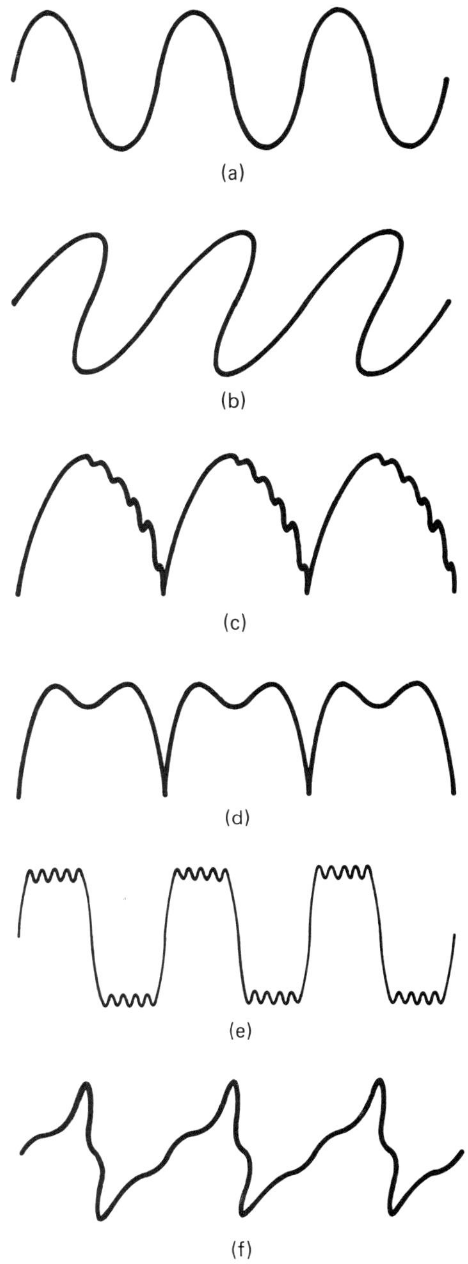

Figure 8–6 Relationship of harmonics to sound: (a) pure A (440 Hz); (b) flute; (c) saxophone; (d) trumpet; (e) clarinet; (f) violin.

TABLE 8-3
Audio-Frequency Range Comparison

Source of Sound	*Frequency Range*
Male voice	100 Hz to 8 kHz
Female voice	150 Hz to 10 kHz
Piano	60 Hz to 10 kHz
Piccolo	500 Hz to 12 kHz
Violin	180 Hz to 15 kHz
Trumpet	180 Hz to 12 kHz
Clarinet	160 Hz to 12 kHz
String bass	40 Hz to 8 kHz
AM radio	50 Hz to 8 kHz
FM radio	50 Hz to 16 kHz

of a sound is increased by 10, its loudness increases by 1. For example, notice the following relationships:

Power Increase	*Loudness Increase*
1	1
10	2
100	3
1000	4

This relationship, based on increments of 10, is referred to as *logarithmic;* therefore, the human ear responds to sound loudness on a logarithmic scale.

The loudness of sound is compared as a ratio of two amounts of power and is given a unit of comparison called the bel. The *bel* was named after Alexander Graham Bell, the American inventor who developed the telephone. Normally, a tenth of a bel, called a *decibel,* is used in sound-level measurements. The decibel for power is expressed by the formula

$$\text{dB} = 10 \log \frac{P_2}{P_1}$$

where P_1 is the original power and P_2 is the new power. The abbreviation for "logarithm" is *log.*

The ratio of P_2 to P_1 is expressed as a logarithmic function which must be looked up in a logarithmic table or solved for with a scientific calculator. As an example, if P_1 equals 1 W and P_2 equals 2 W, the decibel unit can be found, thus:

$$\begin{aligned} \text{dB} &= 10 \log \frac{P_2}{P_1} \\ &= 10 \log \frac{2\text{ W}}{1\text{ W}} \\ &= 10 \log 2 \text{ (at this point you must find the log of 2)} \\ &= 10 \times 0.30103 \\ &= 3.0103 \end{aligned}$$

This problem shows a gain of about 3 dB between the two powers. If the answer had been negative, it would have indicated a 3-dB loss between the powers. Remember, the decibel is not a unit or measurement, but only shows a relationship of the ratio between powers. Table 8-4 shows a decibel-level comparison of some common sounds.

8-1f NONSINUSOIDAL VOLTAGE WAVEFORMS

A *nonsinusoidal waveform* is any waveform that is not a sine wave and, therefore, contains harmonics. There are other nonsinusoidal voltage waveforms that you should become familiar with to enable you to troubleshoot circuits and to facilitate further learning in electronics. Nonsinusoidal voltage waveforms can be ac, where current flows in two directions, or pulsating dc, where the current increases to some maximum amplitude and then decreases to the reference level. Figure 8-7 shows several nonsinusoidal voltage waveforms.

TABLE 8-4
Decibel-Level Comparison of Some Common Sounds

Type of Sound	*Relative Intensity (dB)*
Reference level	0
Soft whisper	20
Normal conversation, radio	60
Group conversation	75
Loud rock band	100
Jet airplane, nearby	120
Threshold of pain from abnormally loud sounds	130

Figure 8–7 Nonsinusoidal voltage waveforms: (a) square wave; (b) ± voltage square wave; (c) square-edge pulse/nonsymetrical; (d) pulse nomenclature; (e) sawtooth wave; (f) reverse sawtooth wave; (g) triangle wave; (h) ± voltage triangle wave.

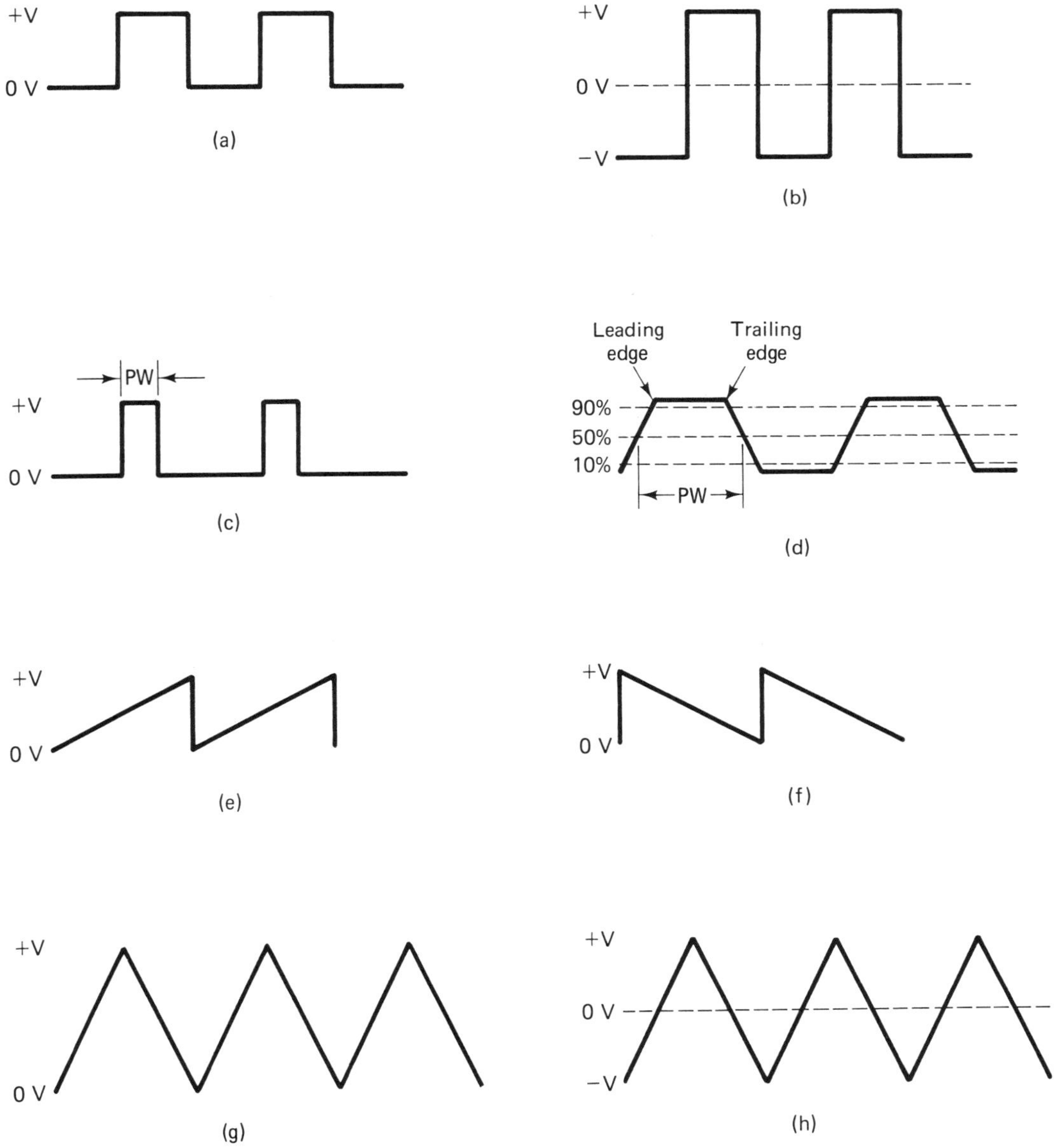

The square wave is shown in Figure 8-7a. This *square-wave* voltage is dc and goes from 0 V to +V, and then back to 0 V. The current increases and decreases but always flows in the same direction. The time that it is high at +V is equal to the time it is low at 0 V, and the waveform is said to be *symmetrical.*

A square wave may be ac and go from positive to negative, as shown in Figure 8-7b. In this case the current would flow in two directions. This waveform is also symmetrical.

Figure 8-7c shows a *nonsymmetrical square-edge pulse.* This pulse goes from 0 V to +V and back to 0 V; however, the time that it is high at +V is much less than the time it is low or at 0 V. The time that it is high is called the *pulse width* (PW).

Voltage waveforms have designated parts that are standardized in the electronics industry, which enables a person to understand written descriptions and be able to communicate effectively with other persons working in electronics. Figure 8-7d shows the nomenclature for a nonsinusoidal waveform. Since in this part of the world we read from left to right, we assume that a representation of lapsed time is also in the same direction. Remember that a voltage waveform shows a relationship of changes in amplitude to time. An increase from a given reference (such as 0 V or ground), whether it be positive or negative, is called the leading edge of a waveform. The *leading edge* is more precisely defined as an increase from 10% to 90% of the total voltage amplitude. The voltage remaining above the reference level a certain period of time and then returning to the reference level is considered the pulse width. A specific definition of *pulse width* is that the voltage exists from 50% of the leading edge to 50% of the trailing edge. The trailing edge is, of course, when the voltage decreases or returns to the reference level. A precise definition of the *trailing edge* is that the voltage is from 90% to 10% of the total voltage.

A *sawtooth voltage waveform* is shown in Figure 8-7e. This voltage increases at a linear rate to its maximum amplitude and then quickly returns to the reference level. Often, it is called a "ramp" voltage. A *reverse sawtooth waveform* is shown in Figure 8-7f. This voltage increases quickly to its maximum amplitude and then decreases at a linear rate to the reference level.

Figure 8-7g shows a *triangle waveform.* This voltage begins at the reference level, increases at a linear rate to its maximum amplitude, and then decreases at a linear rate back to the reference level. A triangle voltage waveform that swings positive and negative is shown in Figure 8-7h. Current will flow in two directions with this type of waveform.

One cycle of a nonsinusoidal waveform is usually measured from the leading edge of one pulse to the leading edge of the following pulse, or from the trailing edge of one pulse to the trailing edge of the following pulse. The frequency of a nonsinusoidal waveform is referred to as the *pulse repetition rate* (PRR). A group of pulses is often referred to as a *pulse train* and its PRR can be found in the same manner as a normal frequency with the formula

$$\text{PRR} = \frac{1}{p}$$

19. Decibel (dB) ______________________________

____

20. Logarithmic (log) ______________________________

____

21. Nonsinusoidal voltage waveform ______________________________

____

22. Square wave ______________________________

____

23. Symmetrical ______________________________

____

24. Nonsymmetrical ______________________________

____

25. Leading edge ______________________________

____

26. Trailing edge ______________________________

____

27. Pulse width ______________________________

____

28. Sawtooth waveform ______________________________

____

29. Triangle waveform ______________________________

____

30. Pulse repetition rate (PRR) ______________________________

____

31. Function generator ______________________________

____

32. Frequency counter ______________________________

____

33. Oscilloscope ______________________________

____

34. Cathode ray tube (CRT) ______________________________

____

35. Trace ______________________________

____

36. *X*-axis ______________________________

____

37. *Y*-axis ______________________________

____

38. Sweep frequency voltage ______________________________

____

SECTION 8-3 EXERCISES AND PROBLEMS

Complete this section before beginning the next section.

1. Draw one cycle of a sine wave and indicate its nomenclature (refer to Figure 8-1).

2. Using the formula $v = \sin\theta \times V_p$, find the instantaneous voltage from the information given.

a. $V_p = 50$ V
$\theta = 60°$

$v =$ ________

b. $V_p = 150$ V
$\theta = 15°$

$v =$ ________

c. $V_{p\text{-}p} = 120$ V
$\theta = 80°$

$v =$ ________

d. $V_{p\text{-}p} = 60$ V
$\theta = 90°$

$v =$ ________

3. Calculate the effective voltage of the data given using the formula $V_{eff} = 0.707\ V_p$.

a. $V_p = 70$ V

$V_{eff} =$ ________

b. $V_p = 140$ V

$V_{eff} =$ ________

c. $V_{p\text{-}p} = 125$ V

$V_{eff} =$ ________

d. $V_{p\text{-}p} = 300$ V

$V_{eff} =$ ________

4. Calculate the average current and voltage for the data given using the formula $V_{av} = 0.637V_p$ or $I_{av} = 0.637I_p$.

a. $V_{p\text{-}p} = 180$ V

$V_{av} =$ ________

b. $I_p = 50$ mA

$I_{av} =$ ________

5. Draw three cycles of a sine wave where the frqeuency is 3 Hz.

6. Calculate the frequency for the given periods using the formula $f = 1/p$.

a. $p = 1$ ms

$f =$ ________Hz

b. $p = 0.005$ ms

$f =$ ________Hz

c. $p = 10\ \mu$s

$f =$ ________Hz

d. $p = 40\ \mu$s

$f =$ ________Hz

7. Calculate the period for the given frequencies using the formula $p = 1/f$.

a. $f = 10$ kHz

$p =$ __________ s

b. $f = 125$ kHz

$p =$ __________ s

c. $f = 20$ MHz

$p =$ __________ s

d. $f = 45$ kHz

$p =$ __________ s

8. List the frequency range for the following designations.

Designation	*Frequency Range*
Power	______
Audio	______
AM radio	______
FM radio	______
Standard television	______
Visible light	______
X-rays	______
Cosmic rays	______

9. Draw a complex ac waveform that contains the fundamental and fourth harmonic frequencies.

10. Referring to Table 8-3, list the instruments that have the respective frequency range.

a. Lowest audio frequencies: ____________

b. Highest audio frequencies: ____________

11. Find the dB gain or loss for the following problems using the formula dB = 10 log P_2/P_1.

a. $P_1 = 10$ W
$P_2 = 20$ W

dB = __________

b. $P_1 = 20$ W
$P_2 = 60$ W

dB = __________

c. $P_1 = 50$ W
$P_2 = 30$ W

dB = __________

d. $P_1 = 15$ W
$P_2 = 5$ W

dB = __________

Refer to Figure 8-7 for Questions 12 through 15.

12. Draw the waveform of a square wave.

13. Draw the waveform of a sawtooth wave.

14. Draw the waveform of a triangle wave.

15. Draw a waveform showing the values indicating leading edge, trailing edge, and pulse width. Label all parts.

16. Match the description in column A with the proper term or symbol in column B.

Column A

____ **a.** Number of cycles per second

____ **b.** Maximum voltage of one alternation

____ **c.** Time to complete one cycle

____ **d.** Total amplitude of a waveform

____ **e.** 0.707 times the peak value

____ **f.** 0.637 times the peak value

____ **g.** Instantaneous voltage

____ **h.** Abbreviation for cycles per second

____ **i.** Sound frequencies

____ **j.** A frequency three times the fundamental frequency

____ **k.** Used for power relationships

____ **l.** A device that produces sine, square, and triangle waves

____ **m.** A device used to count frequency

____ **n.** Produces a replica of an actual voltage waveform

____ **o.** A waveform that is not a sine wave

Column B

1. oscilloscope
2. audio
3. Hz
4. nonsinusoidal
5. harmonic
6. dB
7. $V_{\text{p-p}}$
8. period
9. V_{av}
10. frequency
11. V_p
12. function generator
13. v
14. frequency counter
15. V_{eff}

17. Calculate what is needed for the amplitude, frequency, and phase difference for the waveforms shown in Figure 8-14.

Figure 8–14

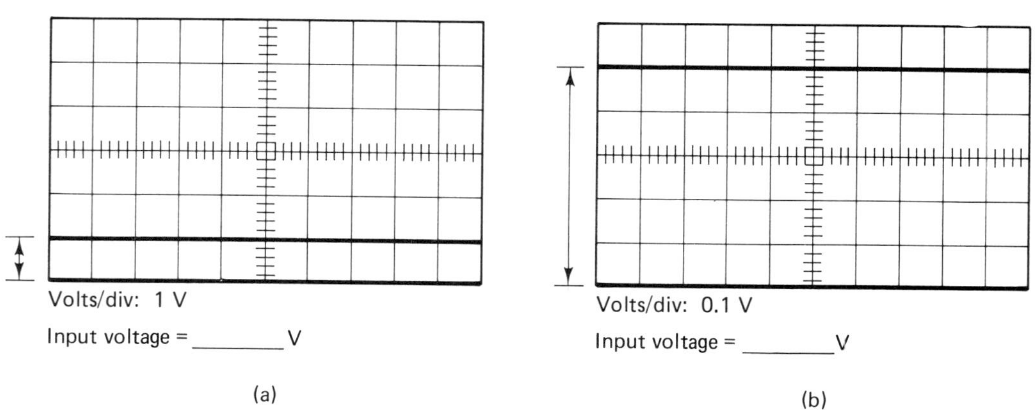

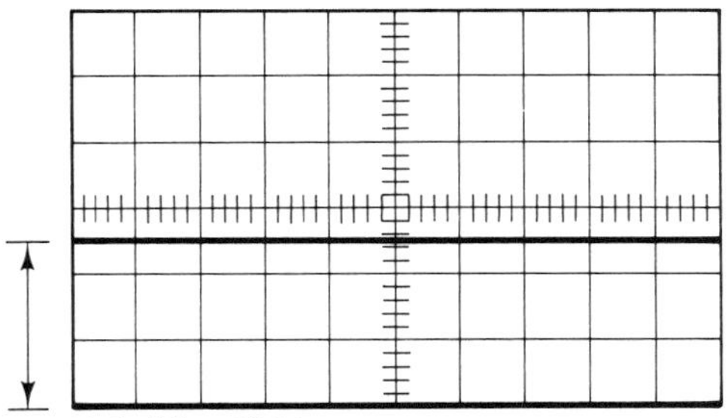

Volts/div: 10 V

Input voltage = ________ V

(c)

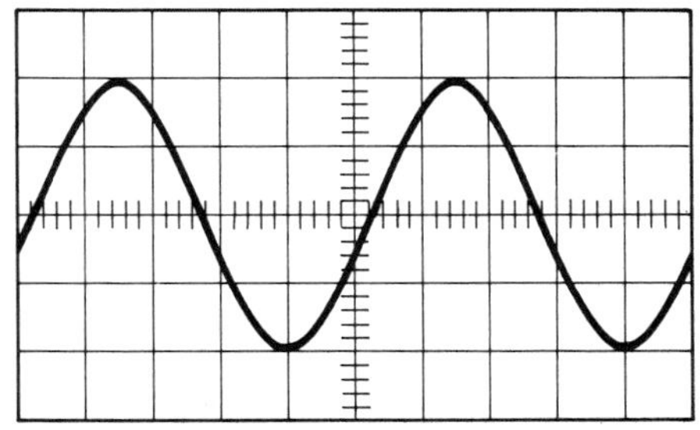

Volts/div: 2 V Time/div: 0.1 ms

v_{in} = ________ $V_{p\text{-}p}$ f = ________ Hz

(d)

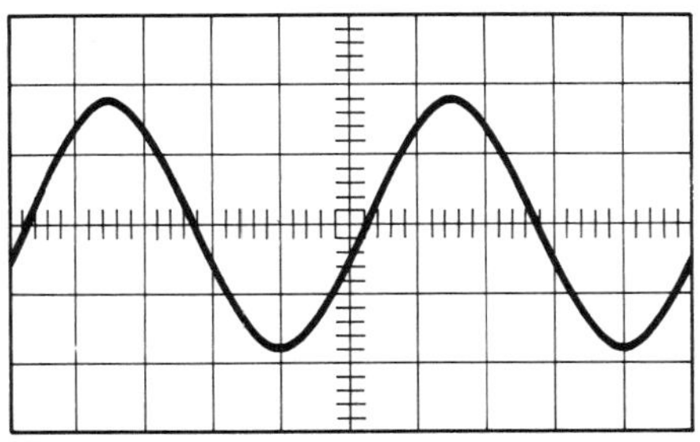

Volts/div: 10 V Time/div: 10 μs

v_{in} = ________ $V_{p\text{-}p}$ f = ________ Hz

(e)

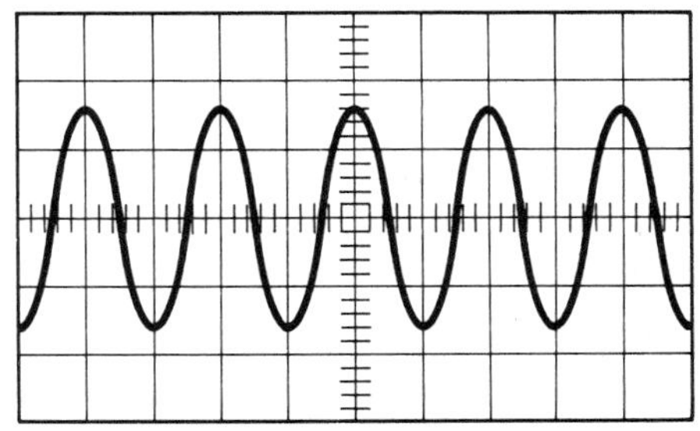

Volts/div: 0.05 V Time/div: 1 μs

v_{in} = ________ $V_{p\text{-}p}$ f = ________ Hz

(f)

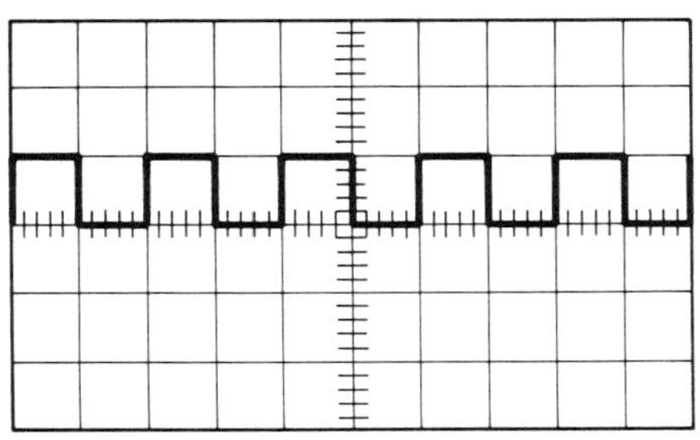

Volts/div: 5 V Time/div: 50 μs

v_{in} = ________ $V_{p\text{-}p}$ PRR = ________ Hz

PW = ________ s

(g)

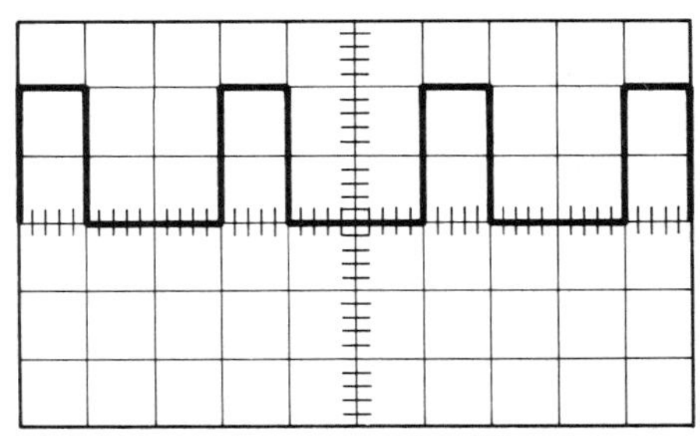

Volts/div: 1 V Time/div: 0.2 μs

v_{in} = ________ $V_{p\text{-}p}$ PRR = ________ Hz

PW = ________ s

(h)

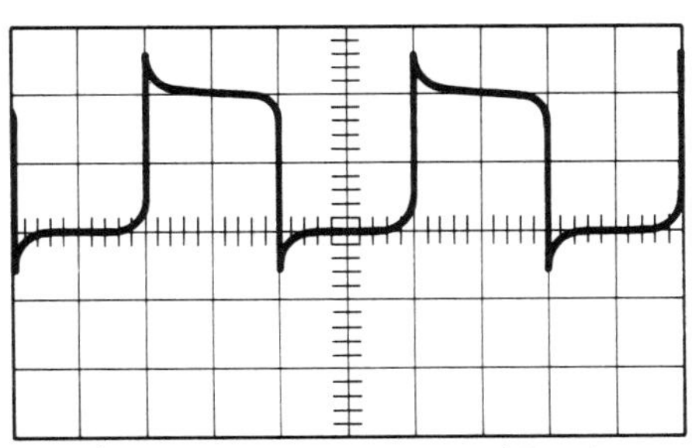

Volts/div: 5 V Time/div: 1 ms

v_{in} = ________ $V_{p\text{-}p}$ PRR = ________ Hz

PW = ________ s

(i)

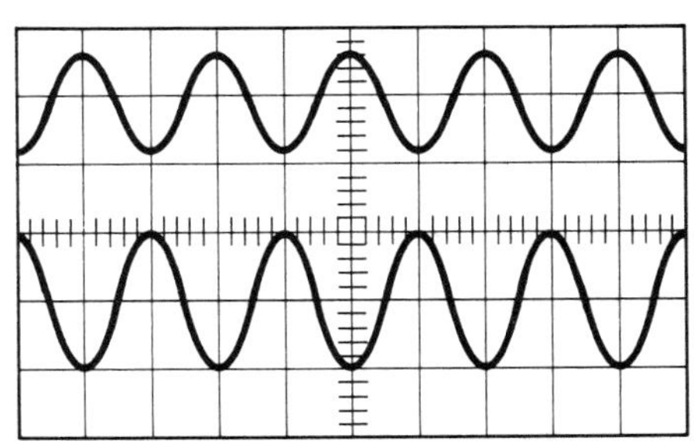

Time/div: 0.5 ms

f = ________ Hz θ = ________ °

(j)

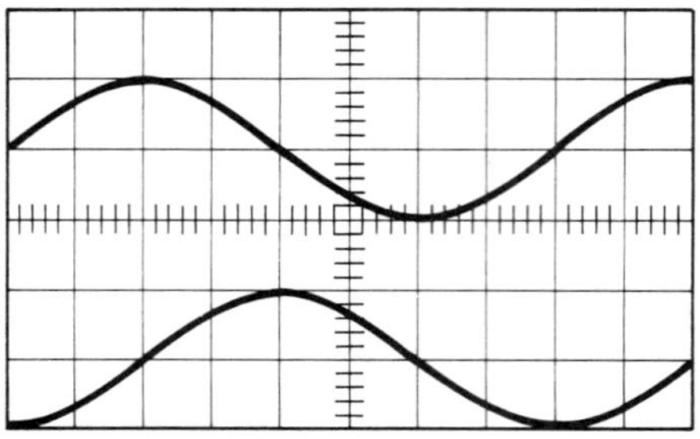
Time/div: 0.2 ms

f = ________ Hz θ = ________ °

(k)

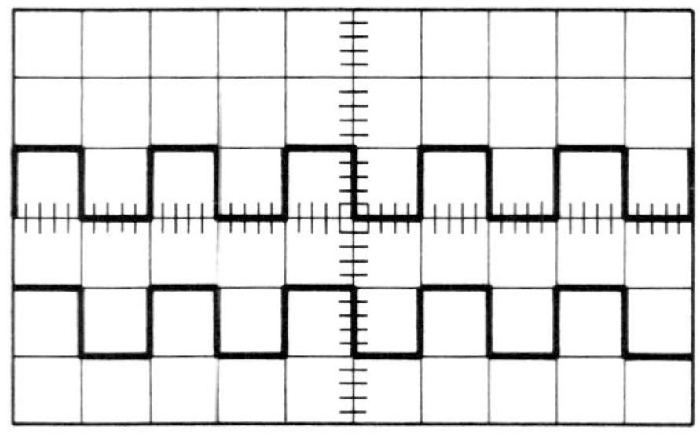
Time/div: 0.5 μs PW = ________ s

f = ________ Hz θ = ________ °

(ℓ)

SECTION 8-4 EXPERIMENTS

EXPERIMENT 1. READING VOLTAGE WAVEFORMS

Objective: To develop the skills necessary for reading and interpreting voltage waveforms with the use of the oscilloscope.

Introduction: This experiment will show you the proper procedures for connecting an oscilloscope to a circuit and develop your abilities in calculating amplitude and frequency for sine waves, square waves, and triangle waves.

Materials Needed:
1 10-kΩ resistor at 0.5 W
2 Clip leads

Equipment Required:
1 15-V dc power supply
1 Function generator
1 Oscilloscope
1 Frequency counter (optional, but enhances the practice to frequency measurements)

Procedure:

1. Perform the setup procedures for using the oscilloscope given in Section 8-1i.2.
2. Set the voltage select switch to ground and move the straight-line trace to the bottom horizontal line on the screen.
3. Set the V/cm control to 1.
4. Place the voltage select switch to dc.
5. Adjust the dc power supply to +5 V.
6. Connect the ground lead of the oscilloscope to the ground terminal of the power supply.
7. Place the signal lead of the oscilloscope on the +*V* terminal of the power supply. How many centimeters did the trace increase? ____________
8. Decrease the voltage on the power supply until the trace is 2 cm from the bottom line. How many volts are you reading now? ____________

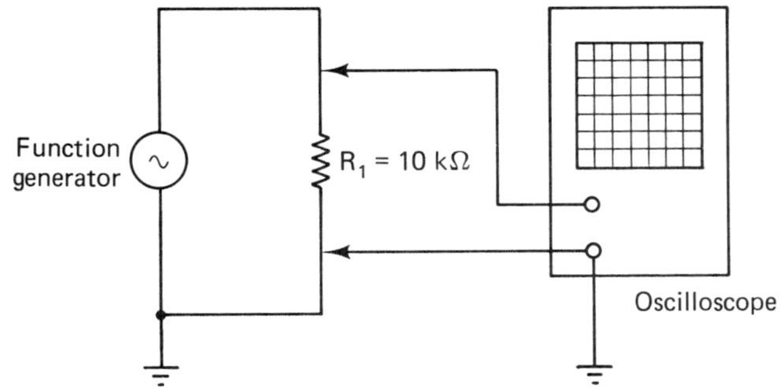

Figure 8–15 Viewing signals.

9. Construct the circuit shown in Figure 8-15.
10. Set the voltage select switch on the oscilloscope to ac.
11. On the oscilloscope, set the V/cm control for 1 and the time/cm control for 1 ms.
12. Set the generator for a sine-wave output. If a frequency counter is available, connect it to the output of the generator at this time and adjust it to read 1 kHz. You may leave the frequency counter connected for the remainder of the experiment and monitor the output of the generator for different settings.
13. Adjust the generator output for a frequency of 1 kHz with an amplitude of 4 $V_{p\text{-}p}$. The waveform should be 4 cm high. (Remember, you may have to adjust the trigger control to "lock in" the trace.)
14. Notice that one cycle of the waveform is 1 cm long on the screen. Set the time/cm control to 0.5 ms. How many centimeters does one cycle now occupy? ____________
15. Set the generator range to the next position (the frequency should now be ×10 greater). How many cycles are there for 1 cm now? ____________. This may be difficult to read.
16. Set the time/cm control to 0.1. Can you read the cycles per centimeter more clearly? ____________. Calculate the frequency of this waveform and enter it here: $f = 1/p =$ ____________ Hz.
17. Set the time/cm control on the oscilloscope to 10 μs.
18. Set the generator to the square-wave output.
19. Adjust the frequency (PRR) of the generator so that one cycle of the square wave is 1 cm long. Calculate the frequency of the waveform and enter it here: $f = 1/p =$ ____________ Hz.
20. Set the time/cm control on the oscilloscope to 2 ms.
21. Set the generator to a triangle-wave output.
22. Adjust the frequency of the generator so that one cycle of the triangle wave is 4 cm long. Calculate the frequency of the waveform and enter it here: $f = 1/p =$ ____________ Hz.

Fill-In Questions:

1. If the V/cm control is set at 0.5 and the waveform on the screen is 3 cm peak to peak, $V_{p\text{-}p} =$ ____________ V.

2. The time/cm control is set at 0.2 ms and the period of the waveform is 3 cm long; therefore, the frequency of waveform is ____________ ____________ kHz.
3. The frequency for a train of pulses is sometimes referred to as the ____________ ____________ ____________ .
4. If the V/cm is set at 2 and the dc voltage shows an increase of 3.6 cm, the dc voltage indicated on the screen is ____________ V.
5. If the trace on the screen of the oscilloscope is too crowded to read its period, you can set the time/cm control to a ____________ value.
6. If the trace on the screen is too high and exceeds the limits of the top and bottom, you can set the V/cm control to a ____________ value.

EXPERIMENT 2. COMPLEX VOLTAGE WAVEFORMS

Objective: To demonstrate harmonic frequencies and their effect on complex voltage waveforms.

Introduction: In this experiment you will place the fundamental frequency on one channel of a dual-trace oscilloscope and its harmonic frequencies on the other channel. Combining these frequencies with the trace-add feature on the oscilloscope produces a complex waveform. Square wave and sawtooth wave pulses contain many harmonics.

Materials Needed: None

Equipment Required:

2 Audio generators
1 Dual-trace oscilloscope with the add channel feature

Procedure:

1. Connect the test instruments as shown in Figure 8-16.
2. Set generator 1 for a frequency of 1 kHz at 4 $V_{p\text{-}p}$.
3. Set generator 2 for a frequency of 2 kHz at 4 $V_{p\text{-}p}$.
4. Set the time/cm control on the oscilloscope to 0.2 ms.
5. Set the trace mode switch to dual and superimpose the two waveforms.
6. Set the trigger source switch to channel 2 and adjust the frequency of generator 1 very carefully until the two waveforms are as stationary as possible.
7. Set the trace mode switch to add and observe the complex waveform that is produced.
8. Set generator 2 for a frequency of 3 kHz and perform steps 5 through 7.

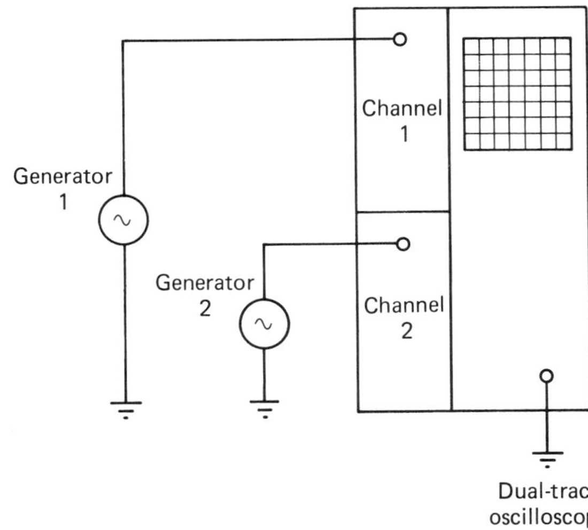

Figure 8–16 Viewing complex waveforms.

Fill-In Questions:

1. A harmonic frequency is an integral multiple of the ____________ frequency.

2. Square-wave and sawtooth-wave pulses contain many ____________ ____________ frequencies.

SECTION 8-5 TROUBLESHOOTING APPLICATION: RECOGNIZING BASIC VOLTAGE WAVEFORM DISTORTION

The most important reason for using an oscilloscope to troubleshoot problems in electronic circuits is that you are able to see the distortion of the waveform that other types of test instruments are not capable of detecting. Figure 8-17 shows some basic types of voltage waveform distortion. Becoming familiar with these distortions will enable you to move ahead faster in developing troubleshooting skills as well as performing better on experiments, because you will be able to clear some basic problems.

A normal desired signal or waveform is shown in Figure 8-17a. Figure 8-17b shows a waveform with extreme distortion that could have many causes. One of the most common problems is where there is a loose or open ground connection. A voltage waveform is present, but without a solid ground connection the waveform may float about the screen and show extreme distortion.

Remember! Always check your connections to all circuits and test equipment.

Figure 8-17c shows a waveform with nonsymmetrical alternations. This is usually caused by an amplifier that does not have linear amplifying characteristics or an overdriven condition where the input signal is too high in amplitude. If you encounter this problem, reduce the amplitude of your input signal to see if the problem can be corrected.

Noise that is present on signal voltage can cause a "scratchy" sound in loudspeakers, as shown in Figure 8-17d. Noise usually attaches itself to the peaks of the voltage waveform. When you work with

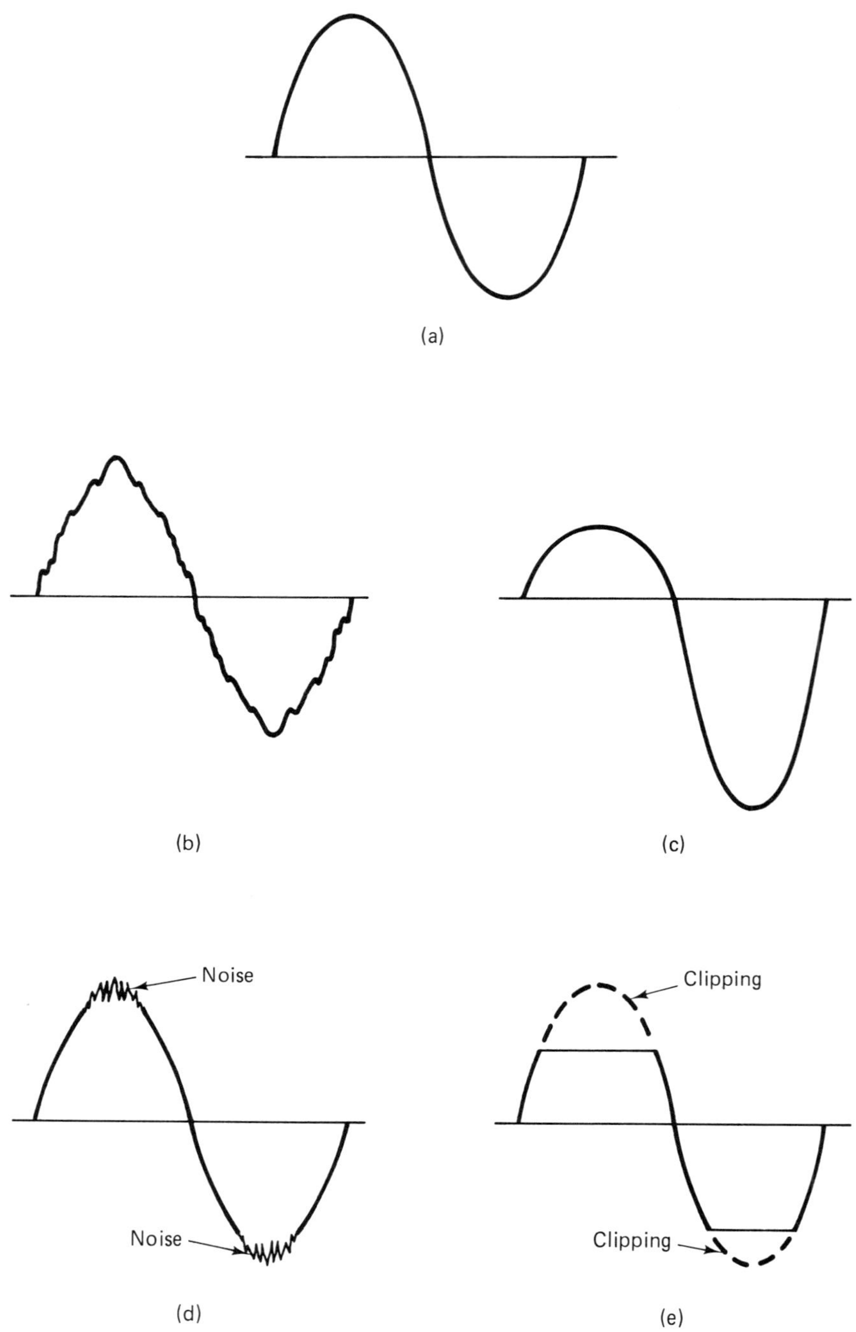

Figure 8–17 Voltage waveform distortion: (a) desired signal; (b) extreme distortion; (c) nonsymetrical; (d) noise; (e) peak clipping.

experiments, again, remember to make secure conenctions that will reduce noise problems.

Clipping the peaks of voltage waveforms "flattens" the peaks, as shown in Figure 8-17e. Usually, this is a condition where an amplifier is overdriven or its operating characteristics have changed. You may encounter this problem in working with experiments. Very often, if the input signal to a circuit is reduced in amplitude, the problem can be corrected.

SECTION 8-6 INSTANT REVIEW

- One cycle or period has 360°.
- A sine wave has two alternations of 180° each.
- V_{max} or V_{peak} (V_p) is the maximum voltage for one alternation.
- $V_{peak\ to\ peak}$ ($V_{p\text{-}p}$) is the total voltage of the maximum points of a waveform.
- Amplitude is the amount of voltage of a waveform.
- Instantaneous voltage $v = \sin \theta \times V_p$.
- A vector shows magnitude and direction.
- A phasor is a rotating vector.
- $V_{eff} = 0.707V_p$; $I_{eff} = 0.707I_p$.
- $V_{av} = 0.637V_p$; $I_{av} = 0.637I_p$.
- Frequency is the number of cycles that occur in a specified time.
- Hertz (Hz) stands for cycles per second.
- $f = 1/p$ or $f = 1/t$; $p = 1/f$ or $t = 1/f$.
- A harmonic is a sine wave that is an integral multiple frequency of the fundamental frequency.
- The characteristics of different sounds, even when producing the same fundamental frequency, are the result of harmonics.
- The loudness of a sound depends on the amplitude of the waveform.
- The human ear responds to sound based on increments of 10, which is called logarithmic.
- The decibel is a comparison of power levels.
- $dB = 10 \log P_2/P_1$.
- A nonsinusoidal waveform is any waveform that is not a sine wave and, therefore, contains harmonics.
- Nonsinusoidal waveforms are square waves, sawtooth waves, triangle waves, and so on.
- The leading edge of a pulse is defined as an increase from 10% to 90% of the total voltage amplitude.
- The trailing edge of a pulse is defined as a decrease from 90% to 10% of the total voltage amplitude.
- The pulse width exists from 50% of the leading edge to 50% of the trailing edge.
- The pulse repetition rate (PRR) is similar to frequency and is used interchangeably.
- A function generator produces several various voltage waveform outputs.
- A frequency counter is a device for measuring the frequency of a voltage waveform.
- The oscilloscope is a test instrument that electronically "paints" a picture on the screen of a CRT of a voltage waveform in a circuit. It can also measure dc voltage.
- An electron beam is focused and aimed at the phosphorus coating on the inside face of the CRT.

- The horizontal sweep frequency causes the trace to go from left to right (x-axis) on the screen of the CRT.
- The input signal voltage placed at the vertical input causes the trace to go from top to bottom (y-axis) on the screen.

SECTION 8-7 SELF-CHECKING QUIZ

Circle the most correct answer for each question.

1. The period of a sine wave has:

a. 90° **b.** 180°
c. 360° **d.** 270°

2. The maximum voltage of a waveform is designated:

a. $V_{p\text{-}p}$ **b.** V_{eff}
c. V_p **d.** V_{av}

3. A waveform has a peak-to-peak voltage of 260 V. Its effective voltage is:

a. 183.82 V **b.** 91.91 V
c. 165.62 V **d.** 82.81 V

4. A straight line with an arrowhead that indicates magnitude and direction is called a:

a. trace **b.** phasor
c. decibel **d.** vector

5. The sin of 20° is 0.342020. The instantaneous voltage at 20° for a sine wave with a peak-to-peak voltage of 160 V is about:

a. 27.36 V **b.** 54.72 V
c 45.67 V **d.** 128 V

6. The average voltage for a sine wave with a maximum voltage of 90 V is about:

a. 31.82 V **b.** 57.33 V
c. 28.67 V **d.** none of the above

7. The period of a particular sine wave is 0.032 ms. Its frequency is about:

a. 31.25 kHz **b.** 3.3 kHz
c. 2.8 kHz **d.** 1.25 kHz

8. Bass sounds are associated with frequencies that are:

a. low **b.** high
c. midrange **d.** in the UHF range

9. The following term *not* used tc describe pulses is:

a. leading edge **b.** trailing edge
c. rms **d.** V_p

10. A symmetrical square wave has a pulse width of 2.5 μs. Its PRR is:

a. 200 kHz **b.** 400 kHz
c. 40 kHz **d.** none of the above

11. A sine wave has an effective voltage of 120 V. Its peak-to-peak voltage is approximately:

a. 170 V **b.** 339 V
c. 85 V **d.** 42 V

12. To obtain the most accurate frequency reading of a specific circuit, you would use:

a. a DMM
b. a frequency generator
c. an oscilloscope
d. a frequency counter

13. Two sine waves are displayed on a CRT. When one is at maximum positive voltage, the other is at maximum negative voltage, and vice versa. The sine waves are out of phase by:

a. 90° **b.** 270°
c. 180° **d.** 45°

14. The horizontal (x-axis) trace on the screen of the oscilloscope is produced by the:

a. sweep voltage
b. input signal
c. intensity control
d. focus control

15. The amplitude of the waveform on the screen of the CRT is produced by the:

a. sweep voltage
b. input signal
c. intensity control
d. voltage select switch

16. To stop or lock in a moving waveform on the screen of an oscilloscope, you have to adjust the:

a. time/cm control
b. focus control
c. trigger control
d. V/cm control

17. The time/cm on an oscilloscope is set at 0.5 ms. The period of the sine wave displayed is 1.6 cm long. The frequency of the signal is:

a. 12.5 kHz **b.** 6.25 kHz
c. 3.125 kHz **d.** 1.25 kHz

18. A term used to show the relationship of power levels is:

a. phasor **b.** log
c. trace **d.** decibel

19. The audio-frequency range is from:

a. 60 to 10 kHz
b. 536 to 160 kHz
c. 20 to 20 kHz
d. 30 to 535 kHz

20. The V/cm control on an oscilloscope is set at 0.2. The amplitude of the waveform on the screen is 4.8 cm; therefore, its $V_{p\text{-}p}$ is:

a. 4.8 V **b.** 0.96 V
c. 2.4 V **d.** none of the above

ANSWERS TO EXPERIMENTS AND SELF-CHECKING QUIZ

Experiment 1: **(1)** 1.5 **(2)** 1.67 **(3)** pulse repetition rate **(4)** 7.2 **(5)** higher **(6)** higher

Experiment 2: **(1)** fundamental **(2)** harmonic

Self-Checking Quiz: **(1)** c **(2)** c **(3)** b **(4)** d **(5)** a **(6)** b **(7)** a **(8)** a **(9)** c **(10)** a **(11)** b **(12)** d **(13)** c **(14)** a **(15)** b **(16)** c **(17)** d **(18)** d **(19)** c **(20)** b

Unit 9

Inductance

INTRODUCTION

In Unit 6 you were introduced to electromagnetism and some of its applications. Most of the circuits described used dc voltage and were based on their magnetic principles, such as the solenoid, speaker, motor, and generator. When the coil is used in ac circuits, its qualities become somewhat more complex. However, it is these qualities that make possible telephone and satellite communication, audio and video recordings, radio and television communication, microwave cooking, and many other devices that make life on this planet more enjoyable.

UNIT OBJECTIVES

Upon completion of this unit, you will be able to:

1. Define and explain *inductance, counter-electromotive force, self-inductance, mutual inductance, henry,* and other terms associated with inductors.
2. List the factors that determine inductance.
3. Identify various types of inductors.
4. Calculate the total inductance of inductors in series and parallel.
5. Describe the effect of a time constant and calculate different values for time constants.
6. Explain the effects of time constants and square-wave pulse-shaping circuits.
7. Define and show how to calculate values of inductive reactance.
8. Demonstrate the relationship of inductive reactance to frequency.
9. Calculate ac formulas for inductive circuits.
10. Show the effect of inductive reactance in an ac circuit with the use of phasors.
11. Describe the relationship of apparent power, true power, and power factor in an inductive circuit.
12. Explain the use of an inductor in a frequency filter.

SECTION 9-1 FUNDAMENTAL CONCEPTS

9-1a DEFINING INDUCTANCE

When a coil is connected to a battery as shown in Figure 9-1a, and the switch S_1 is closed, current begins to flow in the circuit. The flow of current develops a magnetic flux that cuts across the windings of the coil. If you remember from Unit 6, the cutting of flux lines by conductors, or vice versa, induces a current in the conductors. This induced current, in turn, produces a voltage that opposes an increase in current and it is called *counter-electromotive force* (CEMF). CEMF appears to be the same value of the applied voltage that opposes the current trying to rise in the coil. However, the current overcomes this and will eventually increase to its maximum value, depending on the resistance of the wire of the coil. As long as the current is flowing, an electromagnetic force is present around the coil. This force is stored energy.

Figure 9–1 Effects of inductance: (a) increasing field; (b) collapsing field.

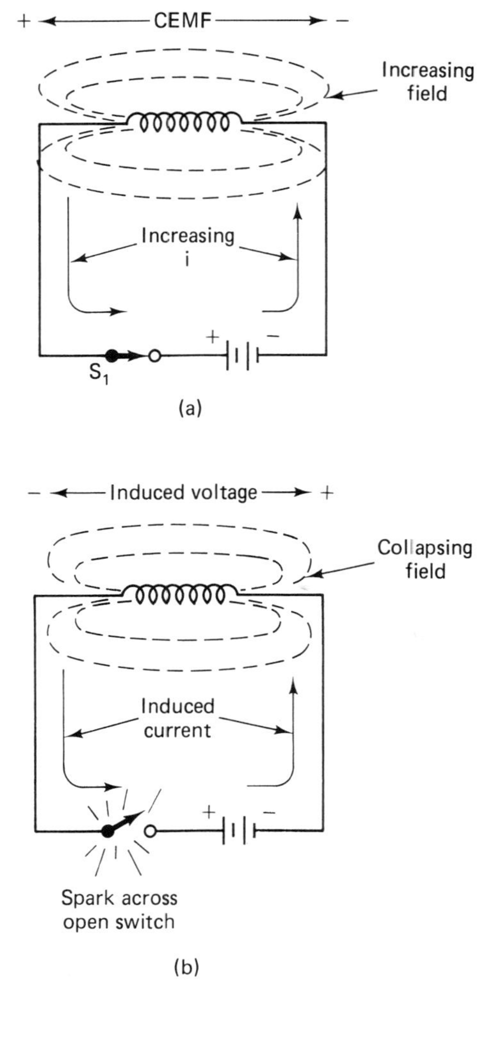

When the switch is opened as shown in Figure 9-1b, the magnetic field around the coil collapses and the stored energy induces a current in the coil. This induced current causes an induced voltage across the coil which tends to keep the current in the circuit flowing in the same direction. There is always somewhat of an electrical spark when a switch is opened; however, in a circuit with a coil the rate of the falling current causes much more current to flow and the spark will be much greater.

The opposition to a change of the current flow in a circuit is called *inductance. Self-inductance* is the action that produces the CEMF of a coil when the current in the coil changes. Since a coil of wire has inductive properties, it is referred to as an *inductor.*

9-1a.1 The Henry

The unit of inductance is called the *henry,* named after Joseph Henry, an American physicist who helped develop the concepts of inductance. Technically, a henry is defined as 1 volt being induced by a current change of 1 ampere per 1 second. It is expressed as a formula:

$$L = \frac{v_L}{di/dt}$$

where L = inductance (henrys)
v_L = induced voltage (volts) di = change in current (amperes)
dt = change in time (seconds)

Figure 9-2b shows a graphical representation of the rise of current in an inductive circuit. Voltage is first applied to an inductor and then the current rises to maximum. In a theoretically pure inductive circuit with no resistance the voltage leads the current by 90°. As a reminder of this action for inductors the name *ELI* is used, where E (for electromotive force) leads or comes before current I in an inductive circuit L.

Schematic symbols for various inductors are shown in Figure 9-2c. An absence of lines beside the coil indicates that it has an air core. Two lines beside the coil indicates that it has an iron core. An arrow through the coil means that it is variable. An inductor may also have taps for providing different amounts of inductance.

Figure 9-2d–i show the various physical types of inductors used in electrical/electronic circuits. As you progress in your study of electronics you will recognize these inductors used in many and varying circuits, ranging from communication circuits to computers and other devices.

9-1a.2 Factors Determining the Amount of Inductance

The physical characteristics of an inductor determine its inductance. There are many factors to consider with the complexity of inductor design; however, there are a few basic ones that should be remembered by the technician. The *inductance* is directly proportional to the number of turns and the diameter of the coil, and inversely proportional to the spacing of the turns and the length of the coil. The *permeability* (μ) of the core also influences the inductance. The permeability of air is 1,

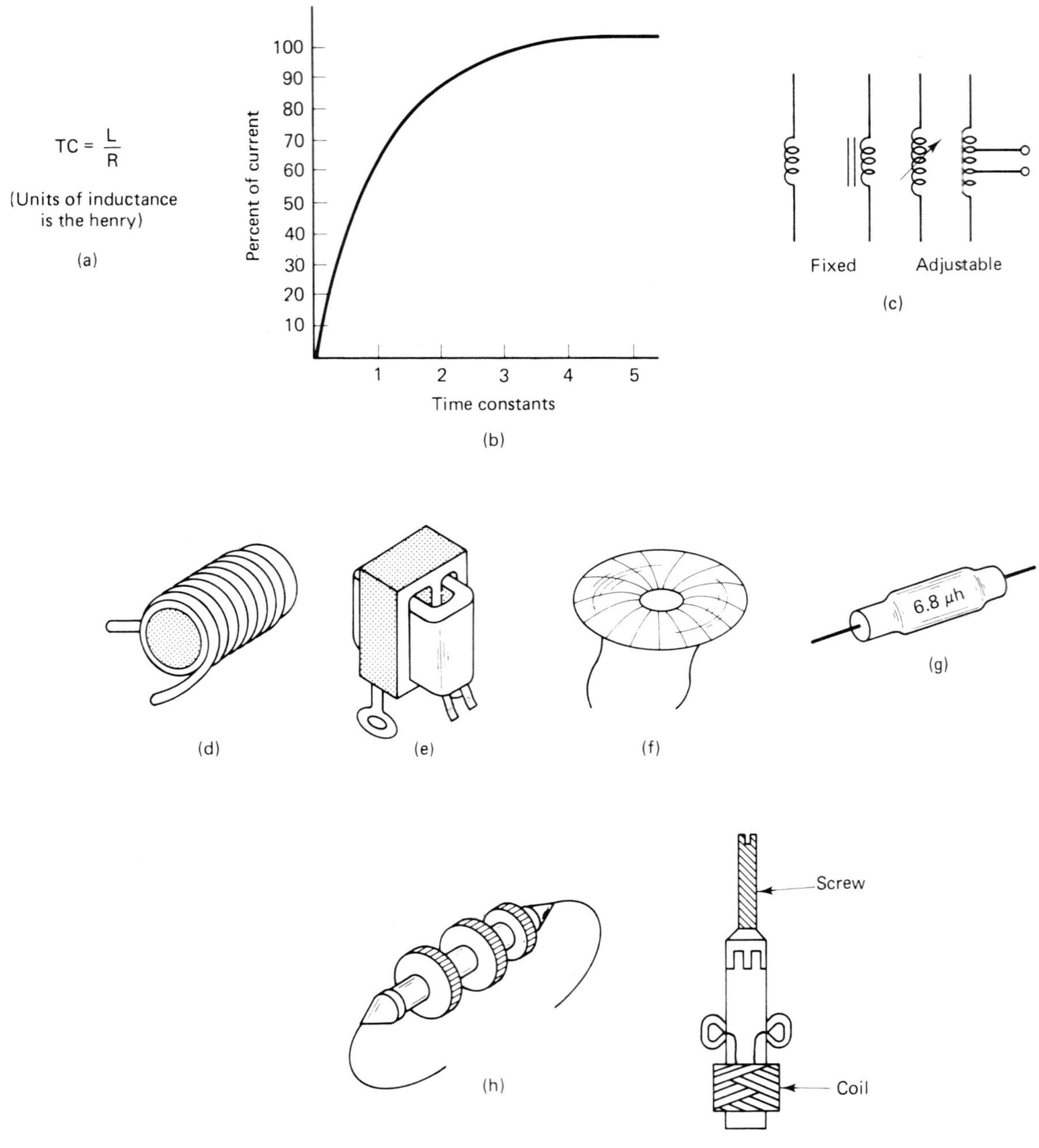

Figure 9–2 Inductors: (a) time-constant formula: (b) current-rise characteristic curve; (c) symbols; (d) air core; (e) iron core; (f) toroidal; (g) tubular; (h) RF choke; (i) turnable RF coil.

but iron has a higher permeability and will create more inductance. Table 9-1 shows these relationships.

Sometimes a current rating is given on inductors. This does not mean the maximum safe current that can flow through the coil, but rather, the amount of current required to develop the given inductance. Remember, inductance results as a change in current; however, in practical applications the amount of current is also important in creating the rated inductance.

TABLE 9-1
Factors Affecting Inductance

	Inductance	
Factor	*Low*	*High*
Number of turns	Few	Many
Diameter	Small	Large
Turns spacing	Wide	Close
Length of coil with same turns	Long	Short
Core material	Low μ	High μ

9-1b INDUCTORS IN SERIES AND PARALLEL CIRCUITS

Calculating the total inductance of inductors in a circuit is accomplished in the manner as was done with resistors. When the inductors are in series as shown in Figure 9-3a, the total inductance is the sum of all the individual inductances and can be expressed with the formula

$$L_T = L_1 + L_2 + L_3$$

When two inductors are in parallel (Figure 9-3b), you would use the formula

$$L_T = \frac{L_1 \times L_2}{L_1 + L_2}$$

If there are three or more inductors in parallel (Figure 9-3c), you would use the reciprocal formula:

$$L_T = \frac{1}{\dfrac{1}{L_1} + \dfrac{1}{L_2} + \dfrac{1}{L_3}}$$

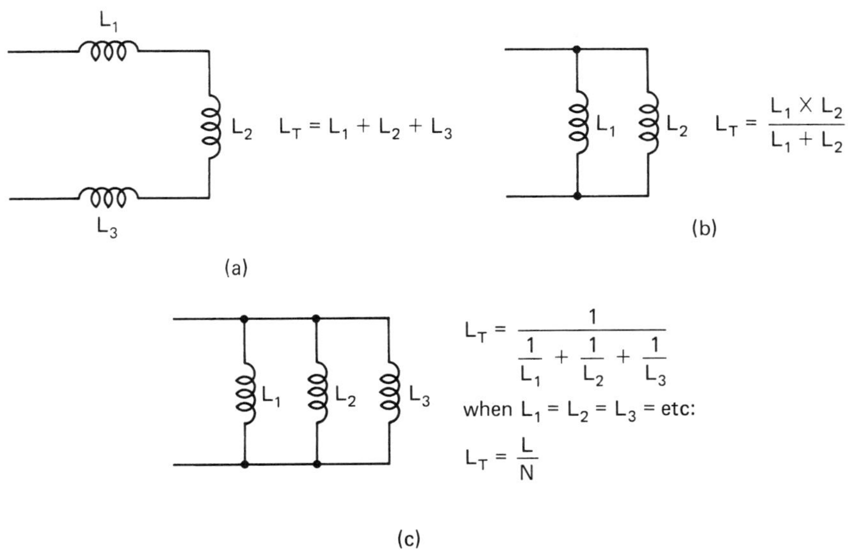

Figure 9–3 Inductor combinations: (a) in series; (b) two in parallel; (c) three in parallel.

When all the inductors in parallel are of the same value, you simply divide the value of one inductor by the total number of inductors.

9-1b.1 Effects of Mutual Inductance

When inductors are placed close enough together in a circuit, the electromagnetic fields about them can affect one another. This effect is known as *mutual induction.* Technically speaking, mutual induction is defined as the condition where a change of 1 ampere per second of current in one coil induces 1 volt into another coil. Mutual induction is designated by the symbol L_m, and its unit of measure is the henry.

The amount of mutual induction that exists between two inductors depends on the number of flux lines that are cut, referred to as the *degree of coupling.* The degree of coupling is expressed by a factor called the coefficient of coupling and is given the symbol k. When $k = 1$, there exists 100% coupling of the flux between two inductors, or in other words, the maximum number of flux lines is cut. When there is no coupling, $k = 0$, and no lines of flux are cut. Therefore, in most cases involving coils, k will be a fractional number indicating that some mutual induction is present.

Mutual induction is expressed as a function of the coefficient of coupling (k) and the values of inductance of two coils (L_1 and L_2). It can be shown by the formula

$$L_m = k\sqrt{L_1 L_2}$$

The mutual induction between two inductors can be aiding or opposing. When two inductors are connected in such a way that their flux fields aid each other, the total inductance can be calculated with the formula

$$L_T = L_1 + L_2 + 2L_m$$

When two inductors are connected in such a way that their flux fields oppose each other, the total inductance can be calculated with the formula

$$L_T = L_1 + L_2 - 2L_m$$

As an example, let $L_1 = 10$ mH, $L_2 = 20$ mH, and $k = 0.5$. Solving for L_m yields

$$\begin{aligned}
L_m &= k\sqrt{L_1 L_2} \\
&= 0.5\sqrt{10\text{ mH} \times 20\text{ mH}} \\
&= 0.5\sqrt{10 \times 10^{-3} \times 20 \times 10^{-3}} \\
&= 0.5\sqrt{200 \times 10^{-8}} \\
&= 0.5 \times 14.14 \times 10^{-3} \\
&= 7.07 \times 10^{-3} \\
&= 7.07\text{ mH}
\end{aligned}$$

If the two inductors were connected series-aiding, the total inductance would be found thus:

$$\begin{aligned} L_T &= L_1 + L_2 + 2L_m \\ &= 10\,\text{mH} + 20\,\text{mH} + 2 \times 7.07\,\text{mH} \\ &= 10 \times 10^{-3} + 20 \times 10^{-3} + 14.14 \times 10^{-3} \\ &= 44.14 \times 10^{-3} \\ &= 44.14\,\text{mH} \end{aligned}$$

If the two inductors were connected in series-opposing, the total inductance would be found thus:

$$\begin{aligned} L_T &= L_1 + L_2 - 2L_m \\ &= 10\,\text{mH} + 20\,\text{mH} - 2 \times 7.07\,\text{mH} \\ &= 10 \times 10^{-3} + 20 \times 10^{3} - 14.14 \times 10^{-3} \\ &= 15.86 \times 10^{-3} \\ &= 15.86\,\text{mH} \end{aligned}$$

Mutual inductance is also a factor in circuits where inductors are connected in parallel. However, it is beyond the scope of this book to deal with further complex problems involving mutual induction. The main point here is that the preceding examples of mutual induction show that for some cases of real-life applications, there are other factors that must be considered when calculating circuit parameters.

9-1c INDUCTANCE IN A DC CIRCUIT

In Section 9-1a we saw what occurs when a dc power source is applied to an inductor. The current lags the applied voltage and it takes a certain amount of time for the current to rise to maximum value. When the maximum amount of current is flowing through the inductor, it is said to be "saturated." In other words, *saturation* is a condition where no further increase of current in the circuit is possible given the existing applied voltage.

9-1c.1 Time Constants

It takes a definite period of time for the current to rise or fall in an inductive circuit. This period will always be the same, or constant, for a particular circuit, and is referred to as the *time constant.* One time constant is defined as the time required for the current to rise 63.2% of its maximum value, as shown in Figure 9-4a.

During the second time constant the current will rise 63.2% more of the remaining current. For example, if the maximum current in a circuit is 100 mA, then at the end of the first time constant the current will be 63.2 mA (100 mA × 63.2% = 63.2 mA). The difference of the maximum current and this value is: 100 mA − 63.2 mA = 36.8 mA. During the second time constant the current will rise 63.2% of this difference current: 36.8 mA × 63.2% = 23.26 mA. Therefore, at the end of the second time constant the total current is equal to 86.46 mA (63.2 mA + 23.26 mA = 86.46 mA). This same process occurs for the remaining time constants. After five time constants the current has risen to

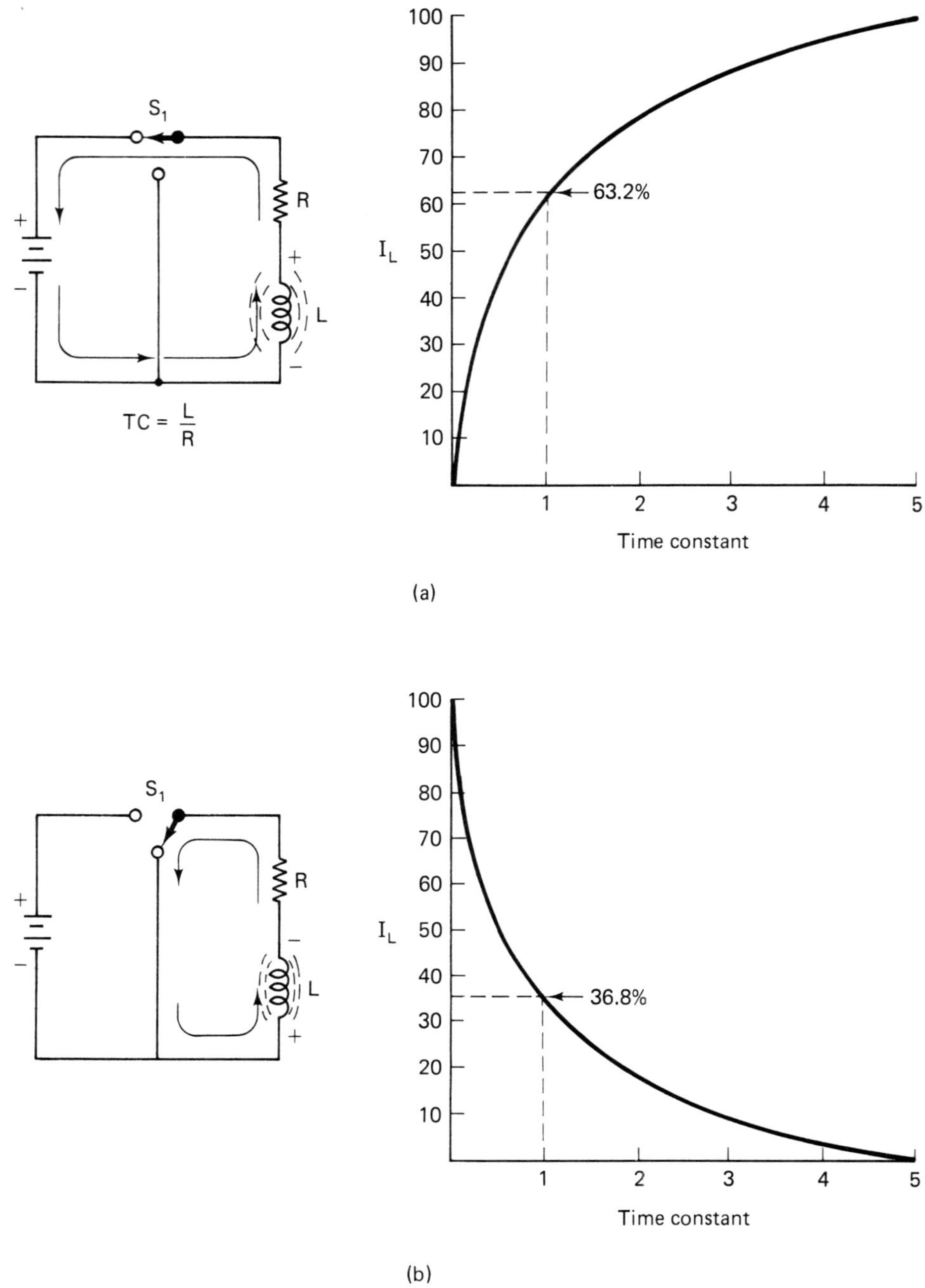

Figure 9–4 Time constants: (a) rising current; (b) falling current.

nearly the maximum amount of current. The total amount of current for each time constant is shown below.

Time Constant	*Amount of Current (mA)*
1	63.20
2	86.26
3	94.94
4	98.14
5	99.32

The current will fall or decrease at the same rate as the rise time, which is shown in Figure 9-4b. In this case the remaining current in the circuit is the reference value. For example, if the maximum current is 100 mA, then during the first time constant the current would have fallen 63.2%:

$$100\text{ mA} \times 63.2\% = 63.2\text{ mA}$$
$$100\text{ mA} - 63.2\text{ mA} = 36.8\text{ mA}$$

After five time constants the current in the circuit will be nearly zero.

The value of one time constant for an inductive circuit is a function of the inductance and resistance. One time constant is directly proportional to the inductance and inversely proportional to the resistance and can be expressed by the formula

$$t = \frac{L}{R}$$

For example, if $L = 10$ mH and $R = 1$ kΩ, then

$$\begin{aligned} t &= \frac{10\text{ mA}}{1\text{ k}\Omega} \\ &= \frac{10 \times 10^{-3}\text{ H}}{1 \times 10^{3}\ \Omega} \\ &= 10 \times 10^{-6}\text{ s} \\ &= 10\ \mu\text{s} \end{aligned}$$

From the values given you can see that it will take 10 μs for the current to rise in one time constant. Since it takes five time constants for the current to reach maximum, it will take 50 μs for this to occur (10 μs × 5 = 50 μs).

9-1c.2 Time Constant and Square-Wave Pulses

The effects of a time constant in an inductive circuit can be seen with the use of square-wave pulses, as shown in Figure 9-5. The time constant of the circuit is 10 μs. If a symmetrical square wave with a PRR of 50 kHz is applied to the circuit, its period can be found thus:

$$p = \frac{1}{f} = \frac{1}{50\text{ kHz}} = \frac{1}{50 \times 10^{3}} = 0.02 \times 10^{-3} = 20\ \mu\text{s}$$

Therefore, the pulse width (the time the pulse is high) is

$$\frac{20\ \mu\text{s}}{2} = 10\ \mu\text{s}$$

The voltage drop V_R across the resistor is in phase with the current in the circuit, and its voltage waveform gives an indication of the action of the current. The pulse width of V_{in} occurs the same amount of time as the time constant; therefore, by the time the current reaches 63.2% of the maximum value, V_{in} changes and goes low. The current now falls

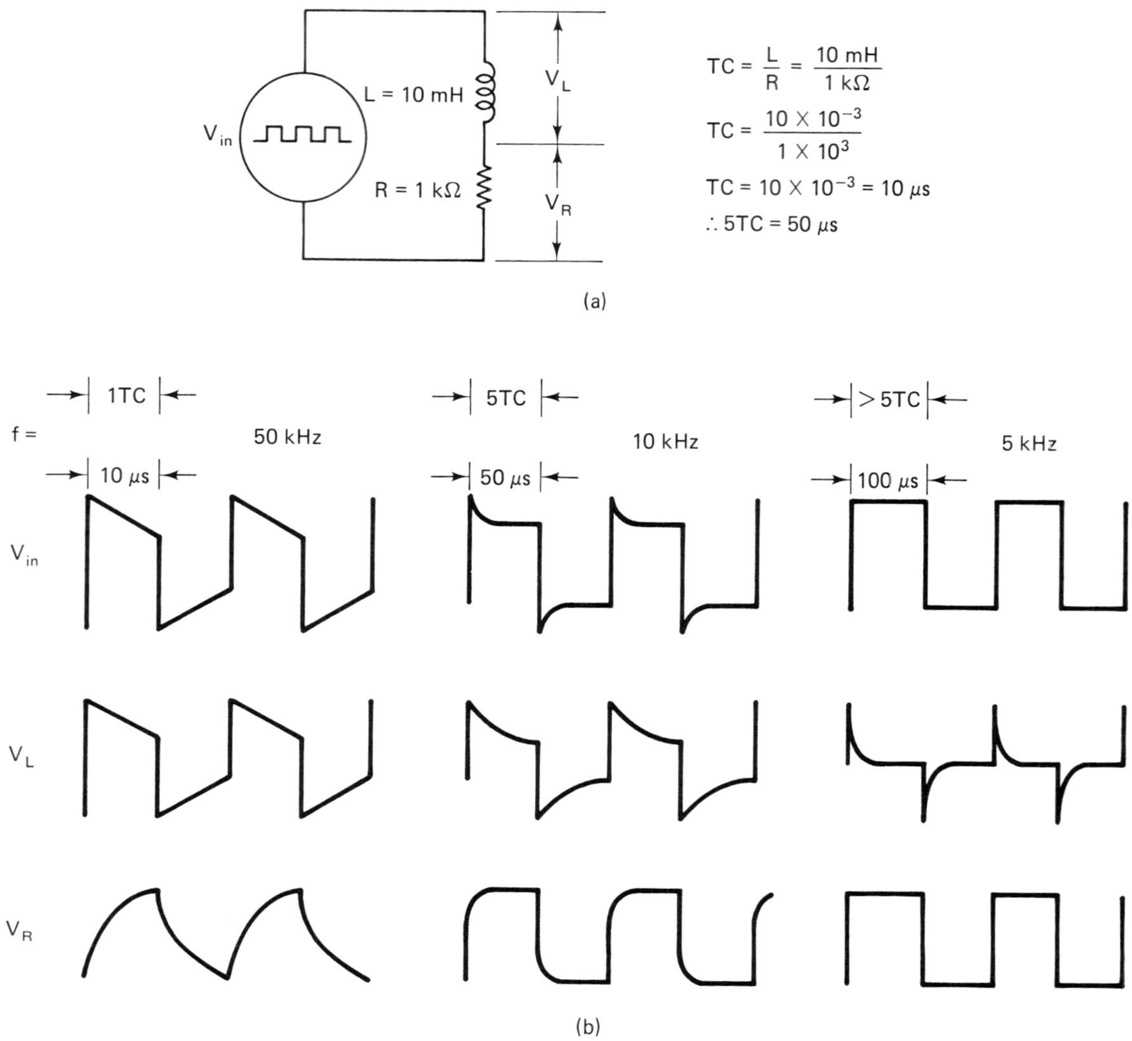

Figure 9–5 Time constant and pulse shaping: (a) schematic diagram; (b) voltage waveforms.

63.2% and V_{in} changes again. The waveform of V_R clearly shows the rise and fall times of the current.

The waveforms of V_L and V_{in} reflect the highly inductive circuit and their voltages show a slope with current changes. When the PRR of V_{in} is decreased to 10 kHz, its pulse width becomes 50 μs, or the same time as five time constants of the circuit. Now the current has sufficient time to rise to the maximum level before V_{in} changes. The voltage waveform V_R resembles more the waveform of V_{in}. The V_L waveform has more of a curve to its slope since the current change was less.

If the PRR of V_{in} is decreased to 5 kHz, its pulse width becomes 100 μs. The input pulse width is now 10 times longer than the time constant of the circuit. Voltage V_R is nearly a perfect square wave as the input waveform, but the V_L waveform resembles a very narrow pulse. This shows that the voltage across the inductor is present for a very short time in relation to the time of the input pulse width.

The output across the resistor is referred to as an integrator. An integrator circuit continuously adds up a quantity being measured over a period of time. The waveform is proportional to the time interval of the input signal. When the output is across the inductor, we are dealing with a *differentiator,* a concept opposite to that of an integrator. The output of a differentiator circuit is proportional to the rate of change of the input voltage.

9-1d INDUCTIVE REACTANCE

Inductance is opposition to a change in the current in a circuit. A change in the current through an inductor creates a CEMF, which in turn offers a resistance to the changing current. When a dc current flowing through an inductor reaches saturation, there is no longer any change in the current and it is in the steady state. The only opposition to the current flow is the resistance of the wire.

The situation is much different when an inductor is placed into an ac circuit. In this case the applied voltage is constantly changing; therefore, the opposition to current flow that the inductance causes is always present. This opposition to ac voltage caused by the inductance of a circuit is called *inductive reactance,* with the symbol X_L. The amount of inductive reactance offered by an inductor is directly proportional to the frequency and its inductance. Its unit of measurement is the ohm.

Inductive reactance is represented by the formula

$$X_L = 2\pi fL$$

where 2π is equal to 6.28. For example, an inductor with 10 mH is placed into an ac circuit with a frequency of 1 kHz. The inductive reactance at this time can be calculated thus:

$$\begin{aligned} X_L &= 2\pi fL \\ &= 6.28 \times 1\ \text{kHz} \times 10\ \text{mH} \\ &= 6.28 \times 1 \times 10^3 \times 10 \times 10^{-3} \\ &= 62.8 \times 10^{\circ}\ \Omega \\ &= 62.8\ \Omega \end{aligned}$$

If the frequency increases, the inductive reactance will increase. For example, if $f = 5$ kHz, then

$$\begin{aligned} X_L &= 2\pi fL \\ &= 6.28 \times 5\ \text{kHz} \times 10\ \text{mH} \\ &= 6.28 \times 5 \times 10^3 \times 10 \times 10^{-3} \\ &= 314 \times 10^{\circ}\ \Omega \\ &= 314\ \Omega \end{aligned}$$

By simple algebra, the basic inductive reactance formula can be rearranged to find f and L when they are unknown:

$$f = \frac{X_L}{2\pi L} \qquad L = \frac{X_L}{2\pi f}$$

The current through an inductor can be found using Ohm's law, thus:

$$I_L = \frac{V_L}{X_L}$$

9-1e INDUCTANCE IN AN AC CIRCUIT

In any electrical circuit, Ohm's law is used to find current, voltage drops, and power. With ac circuits the calculations are a little more complex because of the phase difference between voltage and current caused by the inductance in the circuit.

9-1e.1 Voltage and Current Phase Angles

In a simple series *RL* circuit as shown in Figure 9-6a, the voltage drop across the resistor V_R, and the current i, of the circuit are in phase. The voltage across a resistor and the current through it are always in phase. For this reason, they are often used as the reference point for analysis, as shown in Figure 9-6b. The phase angle difference between i and the voltage drop across the inductor V_L will be a 90° difference. The applied or total voltage V_T is the algebraic sum of the instantaneous values of V_R and V_L and will be out of phase with these two voltages. Trying to analyze the waveforms shown in Figure 9-6b can become very difficult.

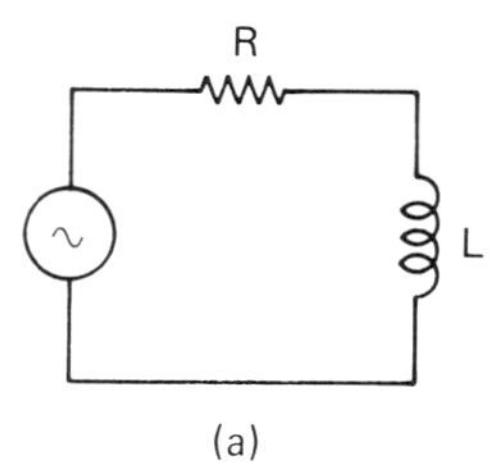

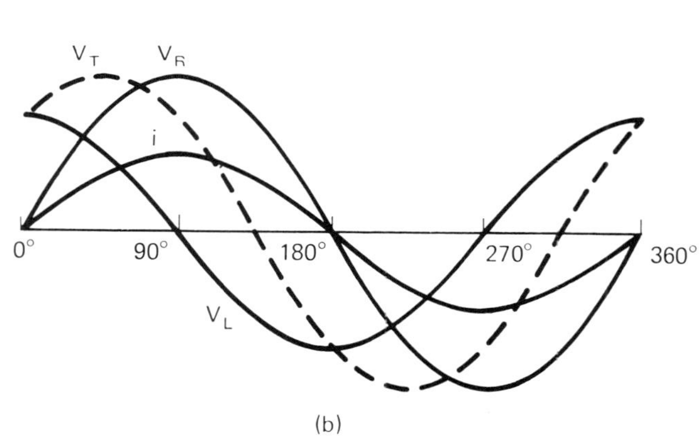

Figure 9–6 Voltage and current relationships in a series *RL* circuit: (a) schematic diagram; (b) waveform phase relationships.

9-1e.2 Voltage Phasor Analysis

It is much more convenient and easier to analyze voltage phase differences by using phasors, as shown in Figure 9-7. Generally, the voltage phase difference between V_L and V_R will be 90°, as shown in Figure 9-7a. The small arrow marked I_R has no value and is shown only to indicate that the circuit current is in phase with V_R. The length of the arrows are directly proportional to the voltages across these devices. Since V_R and I_R are in phase, this phasor becomes the 0° reference point, which extends horizontally to the right. If the V_L phasor is projected to the right as indicated by the dashed line in Figure 9-7b, V_T can be calculated using the right triangle and the Pythagorean theorem. The hypotenuse of the triangle becomes V_T and its value can be

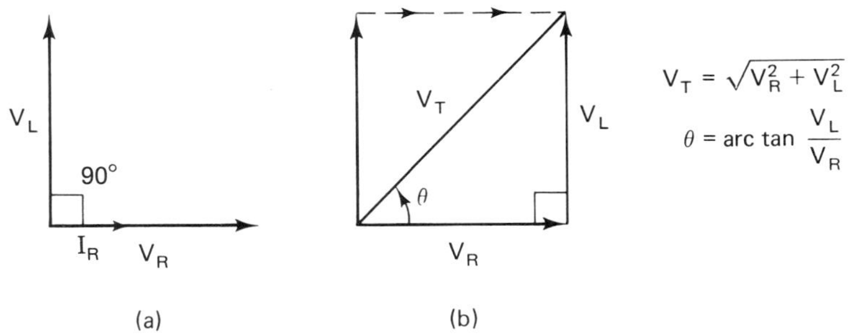

Figure 9–7 Voltage phasor triangle: (a) typical phasor relationship; (b) transposed for easier calculation.

found with the formula $V_T = \sqrt{V_R^2 + V_L^2}$. For example, if $V_R = 25$ V and $V_L = 25$ V, then

$$\begin{aligned} V_T &= \sqrt{V_R^2 + V_L^2} \\ &= \sqrt{25^2\ \text{V} + 25^2\ \text{V}} \\ &= \sqrt{625 + 625} \\ &= \sqrt{1250} \\ &= 35.35\ \text{V} \end{aligned}$$

The voltage values can be expressed as peak to peak, peak, or rms, but in all cases the values must be consistent.

The angle θ between V_R and V_T is found by solving the tangent of the triangle using the formula $\theta = \arctan V_L/V_R$. By definition, it reads "θ is the angle whose tangent is V_L/V_R." If the tangent of an angle is written $\tan \theta = V_L/V_R$, the angle is considered the inverse of the tangent. Since these trigonometric functions are based on the sine wave and angular measurement, the angle is equal to the arctan (or inverse) of the tangent. For example:

$$\begin{aligned} \theta &= \arctan \frac{V_L}{V_R} \\ &= \arctan \frac{25\ \text{V}}{25\ \text{V}} \\ &= \arctan 1 \\ &= 45° \end{aligned}$$

The tangent of 1 can be looked up in a trigonometric table to find the correct angle, or if you have a scientific calculator you may use the following procedure:

1. Divide V_L by V_R.
2. Press the function key.
3. Press the arc, inv, or second function key.
4. Press the tan key.

9-1e.3 Impedance

The current of the series *RL* circuit can be found if the total resistance of the circuit is known. However, the total opposition to current flow in

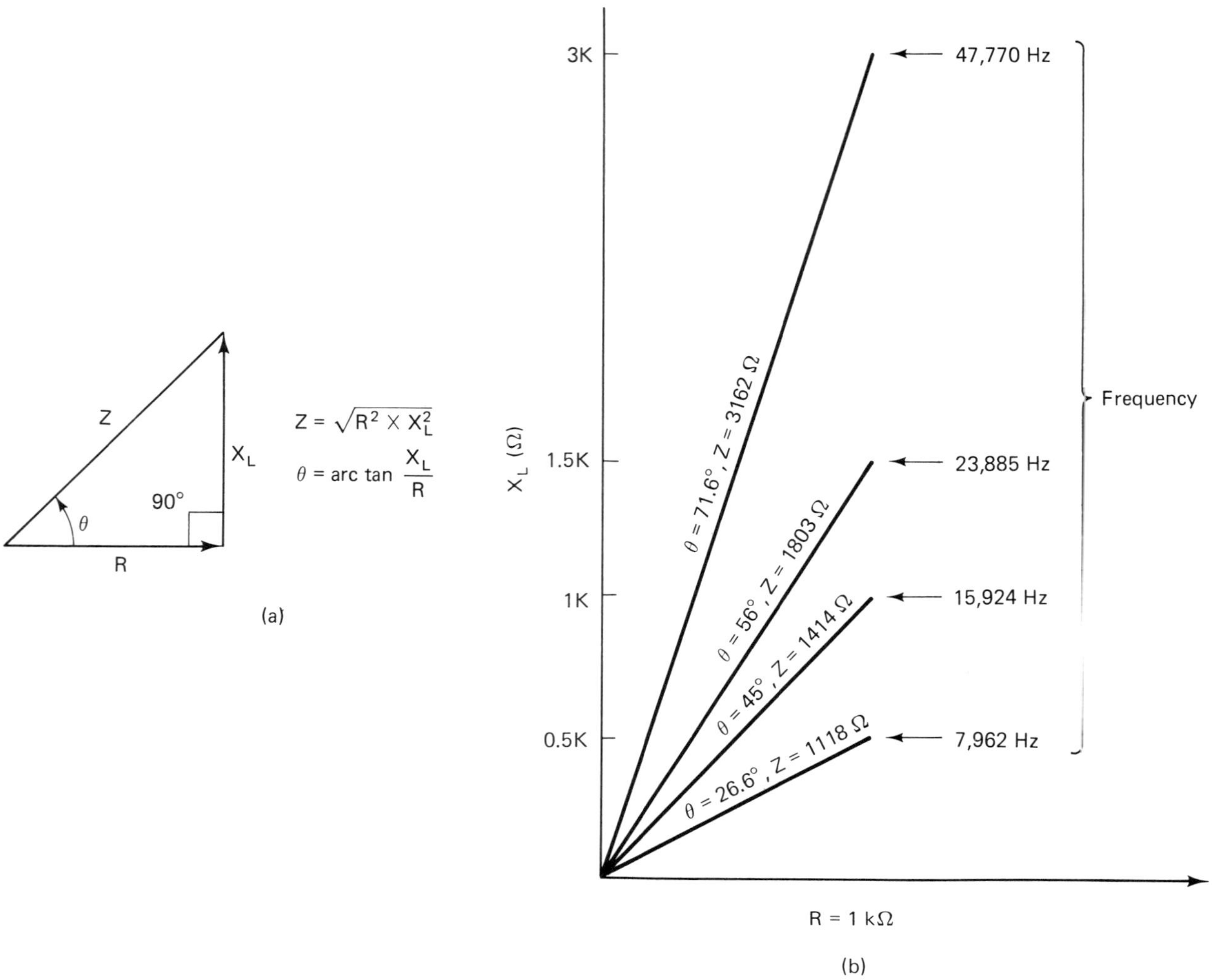

Figure 9–8 Impedance phasor triangle: (a) same relationship as phasor voltage triangle; (b) effects on triangle with frequency.

the circuit is referred to as *impedance* and consists of the resistance and the inductive reactance of the circuit. Since a voltage drop is proportional to the current and the opposition offered, its voltage phasor is similar to the opposition phasor. Therefore, an impedance phasor triangle as shown in Figure 9-8a is similar to the voltage phasor triangle. Phasors R and X_L are 90° out of phase and the impedance phasor z can be found with the formula $Z = \sqrt{R^2 + X_L^2}$. For example, if $R = 1\ \text{k}\Omega$ and $X_L = \text{k}\Omega$, then

$$\begin{aligned}
Z &= \sqrt{R^2 + X_L^2} \\
&= \sqrt{1 \times (10^3)^2 + 1 \times (10^3)^2} \\
&= \sqrt{1 \times 10^6 + 1 \times 10^6} \\
&= \sqrt{2 \times 10^6} \\
&= 1.414 \times 10^3 \\
&= 1.414\ \text{k}\Omega
\end{aligned}$$

The current in the circuit can be found with the formula

$$I_T = \frac{V_T}{Z}$$
$$= \frac{35.35\ \text{V}}{1.414\ \text{k}\Omega}$$
$$= \frac{35.35}{1.414 \times 10^3\ \Omega}$$
$$= 25 \times 10^{-3}\ \text{A}$$
$$= 25\ \text{mA}$$

If you solve for the voltage drops across R and L, you will find that they equal the voltage phasors used in the voltage phasor triangle, thus:

$$V_R = I_T R$$
$$= 25\ \text{mA} \times 1\ \text{k}\Omega$$
$$= 25 \times 10^{-3} \times 1 \times 10^3$$
$$= 25\ \text{V}$$

$$V_L = I_T X_L$$
$$= 25\ \text{mA} \times 1\ \text{k}\Omega$$
$$= 25 \times 10^{-3} \times 1 \times 10^3$$
$$= 25\ \text{V}$$

The impedance phase angle is found with the formula

$$\theta = \arctan \frac{X_L}{R}$$
$$= \arctan \frac{1\ \text{k}\Omega}{1\ \text{k}\Omega}$$
$$= \arctan 1$$
$$= 45^\circ$$

When the frequency of the applied voltage increases, X_L also increases, as shown in Figure 9-8b. Notice that the impedance phase angle increases with a frequency increase and it moves toward 90°. If the angle were 90°, the circuit would be purely inductive. Of course, this is impossible since even the wire in the coil has some resistance. If the angle is equal to 0°, the circuit would be purely resistive, which means that there are no inductive properties in the circuit.

9-1e.4 *RL* Parallel AC Circuit

When a resistor and an inductor are in parallel with ac voltage applied, the voltage across both devices are in phase, as shown in Figure 9-9a. The current through the resistor, i_R, is in phase with the applied voltage, whereas the current through the inductor, i_L, is out of phase with the applied voltage. Both currents combine to form the total current, which is out of phase with the applied voltage. The current phasor triangle shown in Figure 9-9b indicates i_L in the downward position because the current lags the voltage in an inductive circuit.

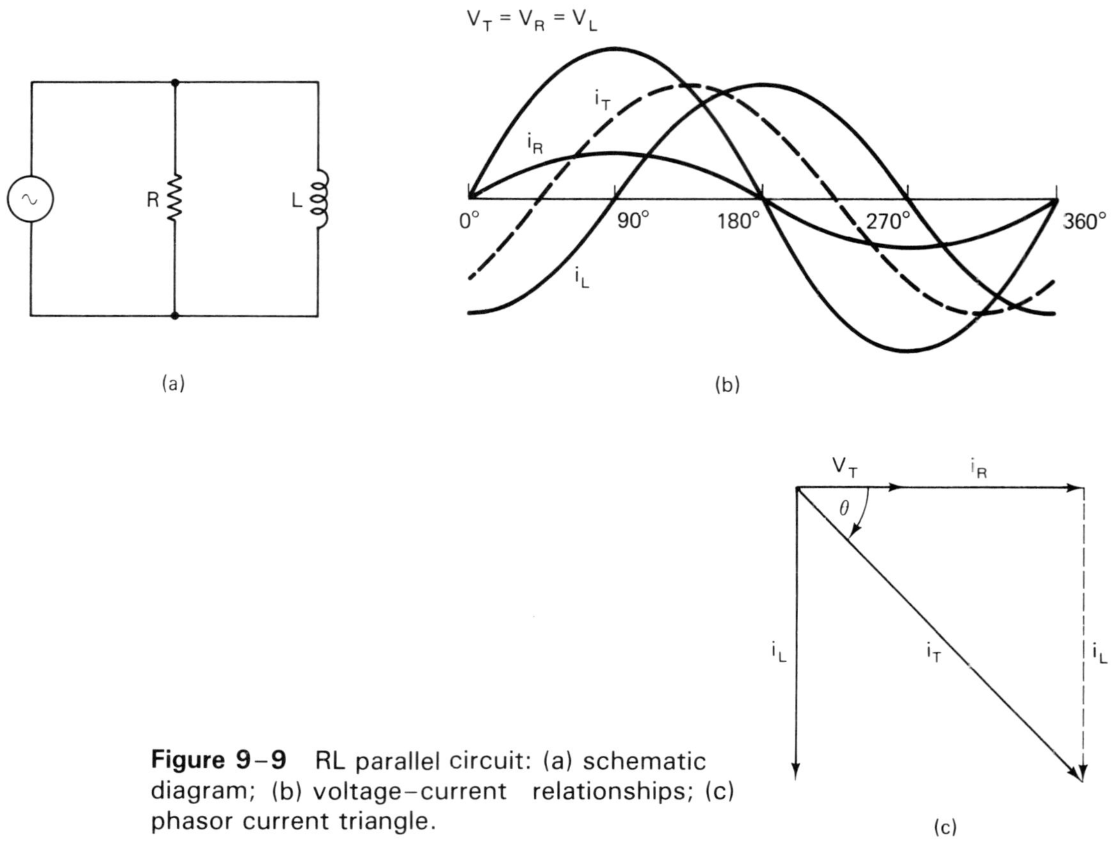

Figure 9–9 RL parallel circuit: (a) schematic diagram; (b) voltage–current relationships; (c) phasor current triangle.

To find the total current in the circuit you must first calculate the current through each branch using Ohm's law. For example, let $R = 1$ kΩ, $X_L = 1$ kΩ, and $V_T = 50$ V; then

$$i_R = \frac{V_T}{R} = \frac{50\text{ V}}{1\text{ k}\Omega} = \frac{50}{1 \times 10^3} = 50 \times 10^{-3}\text{ A} = 50\text{ mA}$$

$$i_L = \frac{V_T}{X_L} = \frac{50\text{ V}}{1\text{ k}\Omega} = \frac{50}{1 \times 10^3} = 50 \times 10^{-3}\text{ A} = 50\text{ mA}$$

Now the total current can be solved using the formula

$$\begin{aligned} i_T &= \sqrt{i_R^2 + i_L^2} \\ &= \sqrt{50\text{ mA}^2 + 50\text{ mA}^2} \\ &= \sqrt{(50 \times 10^{-3})^2 + (50 \times 10^{-3})^2} \\ &= \sqrt{2500 \times 10^{-6} + 2500 \times 10^{-6}} \\ &= \sqrt{5000 \times 10^{-6}} \\ &= 70.71 \times 10^{-3}\text{ A} \\ &= 70.71\text{ mA} \end{aligned}$$

The phase angle between i_L and V_T is found by using the formula

$$\theta = \arctan \frac{i_L}{i_R}$$
$$= \arctan \frac{50 \text{ mA}}{50 \text{ mA}}$$
$$= \arctan 1$$
$$= 45^\circ$$

9-1e.5 Power and Inductive Reactance

When current flowing through an inductor increases, the electromagnetic field about it builds up. When this current decreases the field collapses and induces current back into the circuit. Theoretically, no energy is lost, but because of at least some resistance in the circuit, a small amount of heat is generated, resulting in a power loss. Regardless of this power loss, some energy from the collapsing field is put back into the circuit.

The normal formula for finding the power of a circuit is $P = VI$ and is referred to as *apparent power*, P_A. The power in a RL circuit is called the *true power*, P_T and is a function of the true power and the cosine of the phase angle between the V_T and i_T of the circuit. For example, if $V_T = 35.35$ V and $i_T = 25$ mA in a series circuit with a 45° phase angle, then

$$P_A = V_T i_T$$
$$= 35.35 \text{ V} \times 25 \text{ mA}$$
$$= 35.35 \times 25 \times 10^{-3}$$
$$= 883.75 \times 10^{-3} \text{ W}$$
$$= 883.75 \text{ mW}$$

However, the true power would be found thus:

$$P_T = V_T i_T \cos \theta$$
$$= 35.35 \text{ V} \times 25 \text{ mA} \times \cos 45^\circ$$
$$= 35.35 \times 25\ 10^{-3} \times 0.707$$
$$= 624.8 \times 10^{-3} \text{ W}$$
$$= 624.8 \text{ mW}$$

The relationship of the true power to the apparent power is called the *power factor* (PF), and is expressed by the formula, $\text{PF} = P_T/P_A$. Using the previous two examples, we can find the power factor, thus:

$$\text{PF} = \frac{P_T}{P_A}$$
$$= \frac{624.8 \text{ mW}}{883.75 \text{ mW}}$$
$$= 0.707 \quad \text{(no unit)}$$

As you can see, the power factor is the same value as the cosine of the angle; therefore, PF $= \cos\theta$. If the true power and the power factor are known, the apparent power can be found by rearranging the formula:

$$P_A = \frac{P_T}{\text{PF (or } \cos\theta)}$$

The power in a parallel *RL* circuit can be found by finding the power of each device and then combining them using the Pythagorean theorem:

$$\begin{aligned} P_R &= V_T i_R \\ &= 50\text{ V} \times 50\text{ mA} \\ &= 50 \times 50 \times 10^{-3} \\ &= 2500 \times 10^{-3} \\ &= 2.5\text{ W} \end{aligned} \qquad \begin{aligned} P_L &= V_T i_L \\ &= 50\text{ V} \times 50\text{ mA} \\ &= 50 \times 50 \times 10^{-3} \\ &= 2500 \times 10^{-3} \\ &= 2.5\text{ W} \end{aligned} \qquad \begin{aligned} P_T &= \sqrt{P_R^2 + P_L^2} \\ &= \sqrt{2.5^2 + 2.5^2} \\ &= \sqrt{6.25 + 6.25} \\ &= \sqrt{12.50} \\ &= 3.535\text{ W} \end{aligned}$$

9-1f INDUCTIVE FREQUENCY FILTERS

Since the inductive reactance of an inductor is directly proportional with changes in frequency, it is very useful in frequency filtering circuits, as shown in Figure 9-10. A basic low-pass filter circuit as shown in Figure 9-10a is a simple *RL* series circuit used as a voltage divider. The input signal is across both components, with the output of the circuit taken across the resistor. At relatively low frequencies X_L is small and most of the input voltage is dropped across the resistor. At higher frequencies X_L increases and begins to drop more voltage. Remembering Kirchhoff's law for voltage drops in a series circuit, the total voltage is the sum of the individual voltage drops. Therefore, if X_L increases,

Figure 9–10 Inductive filters: (a) low-pass circuit; (b) low-pass graph;

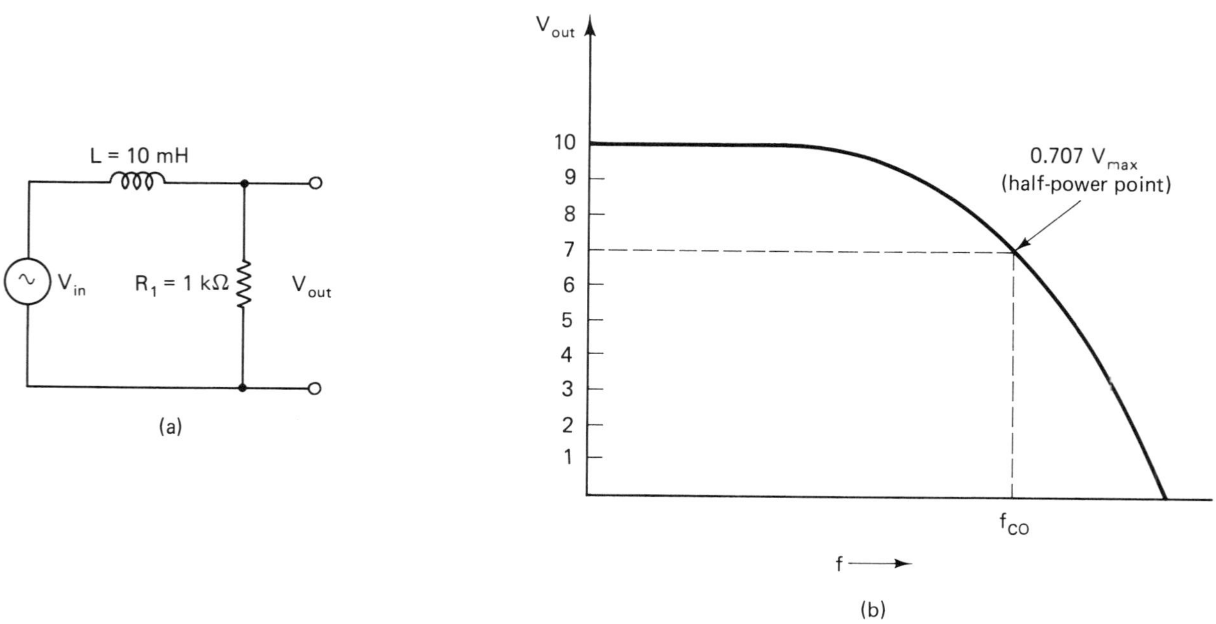

the voltage drop across L will increase and the voltage drop of R will have to decrease. In effect, the output voltage will be less. The amplitude at the output is greater for lower frequencies than it is for higher frequencies, hence the term *low-pass filter.*

A graph of the output voltage versus frequency for a low-pass filter is shown in Figure 9-10b. It is generally accepted that the frequencies that are "passed" by the filter will have an amplitude greater than 70.7% of the input voltage amplitude. The point where the $0.707 V_{max}$ output occurs is called the half-power point. The term "half-power" results from the fact that if the output voltage is $0.707 V_{max}$, then the current is also at $0.707 I_{max}$; therefore, $P = VI = 0.707 V_{max} \times 0.707 I_{max} = 0.4998 VI$, or $0.5 VI$. At the half-power point, X_L equals R and the phase shift is 45°.

The frequency at which the half-power point occurs is called the *cutoff frequency,* f_{co}, and can be calculated by the formula $f_{co} = R/2\pi L$. For example, if $R = 1\ k\Omega$ and $L = 10$ mH, then

$$
\begin{aligned}
f_{co} &= \frac{R}{2\pi L} \\
&= \frac{1\ k\Omega}{6.28 \times 10\ mH} \\
&= \frac{1 \times 10^3}{6.28 \times 10 \times 10^{-3}} \\
&= \frac{1 \times 10^3}{62.8 \times 10^{-3}} \\
&= 0.0159235 \times 10^6\ Hz \\
&= 15{,}923.5\ Hz \\
&= 15.9\ kHz
\end{aligned}
$$

Figure 9–10 Continued (c) high-pass circuit; (d) high-pass graph.

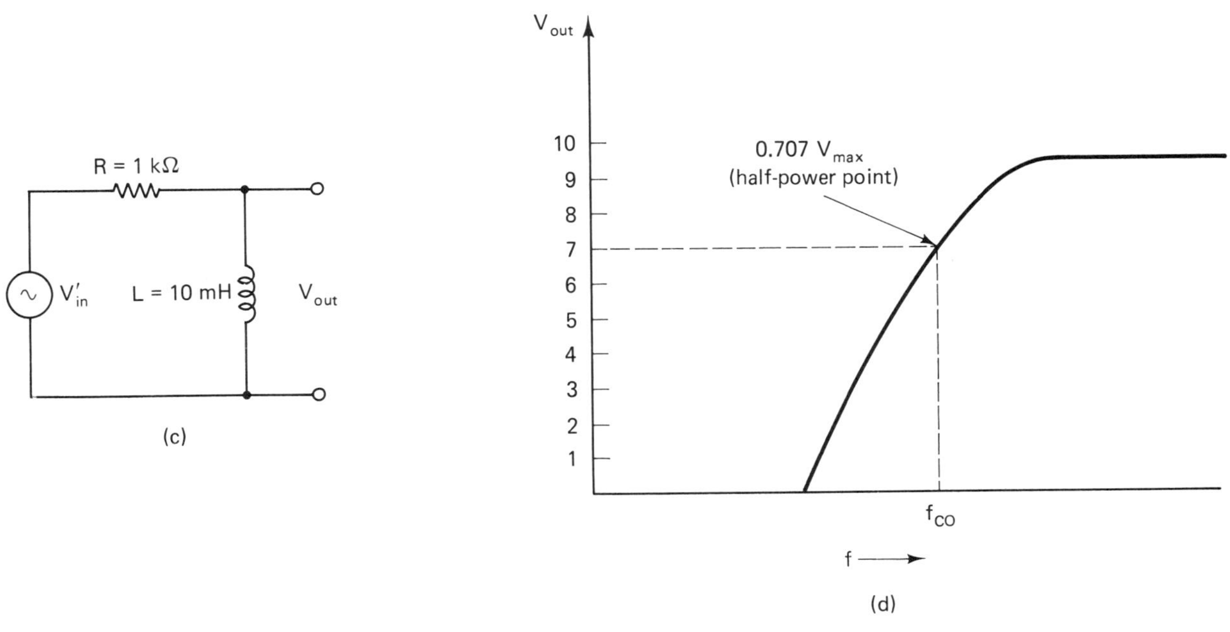

If the input voltage amplitude is 10 $V_{p\text{-}p}$, the output voltage amplitude at the cutoff frequency can be found by using the formula

$$\begin{aligned} V_{fco} &= 0.707\,V_{max} \\ &= 0.707 \times 10\ V_{p\text{-}p} \\ &= 7.07\ V_{p\text{-}p} \end{aligned}$$

If the output is taken across the inductor as shown in Figure 9-10c, the circuit becomes a *high-pass filter.* At low frequencies the value of X_L is very small and very little voltage appears across the inductor. As the frequency increases, so does X_L and the output voltage increases. The graph of the output voltage versus frequency for a high-pass filter is shown in Figure 9-10d. In this case, the cutoff frequency is the point where the output voltage amplitude begins to equal 70.7% and more of the input voltage amplitude.

9-1g SUMMARY OF *RL* CIRCUIT FORMULAS

Series Circuit

$$L_T = L_1 + L_2 + L_3$$

$$X_L = 2\pi fL$$

$$f = \frac{X_L}{2\pi L}$$

$$L = \frac{X_L}{2\pi f}$$

$$Z = \sqrt{R^2 + X_L^2}$$

$$i_T = \frac{V_T}{Z}$$

$$V_R = i_T R$$

$$V_L = i_T X_L$$

$$V_T = \sqrt{V_R^2 + V_L^2}$$

$$P_A = V_T i_T$$

$$P_T = P_A \cos\theta$$

$$\text{PF} = \frac{P_T}{P_A} = \cos\theta$$

$$\text{TC} = \frac{L}{R}$$

$$\theta = \arctan\frac{X_L}{R} = \frac{V_L}{V_R}$$

$$f_{co} = \frac{R}{2\pi L} = \frac{0.159R}{L}$$

$$V_{fco} = 0.707\text{V}_{max}$$

Parallel Circuit

$$L_T = \frac{L_1 \times L_2}{L_1 + L_2}$$

$$L_T = \frac{1}{\dfrac{1}{L_1} + \dfrac{1}{L_2} + \dfrac{1}{L_3}}$$

$$V_T = V_R = V_L$$

$$X_L = 2\pi fL$$

$$i_R = \frac{V_T}{R}$$

$$i_L = \frac{V_T}{X_L}$$

$$i_T = \sqrt{i_R^2 + i_L^2}$$

$$Z = \frac{V_T}{i_T}$$

$$P_R = V_R i_R$$

$$P_L = V_L i_L$$

$$P_T = \sqrt{P_R^2 + P_L^2}$$

$$f_{co} = \frac{R}{2\pi L} = \frac{0.159R}{L}$$

$$V_{fco} = 0.707\text{V}_{max}$$

SECTION 9-2 DEFINITION EXERCISES

Write a brief description for each of the following terms.

1. Counter-electromotive force (CEMF) ____________

2. Inductance ____________

3. Self-inductance ____________

4. Henry ____________

5. Inductor ____________

6. ELI ____________

7. Mutual inductance ____________

8. Saturation ____________

9. Time constant ____________

10. Integrator ____________

11. Differentiator ____________

12. Inductive reactance (X_L) ____________

13. Arctan ____________________

____

14. Impedance ____________________

____

15. Apparent power ____________________

____

16. True power ____________________

____

17. Power factor ____________________

____

18. Low-pass filter ____________________

____

19. High-pass filter ____________________

____

20. Half-power point ____________________

____

21. Cutoff frequency ____________________

____

SECTION 9-3 EXERCISES AND PROBLEMS

Complete this section before beginning the next section.

1. Draw the schematic symbol for the following inductors in the space provided.

   a. air core

   b. iron core

2. List the factors that determine the inductance of a coil.

a.

b.

c.

d.

e.

3. Find the total inductance for the inductors given when they are in series.

a. $L_1 = 1.2$ H, $L_2 = 5.0$ H

L_T = ____________

b. $L_1 = 50$ mH, $L_2 = 20$ mH, $L_3 = 10$ mH

L_T = ____________

c. $L_1 = 100$ mH, $L_2 = 40$ mH, $L_3 = 20$ mH

L_T = ____________

4. Find the total inductance for the inductors given when they are in parallel.

a. $L_1 = 4$ H, $L_2 = 2$ H

L_T = ____________

b. $L_1 = 40$ mH, $L_2 = 20$ mH, $L_3 = 10$ mH

L_T = ____________

c. $L_1 = 60$ mH, $L_2 = 60$ mH, $L_3 = 60$ mH

L_T = ____________

5. Find the total inductance for the data given: $L_1 = 20$ mH, $L_2 = 80$ mH, and $k = 0.3$.

a. Series-aiding circuit: ____________

b. Series-opposing circuit: ____________

6. Calculate the following time constants for the values given.

a. $R = 1$ kΩ, $L = 20$ mH

TC = ____________

b. $R = 2.2$ kΩ, $L = 30$ mH

TC = ____________

c. $R = 10$ kΩ, $L = 60$ mH

TC = ____________

7. A particular *RL* circuit has a maximum dc current of 120 mA. When it is first turned on the current begins to rise. Calculate the amount of current in the circuit for each of the time constants.

Time Constant	*Amount of Current*
1	______
2	______
3	______
4	______
5	______

8. Calculate the inductive reactance for the data given.

a. $f = 1$ kHz
$L = 20$ mH

X_L = ____________ Ω

b. $f = 500$ Hz
$L = 80$ mH

X_L = ____________ Ω

c. $f = 40$ kHz
$L = 50$ mH

X_L = ____________ Ω

9. Find the unknown values from the data given.

a. $X_L = 1256\Omega$
$L = 10$ mH

f = ____________

b. $X_L = 942\Omega$
$L = 30$ mH

f = ____________

c. $X_L = 3140\Omega$
$f = 5$ kHz

L = ____________

d. $X_L = 47$ kΩ
$f = 30$ kHz

L = ____________

10. Draw the *RL* series circuit shown in Figure 9-6a in the space provided. Solve for the unknowns using the data given.

a. $V_{in} = 100$ V at 6 kHz, $R = 2.2$ kΩ, $L = 60$ mH.

X_L = ____________ Ω Z = ____________ Ω i_T = ____________

V_R = ____________ V_L = ____________ V_T = ____________

θ = ____________ P_A = ____________ P_T = ____________

PF = ____________

b. $V_{in} = 50$ V at 16 kHz, $R = 4.7$ kΩ, $L = 100$ mH.

X_L = ____________ Z = ____________ i_T = ____________

V_R = ____________ V_L = ____________ V_T = ____________

θ = ____________ P_A = ____________ P_T = ____________

11. In a *RL* series circuit $R = 10\ \text{k}\Omega$ and $L = 500$ mH. Solve for X_L, Z, and θ with the various frequencies given.

$f = 3$ kHz $X_L =$ ________ $Z =$ ________ $\theta =$ ________

$f = 6$ kHz $X_L =$ ________ $Z =$ ________ $\theta =$ ________

$f = 1.5$ kHz $X_L =$ ________ $Z =$ ________ $\theta =$ ________

12. With the data calculated in Question 10, draw three impedance triangles with all the data shown on the same graph, similar to the one shown in Figure 9-8b (let $\frac{1}{4}$ inch equal 1 kΩ).

13. Draw a *RL* parallel circuit as shown in Figure 9-9a in the space provided. Solve for the unknowns with the data given.

a. V_{in} = 100 V at 12 kHz
R = 2.2 kΩ
L = 60 mH

i_R = ________

i_L = ________

i_T = ________

θ = ________

P_A = ________

P_T = ________

PF = ________

b. V_{in} = 60 V at 25 kHz
R = 10 kΩ
L = 40 mH

i_R = ________

i_L = ________

i_T = ________

θ = ________

P_A = ________

P_T = ________

PF = ________

14. Draw a low-pass filter with its corresponding graph as shown in Figure 9-10a and b.

15. Draw a high-pass filter with its corresponding graph as shown in Figure 9-10c and d.

16. Solve for the cutoff frequency and the voltage at the half-power point for the data given.

a. $V_{in} = 150\ V_{p\text{-}p}$
$R = 3.3\ k\Omega$
$L = 50\ mH$

f_{co} = ____________

V_{out} = ____________

b. $V_{in} = 60\ V_{p\text{-}p}$
$R = 10\ k\Omega$
$L = 8\ H$

f_{co} = ____________

V_{out} = ____________

17. Match the description in column A with the proper term in column B.

Column A

____ a. The opposition to a change of current in a circuit

____ b. The property of a circuit when a CEMF is induced in the same circuit by a change of current

____ c. The unit of inductance

____ d. Opposition to current in a circuit caused by inductance

Column B

1. impedance
2. apparent power
3. ELI
4. inductive reactance
5. henry
6. low-pass filter
7. time constant
8. true power
9. phase angle

____ **e.** Opposition to current, including both resistance and reactance

____ **f.** Guide for remembering that *I* lags *E* in an inductor

____ **g.** The angle that *I* lags *E* in an ac circuit with inductance

____ **h.** The product of *I* and *V* in an ac circuit

____ **i.** The time for current to rise to 63.2% of the maximum current

____ **j.** When the current in one circuit changes and causes a change of current in another circuit

____ **k.** The actual power of a circuit because of inductive reactance

____ **l.** The point where the output voltage equals $0.707V_{max}$

10. mutual inductance
11. inductance
12. cutoff frequency
13. self-inductance

SECTION 9-4 EXPERIMENTS

EXPERIMENT 1. STORED ENERGY IN AN INDUCTOR

Objective: To show the effect of a rising current and a falling current in an inductor.

Introduction: In this experiment you notice how a rising current in an inductor produces a CEMF. When the current is not changing through the inductor, no voltage will be developed. When the current in the inductor falls, a CEMF of opposite polarity is generated.

Materials Needed:
1 Inductor 10 mH or up to 100 mH
1 SPST switch
3 Clip leads for connecting circuit

Equipment Required:
1 12-V power supply
1 Oscilloscope

Procedure:

1. Connect the circuit as shown in Figure 9-11. Be sure to leave S_1 open.
2. Set the input voltage select switch to ac on the oscilloscope.
3. Set the volt/cm control to the highest range.
4. Set the time/cm control to 1 ms.

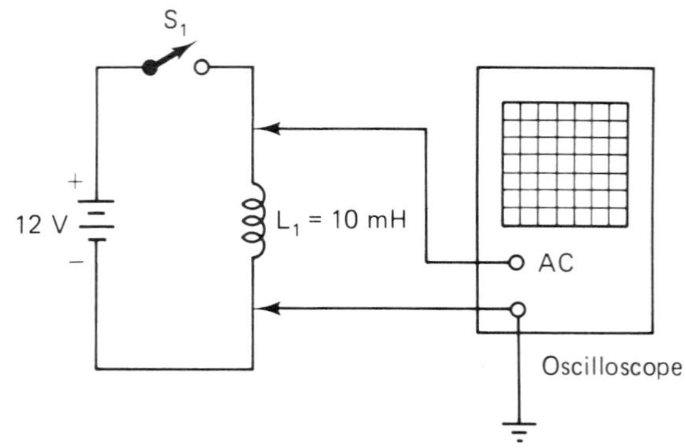

Figure 9-11

5. Adjust the vertical position so that the trace is on the center line of the CRT.
6. Close S_1 and observe the direction in which the trace moves. The trace moved ______________ .
7. Where is the trace now? ______________ There is no change of current through the inductor; therefore, no CEMF is created. The current through the inductor is at saturation.
8. Open S_1 and observe the direction in which the trace moves. The trace moved ______________ . When the circuit is opened, the falling current induces a voltage of the opposite polarity, which is referred to as the *inductive kick.*

Fill-In Questions:

1. When current in an inductor increases, a ______________ is created that tends to oppose the applied voltage.

2. When the current flowing in an inductor is at saturation, no ______________ is created.

3. When the current flowing in an inductor is at saturation, ______________ is stored in the electromagnetic field about the inductor.

4. When the current flowing in an inductor falls, a CEMF is created whose polarity is ______________ to the initial applied voltage.

EXPERIMENT 2. TIME CONSTANTS AND SQUARE-WAVE PULSES

Objective: To demonstrate the effects of various pulse widths with relation to a circuit's time constant.

Introduction: In this experiment you will observe the voltage waveforms across the circuit components R and L to better understand the use of time constants.

Materials Needed:

1 1-kΩ resistor at 0.5 W
1 10-mH inductor

Equipment Required:

1 Square-wave generator
1 Oscilloscope
1 Breadboard with leads or wire for constructing circuit

Procedure:

1. Draw the circuit shown in Figure 9-5a in the space below.
2. What is the time constant of the circuit? ____________ μs
3. Construct the circuit you have drawn.
4. Set the generator for an output of 12 $V_{p\text{-}p}$. Maintain this amplitude for every frequency change that you make.
5. Adjust the generator for 50 kHz. The pulse width is 10 μs.
6. Observe the waveform of V_L and V_R. The voltage across R has only enough time to increase and decrease for the one time constant before the input pulse changes direction. The amplitude of V_R should be about 63.2% of the input voltage. The waveform across L slopes during the increase and decrease of circuit current.
7. Adjust the generator for 10 kHz. The pulse width is 50 μs, five times the amount of the circuit's time constant.
8. Observe the waveforms V_R and V_L. The voltage across R now indicates that there is time for the current to rise to maximum before the input pulse changes direction. The slope of the voltage across L is greater, which indicates a longer rise time for the circuit current.
9. Adjust the generator for 5 kHz. The pulse width is now 100 μs, 10 times greater than the circuit's time constant.
10. Observe the waveforms V_R and V_L. The voltage across R indicates that the current has more than enough time to rise and fall before the pulse changes direction. The voltage across L is a sharp "spike," which indicates that the time required for the current to reach maximum is very small compared to the time the voltage is constant during the pulse width. The rate of current change for this input pulse is very small.

Fill-In Questions:

1. When the input pulse and the circuit's time constant are of the same value, the waveform of V_R shows a ____________ rise and fall time.

2. When the input pulse is 5 to 10 times greater than the circuit's time constant, the waveform of V_L has a __________ appearance.

3. If the square-wave input voltage to a *RL* series circuit is 80 $V_{p\text{-}p}$, after one time constant the amplitude of the voltage across *R* would be __________ $V_{p\text{-}p}$.

EXPERIMENT 3. *RL* LOW- AND HIGH-PASS FILTERS

Objective: To demonstrate the characteristics of low- and high-pass filters and provide the reader with skills in measuring ac reactive circuits.

Introduction: In this experiment you will construct filter circuits and measure the output voltage for various input frequencies. From the data taken, you will then construct frequency response graphs (or curves).

Materials Needed:
1 1-kΩ resistor at 0.5 W
1 10-mH inductor

Equipment Required:
1 Sine-wave generator
1 Dual-trace oscilloscope
1 Breadboard and leads or wire for constructing circuit

Procedure:

1. Construct the circuit shown in Figure 9-12a. This is a low-pass filter circuit. Make sure that both ground leads of the oscilloscope are connected together. If you connect both leads of each channel across each component, you may short out one of the components.
2. Adjust the generator for an output of 0.5 kHz. Observe channel 1 of the oscilloscope for an amplitude of 12 $V_{p\text{-}p}$ reading.
   Important! Every time you change the frequency of the generator be sure that the amplitude is maintained the same to give accurate results.
3. Referring to the following table, calculate the X_L and θ for the frequency given and enter the results in the proper place in the table.

f	*Calculate:*		*Measure:*	
(kHz)	X_L	θ	V_R	V_L
0.5	______	______	______	______
1	______	______	______	______
10	______	______	______	______
16	______	______	______	______
25	______	______	______	______
50	______	______	______	______
100	______	______	______	______
200	______	______	______	______

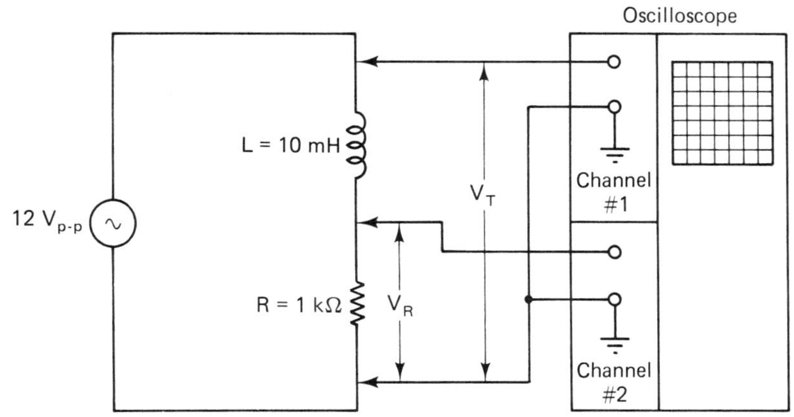

(a)

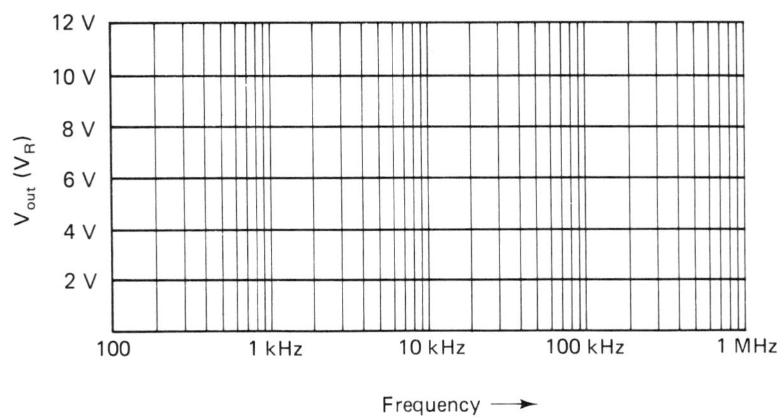

Figure 9–12 (b)

4. Using only one channel of the oscilloscope, measure V_L and V_R and record in the table.
5. Repeat steps 4 and 5 for the other frequency values given in the table.
6. Using the data entries for V_R in the table, construct an output voltage versus frequency curve on the blank graph shown in Figure 9-12b. Notice the type of graph that is used. This is referred to as a four-cycle semilogarithm type of graph. The purpose of using this type of graph is that it condenses the frequencies into a readable scale that can fit on the page in this book. For example, if $\frac{1}{4}$ in. equaled 1 kHz, it would take 250 in. (over 20 ft) on a linear scale to represent the frequency range up to 1 MHz.
7. Turn off the power to the circuit. Exchange the R and L components so that the inductor is now on the bottom. This is a high-pass filter circuit. Apply power to the circuit.
8. Enter the values of V_T and V_L in the following table for the frequencies given.

   Important! Remember to keep the amplitude of the input signal at 12 $V_{p\text{-}p}$ for all frequency changes.

f (kHz)	Measure: V_R	Measure: V_L
0.5	_____	_____
1	_____	_____
10	_____	_____
16	_____	_____
25	_____	_____
50	_____	_____
100	_____	_____
200	_____	_____

9. In the space provided below, construct an output voltage versus frequency graph for the data of V_L in the table.

10. Calculate the cutoff frequency for these circuits using the formula

$$f_{co} = \frac{R}{2\pi L} = \underline{\hspace{3cm}} \text{ Hz}$$

11. Calculate the amplitude of the output voltage at the cutoff frequency using the formula

$$V_{fco} = 0.707\text{V}_{\text{max}} = \underline{\hspace{3cm}} \text{ V}_{\text{p-p}}$$

Fill-In Questions:

1. When the frequency applied to a RL series circuit increases, X_L ______________.

2. When the frequency applied to a RL series circuit decreases, X_L ______________.

3. The cutoff frequency for these circuits is approximately

____________ kHz.

4. The output voltage amplitude of these circuits at the cutoff frequency is about ____________ $V_{p\text{-}p}$.

5. The characteristic that identifies a low-pass filter is that the output voltage is ____________ than $0.707\,V_{max}$ for all frequencies ____________ the f_{co} point.

6. The characteristic that identifies a high-pass filter is that the output voltage is greater than ____________ V_{max} for all frequencies ____________ the f_{co} point.

SECTION 9-5 TROUBLESHOOTING APPLICATION: TESTING INDUCTORS

An inductor can be tested for an open by making a continuity test with an ohmmeter. If the dc resistance of the wire is known, you can make a resistance check for a partial short. In either case, if a problem is found, the inductor will not perform properly in a circuit.

There are cases where the inductance value is not given on an inductor and it may be necessary to find the value. A rather simple test can be performed to determine the value of inductance, as shown in Figure 9-13.

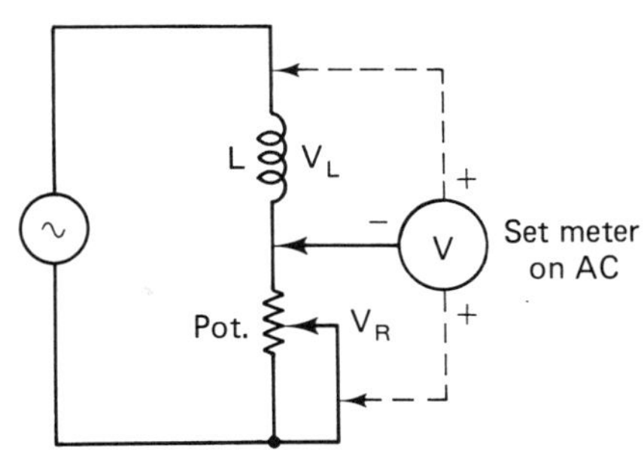

Figure 9–13

Once the circuit is connected, you adjust the generator for a desired frequency, perhaps 1 kHz, and measure the amplitude of its output voltage. Using a voltmeter or oscilloscope, you alternately measure V_L and V_R. You adjust the potentiometer until V_R equals V_L. When these two voltages are equal, it can be assumed that X_L equals R. Now, turn off the generator and disconnect the circuit, being very careful not to disturb the setting of the potentiometer. Measure the resistance of the potentiometer and use these data as the value of X_L, with the following formula to find L:

$$L = \frac{XL}{2\pi f}$$

SECTION 9-6 INSTANT REVIEW

- Inductance is the opposition to a change of the current in a circuit.
- Total inductance for inductors in series is found by adding up the individual inductors.
- Total inductance for inductors in parallel is found with the reciprocal method or product-over-sum method.
- Total inductance for the same-value inductors in parallel is found by dividing the value of one inductor by the total number of inductors.
- Counter-electromotive force (CEMF) is the voltage developed by inductance which opposes the applied voltage.
- Self-inductance is the development of a CEMF within a circuit with no other outside influence.
- Mutual inductance is the development of a CEMF in one circuit by the induction of another circuit. Inductors in close proximity may affect each other by aiding or opposing through mutual inductance.
- An inductor is a coil of wire that exhibits inductive properties.
- The henry is the unit of inductance.
- "ELI" means that the voltage (electromotive force) leads the current in an inductive circuit.
- Saturation occurs in an inductor when the dc current reaches maximum for the given applied voltage.
- A time constant is the required time for the current in a circuit to rise to 63.2% of the maximum current.
- A time constant is also the time required for the current in a circuit to fall to 36.8% of its maximum value.
- Five time constants are required for the current in an inductive circuit to reach maximum or fall to zero.
- An integrator is a circuit that continuously adds up a quantity being measured over a period of time.
- A differentiator is a circuit whose output is proportional to the rate of change of the input voltage.
- Inductive reactance (X_L) is the opposition an inductor offers to ac voltage and is measured in ohms.
- "Arctan" stands for "θ is the angle whose tangent is" and is used to find the phase angle between voltage and current in an inductive circuit.
- Impedance (Z) is the total opposition to current flow in an ac circuit, including resistance and inductive reactance.
- In a series RL circuit, the vectorial sum of V_R and V_L equals the applied voltage.
- In a series RL circuit, the vectorial sum of R and L equals the impedance.
- In a series RL circuit, $I_T = I_R = I_L$ and is found by V_T/Z.
- In a parallel RL circuit, $V_T = V_R = V_L$.
- In a parallel RL circuit, the vectorial sum of i_R and i_L equals i_T.
- Apparent power is V times I.

- A reactive component such as an inductor gives some energy back to the circuit.
- True power is V times I times cos θ, which is the phase angle of the voltage and current in a reactive circuit.
- The power factor (PF) is the true power divided by the apparent power.
- The cutoff frequency (f_{co}) of a filter circuit is where the output voltage reaches $0.707V_{max}$ of the input voltage.
- The output of a low-pass filter will be greater than $0.707V_{max}$ for all frequencies below f_{co}.
- The output of a high-pass filter will be greater than $0.707V_{max}$ for all frequencies above f_{co}.

SECTION 9-7 SELF-CHECKING QUIZ

Circle the most correct answer for each question.

1. The property that opposes a change in current is called:

a. mutual inductance
b. inductive reactance
c. inductance
d. reactive resistance

2. The relationship of voltage and current in a pure inductive circuit is that current:

a. leads voltage by 90°
b. lags voltage by 90°
c. lags voltage by 45°
d. leads voltage by 180°

3. The force that tends to keep current flowing in an inductive circuit when it is opened is called:

a. inductive reactance **b.** apparent power
c. saturation **d.** CEMF

4. The unit of inductance is the:

a. henry **b.** ohm
c. mho **d.** watt

5. Two inductors are in series. One inductor has 45 mH and the other has 60 mH. The total inductance is about:

a. 15 mH **b.** 105 mH
c. 25 mH **d.** none of the above

6. Two inductors are in parallel. One inductor has 30 mH and the other has 60 mH. The total inductance is about:

a. 90 mH **b.** 20 mH
c. 45 mH **d.** 75 mH

7. When a change of current in one circuit induces a voltage in another circuit it is referred to as:

a. saturation **b.** inductive reactance
c. mutual induction **d.** power factor

8. In a series RL circuit, R = 2.2 kΩ and L = 80 mH. The time constant for this circuit is approximately:

a. 1.76 s **b.** 27.5 ms
c. 36 μs **d.** 17.6 μs

9. If the square-wave voltage pulse width applied to an *RL* series circuit is longer than five times the circuit's time constant, the voltage waveform appearing across the inductor would resemble a:

a. triangle waveform

b. square waveform

c. sawtooth waveform

d. spike waveform

10. The impedance of a series *RL* circuit:

a. consists of resistance and inductive reactance

b. is determined by the vectorial sum of R and X_L

c. both answers a and b are correct

d. only answer a is correct

11. Which one of the following is *not* true of inductive reactance:

a. directly proportional to inductance

b. inversely proportional to frequency

c. is measured in ohms

d. has no value for a dc circuit

12. A particular inductor with 50 mH is placed into a circuit with a frequency of 80 kHz. Its X_L is approximately:

a. 25 kΩ **b.** 255 kΩ

c. 4 kΩ **d** 12.5 kΩ

13. In a series *RL* circuit, $R = 4.7$ kΩ and $X_L = 9.5$ kΩ. The impedance of this circuit is approximately:

a. 10.6 kΩ **b.** 112 kΩ

c. 3.8 kΩ **d.** 14.2 kΩ

14. In a series *RL* circuit $R = X_L$; therefore, θ equals:

a. 0° **b.** 30°

c. 45° **d.** 90°

15. In a parallel *RL* circuit $R = 1$ kΩ, $X_L = 750$ Ω, and $V_T = 30$ V. The total current of the circuit is:

a. 10 mA **b.** 50 mA

c. 70 mA **d.** 2.5 A

16. To find the θ of a *RL* circuit, you solve for the:

a. sin **b.** cos

c. tan **d.** arctan

17. The formula to use to find true power is:

a. $V_T I_T$ **b.** P_T/P_T

c. P_A/P_T **d.** $P_A \cos\theta$

18. A circuit that has an output greater than $0.707V_{max}$ for frequencies above the cutoff frequency is called a:

a. differentiator **b.** low-pass filter

c. high-pass filter **d.** voltage divider

19. In a series *RL* circuit, $R = 4.7$ kΩ and $L = 120$ mH. The f_{co} for this circuit is approximately:

a. 3 kHz **b.** 6 kHz

c. 12 kHz **d.** 24 kHz

20. If the frequency of a *RL* circuit increases, the X_L:

a. increases **b.** decreases

c. remains the same **d.** increases by the cos θ

ANSWERS TO EXPERIMENTS AND SELF-CHECKING QUIZ

Experiment 1: **(1)** CEMF **(2)** CEMF **(3)** energy **(4)** opposite

Experiment 2: **(1)** long **(2)** spike **(3)** 50.56

Experiment 3: **(1)** increases **(2)** decreases **(3)** 16 **(4)** 8.5 **(5)** greater, below **(6)** 0.707, above

Self-Checking Quiz: **(1)** c **(2)** b **(3)** d **(4)** a **(5)** b **(6)** b **(7)** c **(8)** c **(9)** d **(10)** c **(11)** b **(12)** a **(13)** a **(14)** c **(15)** b **(16)** d **(17)** d **(18)** c **(19)** b **(20)** a

Unit 10

Transformers

INTRODUCTION

Transformers are a practical application of electromagnetism and are based on the principle of mutual induction. Electricity is efficiently transferred from the power plant to cities and homes because of transformers. Nearly every consumer electronic device found in the home requires the use of a transformer. Undoubtedly, as a technician, you will have to test and replace several transformers during your career in electronics.

UNIT OBJECTIVES

Upon completion of this unit, you will be able to:

1. Define and explain transformer action.
2. Describe transformer turns ratio.
3. Explain the step-up transformer.
4. Explain the step-down transformer.
5. Calculate voltage/turns ratio.
6. State the transformer power ratio.
7. Derive the power/current ratio.
8. Describe the impedance ratio.
9. Demonstrate impedance matching.
10. Explain transformer losses and how they occur.
11. List and explain different types of transformers.
12. Identify transformer ratings.
13. Recognize transformer symbols on a schematic drawing.
14. Measure voltage across transformers.
15. Test transformers with an ohmmeter.

SECTION 10-1 FUNDAMENTAL CONCEPTS

10-1a MUTUAL INDUCTION AND TRANSFORMER ACTION

You were introduced to mutual induction in Section 9-1b.1, where it was explained that the electromagnetic field of one inductor can affect the field of another inductor when they are in close proximity. If one coil is placed very close to another coil as shown in Figure 10-1, the coefficient of coupling can be nearly 1.

The coil to which power is applied is called the *primary winding.* As the current flows in this winding it produces a magnetic flux field, which induces a current in the other winding. The winding in which a current is induced is called the *secondary winding.* If the secondary winding has a complete circuit, a voltage drop will appear across the load resistor *R*.

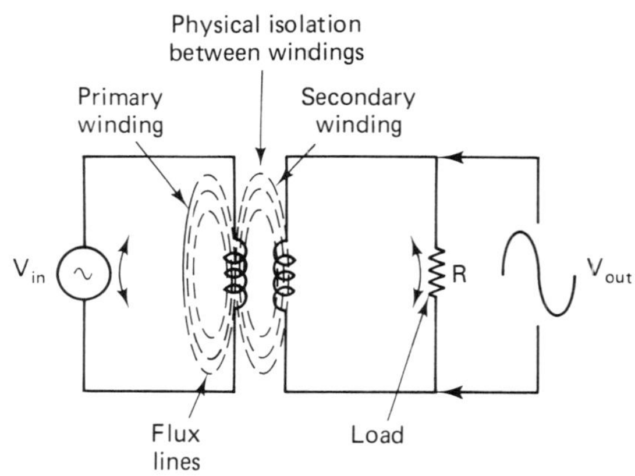

Figure 10–1 Mutual inductance producing transformer action.

The voltage applied to the primary winding must be ac or pulsating dc. Remember that current is induced in a coil only when the flux field is changing. A dc voltage would saturate the primary winding and the flux field would not change; therefore, there would be no current induced in the secondary winding.

A *transformer,* then, is an electrical device that transforms the power applied to one circuit to another circuit without a physical connection. The schematic symbol for the transformer is two coils facing each other. The transformer shown in Figure 10-1 has an air core.

The power transformer is very widely used and is shown in Figure 10-2. The primary winding is wound around a laminated steel core and is insulated by very thin paper, often called "fish" paper. The secondary winding is wound on top of the primary winding. This method produces a high degree of electromagnetic coupling between the coils.

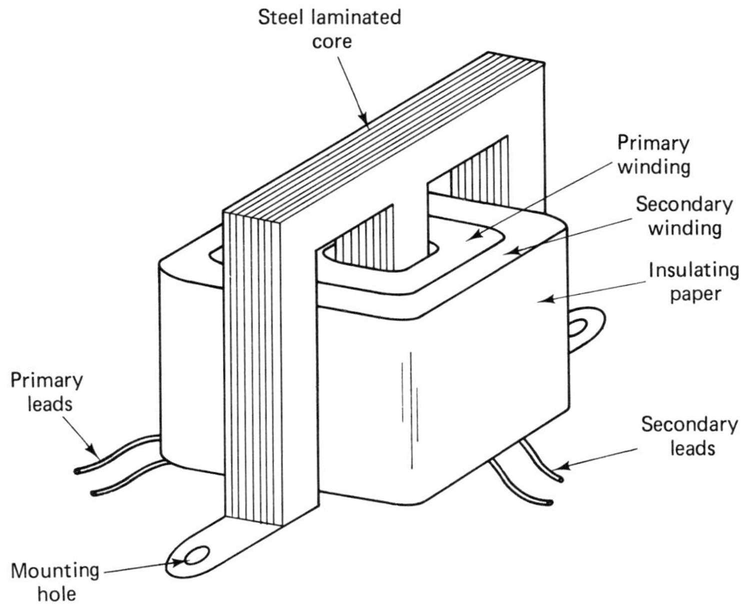

Figure 10–2 Typical power transformer construction.

10-b TRANSFORMER RATIOS

Numerical values associated with the windings of a transformer determine its operation and use. These values are combined into ratios that are used in circuit calculations.

10-b.1 Voltage/Turns Ratio

The number of turns in the windings determine the amount of electrical energy that is transferred. The number of turns in the primary winding, N_p, are directly proportional to the voltage across the primary winding, V_p. The number of turns in the secondary winding, N_s, are directly proportional to the voltage across the secondary winding, V_s. Expressed as a ratio, these values become

$$\frac{N_p}{N_s} = \frac{V_p}{V_s}$$

The *voltage/turns ratio* relationship is shown in Figure 10-3. The lines between the coils indicate that it has an iron core. If the number of turns in the primary winding N_p is 100 and the number of turns of the secondary winding N_s is also 100, the ratio is

$$\frac{N_p}{N_s} = \frac{100}{100} = \frac{1}{1} \quad \text{or} \quad 1:1$$

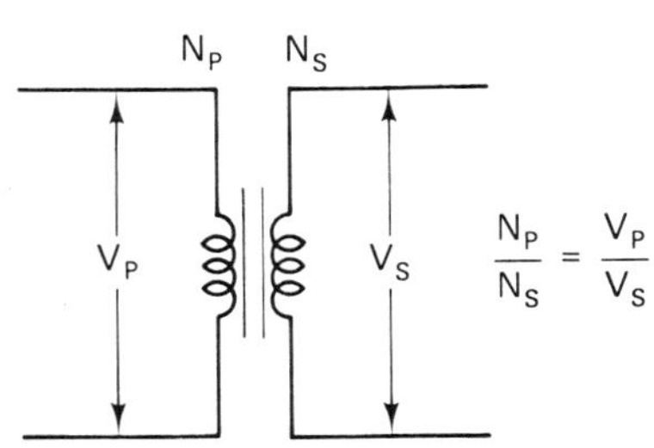

Figure 10–3 Voltage/turns ratio.

The value of the voltage induced into the secondary winding will be the same value of the voltage applied to the primary winding, because the ratio is 1:1 and must remain the same value. For example, if $V_p = 60$ V, then $V_s = 60$ V, and the formula can be rearranged thus:

$$\frac{N_p}{N_s} = \frac{V_p}{V_s} = V_s = \frac{V_p N_s}{N_p} = \frac{100 \times 60 \text{ V}}{100} = \frac{6000}{100} = 60 \text{ V}$$

10-1b.1a The Step-Up Transformer

When the secondary winding has more turns than the primary winding, it is called a *step-up transformer*. For example, if $N_p = 100$ turns and $N_s = 1000$ turns, the step-up ratio is 1:10, as given by the formula

$$\frac{N_p}{N_s} = \frac{100}{1000} = \frac{1}{10} \quad \text{or} \quad 1{:}10$$

The voltage across the secondary winding will be 10 times greater than the voltage applied to the primary winding. For example, if $V_p = 60$ V, then V_s can be found with the formula

$$V_s = \frac{V_p N_s}{N_p} = \frac{60 \times 1000}{100} = \frac{60{,}000}{100} = 600 \text{ V}$$

or simply, $V_s = 10V_p = 10 \times 60 = 600$ V.

10-1b.1b The Step-Down Transformer

When the secondary winding has less turns than the primary winding, it is called a *step-down transformer*. For example, if $N_p = 1000$ turns and $N_s = 500$ turns, the step-down ratio is 2:1 as given by the formula

$$\frac{N_p}{N_s} = \frac{1000}{500} = \frac{10}{5} = \frac{2}{1} \quad \text{or} \quad 2{:}1$$

The voltage across the secondary winding will be one-half the voltage applied to the primary winding. For example, if $V_p = 60$ V, V_s can be found with the formula

$$V_s = \frac{V_p N_s}{N_p} = \frac{60 \text{ V} \times 500}{1000} = \frac{30{,}000}{1000} = 30 \text{ V}$$

or simply, $V_s = V_p/2 = 60 \text{ V}/2 = 30$ V.

Figure 10-4 shows a practical application of step-up and step-down transformers. Most electrical power plants are located very far from cities and factories where the electrical energy is needed. This energy must be transported over great distances by transmission lines. Copper wire is a very good conductor but would be extremely expensive to use as current carriers over these long distances. Aluminum wire is less expensive and is used instead.

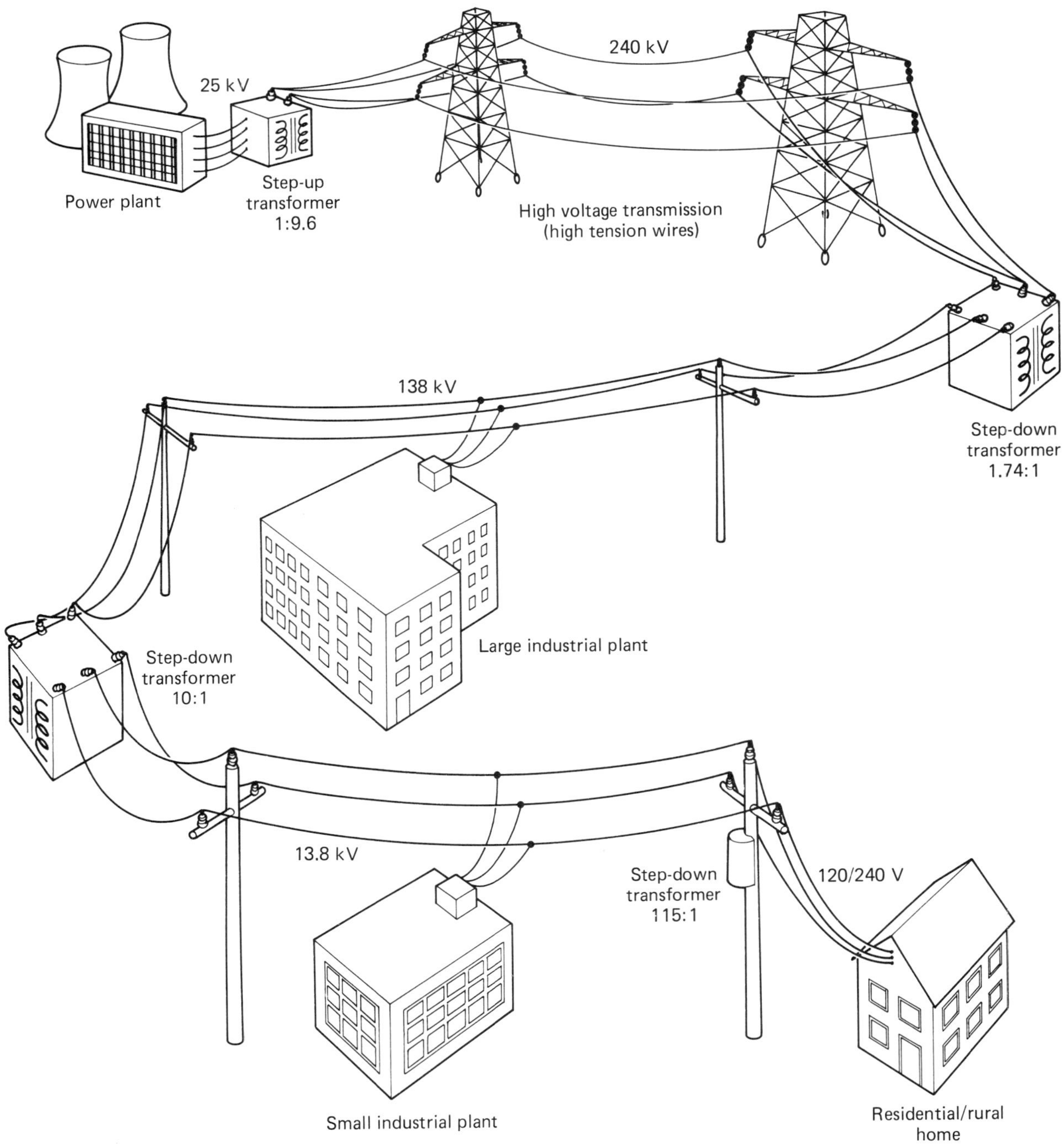

Figure 10–4 General transmission system for electricity.

Aluminum wire has more resistance than copper wire, so that the power loss is greater over long distances. Increasing the voltage to very high voltage levels before transmitting the current over the transmission lines reduces the power loss. At various destinations the voltage is then decreased to more usable levels for factories and homes. This method of transmitting electrical energy is the most efficient at present.

The power plant produces 25 kV of power, which is applied to a step-up transformer with a ratio of 1:9.6. The stepped-up voltage of 240 kV is transferred over high-tension wires. The 240 kV is applied to the step-down transformer with a ratio of 1.74:1. The reduced voltage of 138 kV is supplied to large industrial plants. These plants, in turn, step-down this voltage for other uses. The 138 kV is applied to another step-down transformer, with a ratio of 10:1, producing 13.8 kV, which is used by smaller industrial plants. This 13.8 kV is stepped down by a transformer with a ratio of 115:1 to produce the commonly used 120/240 V used by homes and farms.

10-1b.2 Power Ratio

Another transformer relationship is that the power in the primary winding is equal to the power in the secondary winding, ignoring any loss between the windings. Figure 10-5 shows this power relationship.

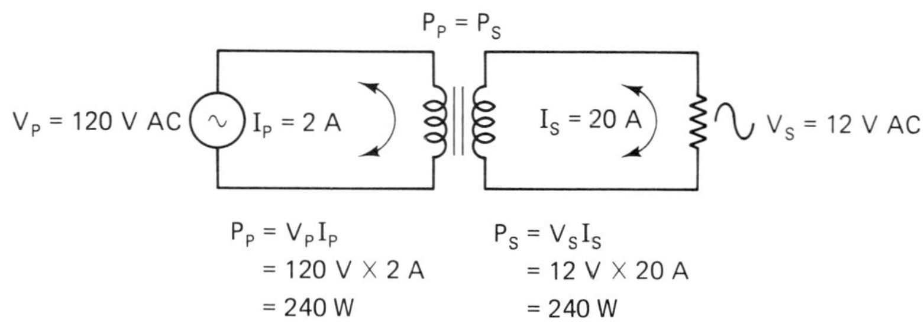

Figure 10–5 Power ratio of a transformer.

If the voltage applied to the primary winding is 120 V and develops a current of 2 A, then the power in the primary, P_p, is equal to 240 W. If the transformer has a step-down ratio of 10:1, the voltage across the secondary will be 12 V. Since the power of each winding must be the same, the current in the secondary I_s can be calculated from the formula

$$I_s = \frac{P_s}{V_s} = \frac{240\text{ W}}{12\text{ V}} = 20\text{ A}$$

Notice that since $P_p = P_s$, the product of $V_p I_p$ must equal the product of $V_s I_s$. In this case, a 10:1 step-down transformer is used and V_s is 10 times less than V_p; therefore, I_s must be 10 times greater than I_p.

10-1b.3 Current Ratio

From the previous discussion of power in a transformer, the relationship of the voltage and current of each winding can be expressed in the formula

$$\frac{V_p}{V_s} = \frac{I_s}{I_p}$$

This formula shows that the current in each winding is inversely proportional to the voltage across each winding. If the voltage ratio is 10:1, the current ratio would have to be 1:10. For example, using the same voltage ratio, if $V_p = 200$ V, then $V_s = 20$ V. If $I_p = 3$ A, then $I_s = 30$ A. The product of each winding would equal 600 W, because $P_p = V_p I_p = 200\text{ V} \times 3\text{ A} = 600\text{ W}$ and $P_s = V_s I_s = 20\text{ V} \times 30\text{ A} = 600\text{ W}$.

10-1b.4 Impedance Ratio

It has been determined that the *impedance ratio* of the windings of a transformer is directly related to the square of the voltage/turns ratio. This relationship can be expressed by the formula

$$\frac{Z_p}{Z_s} = \frac{N_p^2}{N_s^2} \text{ or } \left(\frac{N_p}{N_s}\right)^2 \quad \text{and} \quad \frac{Z_p}{Z_s} = \frac{V_p^2}{V_s^2} \text{ or } \left(\frac{V_p}{V_s}\right)^2$$

where Z_p is the impedance of the primary winding and Z_s is the impedance of the secondary winding. Figure 10-6 shows this relationship.

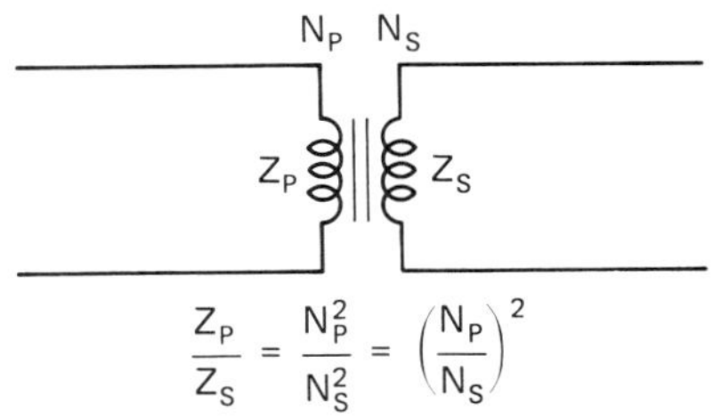

Figure 10–6 Impedance ratio.

10-1b.4a Impedance Matching

An often critical situation in electrical/electronic circuits is the transferring of power from one circuit to another circuit. It has been found that the maximum amount of power is transferred when the impedance of each circuit is the same. If there is a mismatch of impedances in the two circuits, there is a considerable loss of power in the transfer and some waveform distortion may occur. *Impedance matching* means to make the impedance of two circuits as close as possible to the same impedance for a transfer of maximum power.

One of the most important applications of transformers is matching the impedance of one circuit to the impedance of another circuit. For example, an audio-amplifier output circuit with an impedance of 1800 Ω must be connected efficiently to a loudspeaker with 8 Ω. Figure 10-7 shows how impedance matching is accomplished with a transformer.

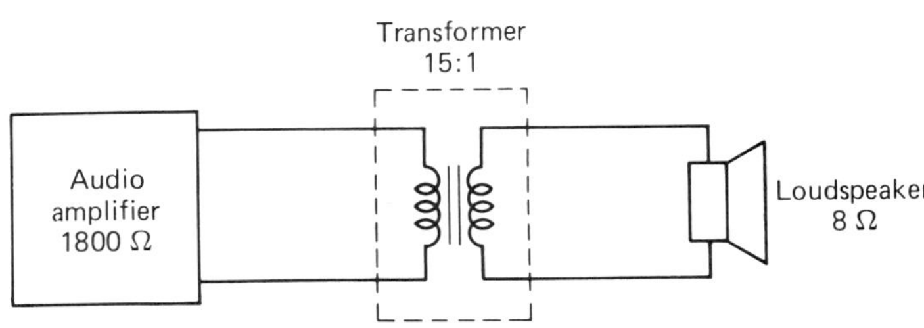

Figure 10–7 Impedance matching for maximum power transfer.

A transformer with the proper turns ratio connected between the amplifier and the loudspeaker will match the impedances and make the loudspeaker appear as 1800 Ω to the amplifier. We can find the turns ratio needed by using the impedance ratio formula:

$$\frac{Z_p}{Z_s} = \left(\frac{N_p}{N_s}\right)^2 = \sqrt{\frac{1800}{8}} = \sqrt{225} = 15 \quad \text{or } 15:1$$

The impedances of 1800 and 8 Ω assumes the values of the turns ratio squared. Since we do not want to find the turns ratio squared, we must take the square root of 1800/8 to get the true turns ratio. A step-down transformer with a turns ratio of 15:1 is required to match these two devices properly.

The true load on the amplifier is 8 Ω, but the apparent impedance seen across the primary winding of the transformer is 1800 Ω. This condition is called reflected impedance, or in other words, the *reflected impedance* is the apparent impedance across the primary winding when current flows in the secondary winding.

We can find the electrical values of the transformer:

Amplifier side:

Given: $V_p = 50$ V
$I_p = 1$ A.
Find: $P_p = V_p \times I_p$
$= 50 \text{ V} \times 1 \text{ A}$
$= 50$ W

Loudspeaker side:

Given: turns ratio = 15:1
Find: $V_s = V_p/15 = 50 \text{ V}/15$
$= 3.33$ V
Since $P_p = P_s$, find: $I_s = P_s/V_s$
$= 50 \text{ W}/3.3 \text{ V} = 15$ A

10-1c TRANSFORMER LOSSES AND EFFICIENCY

Transformers are very efficient in transferring power, but have the same losses as other electromagnetic devices. These losses were mentioned briefly in Unit 7 and are described in more detail at this time.

10-1c.1 Eddy Current Losses

The steel or iron core of a transformer has a high permeability and conducts lines of magnetic force very easily. With such a core a transformer has a greater flux density, which is the reason for its use. However, the alternating current flowing through the windings causes a changing magnetic field in the core, which induces an EMF. This induced EMF produces eddy currents flowing in the core. The eddy currents have a power loss proportional to the current squared, as expressed by the familiar formula: $P = I^2R$. This power loss is in the form of heat dissipation. Therefore, *eddy current losses* are currents produced in magnetic materials from induced voltages that result in power loss.

If the current can be reduced, the power loss can be reduced. A laminated core, shown in Figure 10-2, consists of very thin plates or laminates of steel. These plates are coated with a very thin varnish which acts as an insulator to the eddy currents. The eddy currents are limited to flowing in each plate only and the total current is reduced. Although the laminated core increases flux density and reduces eddy currents, there is still some power loss.

10-1c.2 Hysteresis Loss

When steel or iron is magnetized, the magnetic domains in the material align themselves in a specific direction. The transformer is an ac operating device; therefore, the magnetic domains must change direction periodically. In doing so, they must overcome friction and inertia, which produces a power loss in the form of heat. In other words, *hystere-*

sis loss is a result of changing periodically the direction of alignment of the magnetic domain in a material.

10-1c.3 Copper Losses

Copper losses are the normal power loss of the resistance of the copper wire in the windings. For example, using the power formula $P = I^2R$, if the current is doubled, the power increases by a factor of 4. This loss is reduced by increasing the size of the copper wire or, whenever possible, designing transformers with a lower operating current.

10-1c.4 Transformer Efficiency

The efficiency of any device is described as the amount of work that is accomplished with the minimum of wasted or lost power. The efficiency of an electrical device can be expressed as a ratio as shown by the formula

$$\text{efficiency} = \frac{\text{output power}}{\text{input power}}$$

The efficiency of a transformer can easily be found since the input power is the power applied to the primary winding and the output power is the resulting power in the secondary winding. It is usually expressed as a percentage using the formula

$$\text{efficiency}\ (\%) = \frac{P_s}{P_p} \times 100$$

For example, if $P_p = 50$ W and $P_s = 48$ W, then

$$\begin{aligned}\text{efficiency}\ (\%) &= \frac{p_s}{P_p} \times 100 = \frac{48\text{ W}}{50\text{ W}} \times 100 \\ &= 0.96 \times 100 = 96\%\end{aligned}$$

The efficiency of a transformer and most devices will always be less than 100%.

10-1d TYPES OF TRANSFORMERS

There are many types of transformers used for a variety of applications. Figure 10-8 shows a sample of schematic symbols and physical construction for some transformers. The schematic symbol for an iron-core transformer is shown in Figure 10-8c. The schematic symbol for an air-core transformer is shown in Figure 10-8d.

The transformer shown in Figure 10-8e has dots, which represent voltage phasing. If the dots are opposite to each other as shown, the voltage of the primary and secondary windings are in phase. If the dot appears on the other lead of the secondary winding, the transformer is wound in such a manner that this voltage is 180° out of phase with the voltage at the primary winding.

The schematic symbol for a step-down autotransformer is shown in Figure 10-8f. It has one lead that is common to the primary and

$$\frac{N_p}{N_s} = \frac{V_p}{V_s} \qquad \frac{N_p}{N_s} = \frac{I_s}{I_p}$$

$$P_p = P_s \qquad \frac{Z_p}{Z_s} = \left(\frac{N_p}{N_s}\right)^2$$

$$(V_p I_p = V_s I_s)$$

(a)

Color	Lead
Black	Primary
Red Blue/red Red/yellow	Secondary
Green Yellow	Secondary filament winding

(b)

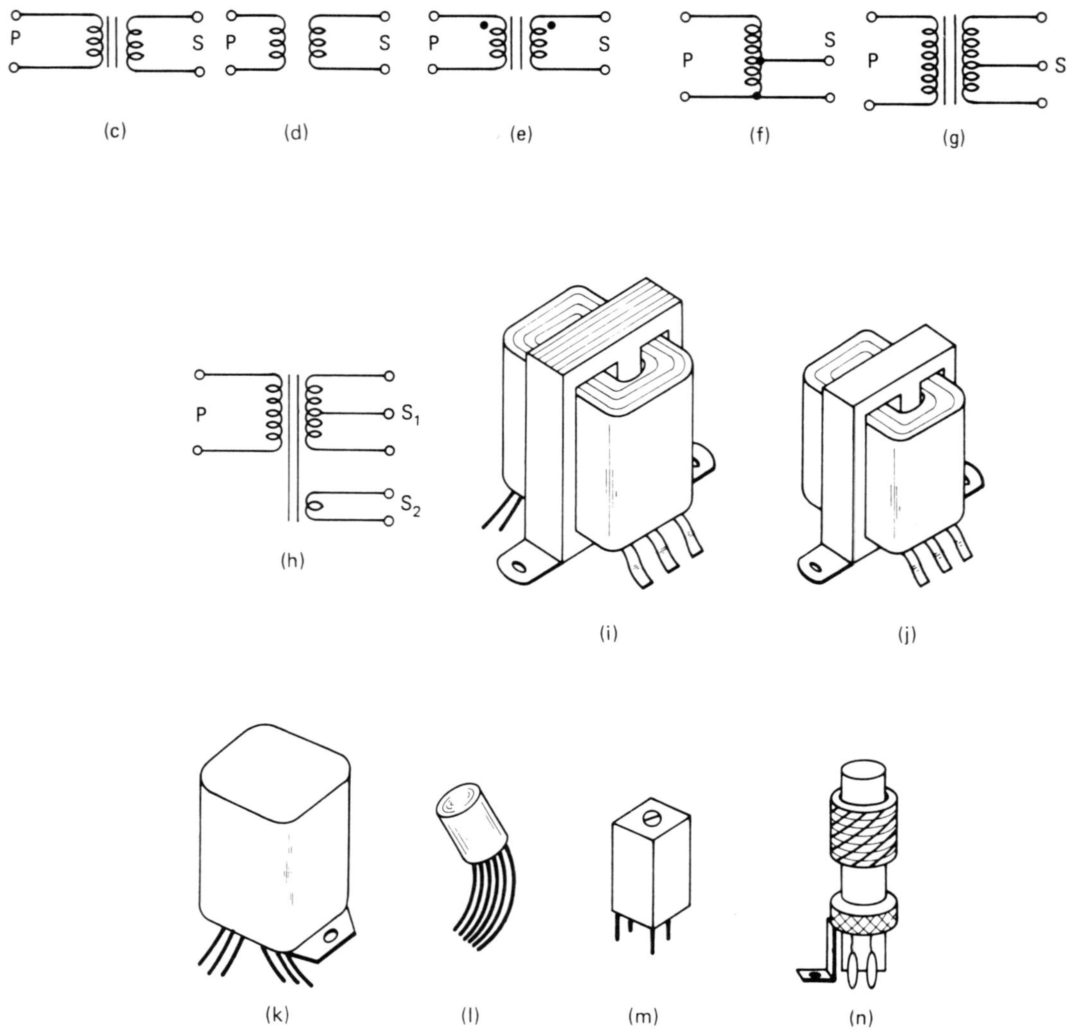

Figure 10–8 Transformers: (a) formulas; (b) lead color code; (c) iron core; (d) air core; (e) phasing symbols; (f) autotransformer; (g) center-tapped secondary; (h) multiple-secondary windings; (i) power with center tap; (j) autotransformer; (k) can-type; (l) miniature; (m) IF; (n) RF.

secondary windings and can usually be recognized as having three leads (Figure 10-8j). This type of transformer can also be found in a step-up version.

Figure 10-8g shows the schematic symbol for a transformer with a center-tapped secondary winding. The voltage between the center tap on either end lead is one-half the total voltage across the secondary winding. The physical construction of this transformer may appear as shown in Figure 10-8i.

Some transformers have more than one winding as the secondary winding, as shown in Figure 10-8h. The center-tapped winding may be a step-up and the other winding may be a step-down, or any other combination. Transformers may be enclosed in metal cans to reduce extraneous electromagnetic radiation, as shown in Figure 10-8k and l.

Small transformers, shown in Figure 10-8m and n, are used in radio and television receivers to tune in or select the various frequencies. They have movable cores made of powered iron particles which are adjusted through the top with a screwdriver. An isolation transformer has a 1:1 ratio and is used to isolate some circuits and equipment from the earth ground to prevent shock hazard to people and the unintentional shorting of equipment.

10-1e TRANSFORMER LEAD COLOR CODE AND RATINGS

There is a general standard for the color of transformer winding leads, as given in Figure 10-8b. The primary leads are usually black or a combination of black and some other color. The main secondary high-voltage leads are commonly red or a combination of red and another color. Other secondary leads may be colored green, yellow, brown, blue, orange, or a combination of these colors.

Transformers have a few ratings, which in some cases are printed on them. Power transformers used in electronic systems may be rated in volt-amperes (VA), which refers to the ac wattage that can safely be handled by the primary winding. This winding may also have printed on it the maximum input voltage amplitude and frequency. The secondary winding side may show the induced voltage and the maximum current that can be used. There may also be an insulation test rating of the maximum voltage that can be applied between the windings.

SECTION 10-2 DEFINITION EXERCISES

Write a brief description for each of the following terms.

1. Mutual induction ______________________________

____

2. Primary winding ______________________________

____

3. Secondary winding ______________________________

____

4. Transformer action ______________________________

____

5. Transformer ______________________________

____

6. Voltage/turns ratio ______________________________

____

7. Step-up transformer ______________________________

____

8. Step-down transformer ______________________________

____

9. Power ratio ______________________________

____

10. Current ratio ______________________________

____

11. Impedance ratio ______________________________

____

12. Impedance matching ______________________________

____

13. Reflected impedance ______________________________

____

14. Eddy current losses ______________________________

____

15. Hysteresis loss ______________________________

____

16. Copper losses ______________________________

____

17. Efficiency formula ______________________________

____

18. Air-core transformer ______________________________

____

19. Iron-core transformer ______________________________

____

20. Transformer winding phasing ______________________________

____

21. Autotransformer ______________________________

____

22. Center-tapped transformer ______________________________

____

23. IF and RF transformers ______________________________

____

24. Transformer ratings ______________________________

____

SECTION 10-3 EXERCISES AND PROBLEMS

Complete this section before beginning the next section.

1. Draw the schematic symbol for the following transformers.

a. air core

b. iron core

c. autotransformer

d. center-tapped secondary

e. multiple secondary windings

f. winding phasing

2. Find the turns ratio for the given data.

a. $N_p = 150$
$N_s = 3000$

Ratio = ___________

b. $N_p = 250$
$N_s = 750$

Ratio = ___________

c. $N_p = 500$
$N_s = 100$

Ratio = ___________

3. Find the secondary voltage for the data given.

a. $N_p = 150$, $N_s = 300$, $V_p = 120$ V, $V_s =$ ___________

b. $N_p = 300$, $N_s = 30$, $V_p = 120$ V, $V_s =$ ___________

c. $N_p = 800$, $N_s = 250$, $V_p = 240$ V, $V_s =$ ___________

4. The power of a transformer is stated as $P_p = P_s$. Using these formulas:

$$P_p = V_p I_p \qquad P_s = V_s I_s$$

$$V_p = \frac{P_p}{I_p} \qquad V_s = \frac{P_s}{I_s}$$

$$I_p = \frac{P_p}{V_p} \qquad I_s = \frac{P}{V_s}$$

find the unknown values for the data given.

a. Ratio = 1:2

$V_p = 120$ V

$I_p = 0.5$ A

$P_p =$ ___________

$P_s =$ ___________

$V_s =$ ___________

$I_s =$ ___________

b. Ratio = 5:2

$V_p = 60$ V

$I_p = 2$ A

$P_p =$ ___________

$P_s =$ ___________

$V_s =$ ___________

$I_s =$ ___________

c. $V_p = 120$ V

$I_p =$ ___________

$P_p =$ ___________

$P_s = 18$ W

$V_s = 6$ V

$I_s =$ ___________

d. $V_p =$ ___________

$I_p = 0.64$ A

$P_p =$ ___________

$P_s = 96$ w

$V_s = 48$ v

$I_s =$ ___________

e. $V_p =$ ___________

$I_p = 0.5$ A

$P_p =$ ___________

$P_s = 6$ W

$V_s =$ ___________

$I_s = 1$ A

f. $V_p = 120$ V

$I_p = 5$ A

$P_p =$ ___________

$P_s =$ ___________

$V_s = 120$ V

$I_s =$ ___________

5. Find the impedance ratio for the data given.

a.	b.	c.
$N_p = 500$ $N_s = 100$	$N_p = 250$ $N_s = 1000$	$V_p = 120$ V $V_s = 12$ V
Ratio = ____________	Ratio = ____________	Ratio = ____________

6. Find the turns ratio for the given data.

a.	b.	c.
$Z_p = 144$ $Z_s = 16$	$Z_p = 50$ $Z_s = 5000$	$Z_p = 3200$ $Z_s = 8$
Ratio = ____________	Ratio = ____________	* Ratio = ____________

*It may be easier to invert these numbers, solve for the answer, and show it has a step-up ratio.

7. List the losses that can occur in a transformer.

a.

b.

c.

8. Find the transformer efficiency for the data given (show the answer as a percentage).

a.	b.	c.
$P_p = 100$ W $P_s = 98$ W	$P_p = 240$ W $P_s = 230$ W	$P_p = 560$ W $P_s = 515$ W
eff = ____________	eff = ____________	eff = ____________

9. List the rating that may be given with a transformer.

a.

b.

c.

d.

10. Explain transformer lead color code and give examples.

11. List several applications where transformers are used.

a.

b.

c.

d.

12. Match the description in column A with the proper term in column B.

Column A

____ **a.** Used to transfer maximum power

____ **b.** To increase voltage

____ **c.** To decrease voltage

____ **d.** The ratio of primary voltage to secondary voltage

____ **e.** The square of the turns ratio

____ **f.** Ratio of output power to input power

____ **g.** Used to reduce eddy currents

____ **h.** Result of polarity switching of magnetic domains

____ **i.** Loss from the resistance of wire

____ **j.** Transformer power ratio

Column B

1. copper losses
2. turns ratio
3. efficiency
4. impedance matching
5. step-down
6. hysteresis loss
7. $P_p = P_s$
8. impedance ratio
9. step-up
10. lamination

SECTION 10-4 EXPERIMENTS

EXPERIMENT 1. VOLTAGES OF A TRANSFORMER

Objective: To demonstrate the safe method for measuring the voltage of a transformer.

Introduction: Most power transformers use 120 V ac (line voltage) at the primary winding. This voltage can produce severe shock in a person and may even be fatal. The voltages on the secondary winding may be greater in voltage and cause more damage. A special power cord is used in this experiment. The cord has a two-prong ac plug at one end and an alligator clip attached to the end of each wire.

Materials Needed:
1 Power transformer
1 Special power cord

Equipment Required:
1 Ac power outlet
1 Multimeter or DMM

Caution! Take your time and follow every instruction precisely. This is a dangerous experiment, but can be performed safely. When you apply power to the circuit, keep one hand behind your back.

Procedure:

1. Draw a schematic symbol of the transformer you are going to use for this experiment in the space below.

2. Be sure that this experiment is performed on an electrically insulated surface. Make sure that no extraneous metal objects are close to any of the components or wires.
3. Identify the primary winding leads and the secondary winding leads of the transformer. Small pieces of masking tape can be attached to the leads for identification.
4. Connect the power cord leads to the primary winding leads. Make sure that all bare wires are not touching each other or some other metal component.
5. Set the meter to the proper ac range and connect it on top of the primary winding connections.
6. Stand back from the experiment and very carefully place the power cord plug into the ac receptacle. Record the meter reading here: V_p = ______________ .
7. *Unplug the power cord from the ac receptacle.*
8. Disconnect the meter from the primary winding.
9. Set the meter to the highest ac voltage.
10. Connect the meter to the secondary winding. *Be sure that no bare wires are touching.*
11. Stand back from the experiment and very carefully place the power cord plug into the ac receptacle. Carefully adjust the meter for a good reading. Record the meter reading here: V_s = ______________. This reading will be higher than the voltage across the secondary winding in an actual circuit because there is no load present.
12. Carefully remove the power cord plug from the ac receptacle.

Fill-In Questions:

1. You must be very careful when working with ac ______________ voltage.

2. According to the voltage readings, a ______________ transformer was used for this experiment.

SECTION 10-5 TROUBLESHOOTING APPLICATION: TESTING A TRANSFORMER WITH AN OHMMETER

Testing a transformer for opens or shorts is very similar to testing an inductor. In some cases, the dc resistance is printed on the transformer or on the schematic diagram of the circuit in which it is used.

To make resistance measurements of a transformer, you place the ohmmeter across the leads of the winding you want to test. If the winding is open, the resistance will be infinite. Generally, the resistance will be rather low for a good winding and will probably never be more than a few thousand ohms. The secondary winding of a step-up transformer should be a higher resistance than the primary winding. The secondary winding of a step-down transformer should be a lower resistance than the primary winding.

Testing for shorts is somewhat more difficult. If you know the dc resistance of a winding, a lower measurement would, of course, indicate a partial short. A shorted winding will usually draw more current in an operating circuit and can be detected several ways: the fuse or circuit breaker may activate, the transformer may be excessively hot to the touch, or there could be smoke or even fire.

Using the transformer from the previous experiment, draw its schematic symbol in the space provided.

Use an ohmmeter to measure the dc resistance of each winding and list the amount next to each winding in the drawing.

SECTION 10-6 INSTANT REVIEW

- Transformer action is based on the mutual induction between two coils.
- The primary winding has ac voltage applied to it.
- The secondary winding has the induced voltage.

- A transformer is an electrical device that transfers power from one circuit to another circuit without a physical connection.
- The transformer provides electrical isolation between circuits.
- An iron-core transformer produces better electromagnetic coupling between the primary and secondary windings.
- The turns ratio of a transformer is directly proportional to the voltage across each winding or voltage ratio.
- A step-up transformer has more turns in the secondary winding than in the primary winding and produces a greater output voltage.
- A step-down transformer has fewer turns in the secondary winding than in the primary winding and produces a lower output voltage.
- The power in the primary winding is equal to the power in the secondary winding for an ideal transformer.
- In a step-up transformer, the current in the secondary will be less than in the primary to maintain the power equality.
- In a step-down transformer, the current in the secondary will be more than in the primary to maintain the power equality.
- Transformers are highly efficient and their percentage of efficiency can be found with the formula

$$\text{efficiency}\ (\%) = \frac{\text{output power}}{\text{input power}}$$

- The impedance ratio of a transformer is the square of the turns ratio or voltage ratio.
- Impedance matching is a method of matching the impedance of two circuits to obtain the maximum transfer of power.
- Transformers are excellent impedance-matching devices.
- To find the proper impedance ratio for impedance matching, you must solve for the square root of the turns or voltage ratio.
- Eddy current losses result in a power loss from the heat of currents induced in magnetic materials. Lamination of the core of a transformer reduces eddy currents.
- Hysteresis loss in a transformer is a result of the periodic changing of the polarity alignment of magnetic domains.
- Copper losses are I^2R loss of the dc resistance in the wire of the windings of a transformer.
- An autotransformer usually has only three leads.
- A transformer may have several secondary windings, which may be step-up and/or step-down.
- Dots next to the windings on a schematic drawing indicate winding phasing.
- Some small transformers have tunable cores and are used in radio and television circuits.
- An isolation transformer has a 1:1 ratio.
- Transformer leads have a color code for identification and also have certain ratings.

SECTION 10-7 SELF-CHECKING QUIZ

Circle the most correct answer for each question.

1. A transformer:

a. transforms power from one circuit to another circuit

b. provides isolation between two circuits

c. matches impedance of two circuits

d. all of the above

2. Transformer action is a result of:

a. self-induction

b. mutual induction

c. hysteresis loss

d. none of the above

3. To construct a 12-V power supply from a 120-V ac line source, you would use:

a. an isolation transformer

b. a step-up transformer

c. a step-down transformer

d. a radio-frequency (RF) transformer

4. The secondary voltage of a transformer is determined by the:

a. power ratio

b. copper losses

c. turns ratio

d. impedance

5. A transformer has a 1:2.5 turns ratio. The primary voltage is 60 V. The secondary voltage is:

a. 60 V **b.** 120 V

c. 150 V **d.** 24 V

6. An iron-core transformer has less electromagnetic coupling than does an air-core transformer.

a. True **b.** False

7. The power in the secondary winding of a transformer is nearly the same as the power in the primary winding.

a. True **b.** False

8. Given for a particular transformer: $P_p = 100$ W, $V_p = 50$ V, $V_s = 10$ V. The I_s is equal to:

a. 2 A **b.** 10 A

c. 20 A **d.** 50 A

9. A method of reducing eddy currents in a transformer is called:

a. isolation **b.** phasing

c. impedance matching **d.** lamination

10. The impedance ratio of a transformer is the square of the turns ratio or voltage ratio.

a. True **b.** False

11. An amplifier with an output impedance of 1296 Ω is going to be connected to a 16-Ω speaker. The turns ratio for the matching transformer should be about:

a. 80:1 **b.** 1:16

c. 9:1 **d.** 1:9

12. The losses in a transformer as a result of the switching of the polarity of the magnetic domains are called:

a. hysteresis losses **b.** copper losses

c. eddy current losses **d.** coupling losses

13. A transformer has a turns ratio of 4:1. The primary voltage is 180 V. The secondary voltage is:

a. 180 V **b.** 90 V
c. 45 V **d.** 720 V

14. A transformer has an input power of 120 W and an output power of 118 W. Its efficiency is about:

a. 100% **b.** 97.2%
c. 89.6% **d.** 98.3%

15. The primary winding leads to a transformer are usually colored:

a. black **b.** red
c. yellow **d.** white

16. Which of the following is *not* a transformer rating?

a. volt-amperes of the primary winding
b. secondary voltage and maximum current
c. impedance ratio
d. insulation test voltage

17. The total voltage across the center-tapped secondary winding of a transformer is 350 V. The voltage across one end lead and the center tap is:

a. 350 V **b.** 175 V
c. 700 V **d.** none of the above

18. If a step-up transformer used the same-size wire for both windings and the resistance of the primary winding was 345 Ω, the resistance of the secondary winding could possibly be:

a. 345 Ω **b.** 1035 Ω
c. 250 Ω **d.** none of the above

19. One method to test for a shorted transformer in an operating circuit is to:

a. measure the resistance of its primary winding
b. measure the resistance of its secondary winding
c. test it for excessive heat
d. none of the above

20. Given for a particular transformer, $V_s = 12$ V, $I_s = 3$ A, and its turns ratio is 5:1. The I_p is equal to:

a. 4 A
b. 1.5 A
c. 0.6 A
d. none of the above

ANSWERS TO EXPERIMENT AND SELF-CHECKING QUIZ

Experiment 1: **(1)** line **(2)** step-up or step-down

Self-Checking Quiz: **(1)** d **(2)** b **(3)** c **(4)** c **(5)** c **(6)** b **(7)** a **(8)** b **(9)** d **(10)** a **(11)** c **(12)** a **(13)** c **(14)** d **(15)** a **(16)** c **(17)** b **(18)** b **(19)** c **(20)** c

Unit 11

Capacitance

INTRODUCTION

An electrical *passive device* is a component that may control but does not create or amplify energy. There are three basic passive devices associated with electrical/electronic circuits: the resistor, the inductor, and the capacitor. You already have an understanding and practical use of resistors and inductors. Capacitance is one of the most important elements of electrical circuits. Nearly every electrical/electronic circuit is affected by capacitance in some manner. The *capacitor* is a physical unit of capacitance and is found in most circuits.

UNIT OBJECTIVES

Upon completion of this unit, you will be able to:

1. Define and explain *capacitance, farad,* and other terms associated with capacitors.
2. List the factors that determine capacitance.
3. Identify various types of capacitors.
4. Calculate the total capacitance of capacitors in series and parallel.
5. Describe the effect of an *RC* time constant and calculate different values for time constants.
6. Explain the effects of *RC* time constants and square wave pulse shaping circuits.
7. Define and show how to calculate values of capacitive reactance.
8. Demonstrate the relationship of capacitive reactance to frequency.
9. Calculate ac formulas for capacitive circuits.
10. Show the effect of capacitive reactance in an ac circuit with the use of phasors.
11. Describe the relationship of apparent power, true power, and power factor in a capacitive circuit.
12. Explain the use of a capacitor in a frequency filter.
13. Test capacitors with an ohmmeter.

SECTION 11-1 FUNDAMENTAL CONCEPTS

11-1a DEFINING CAPACITANCE

Capacitance is usually defined as that property which permits the storage of an electrical charge that is separated by two conductors of a different potential. Two parallel wires in close proximity with a difference of potential will exhibit a slight amount of capacitance. Generally, capacitance is associated with the capacitor as shown in Figure 11-1.

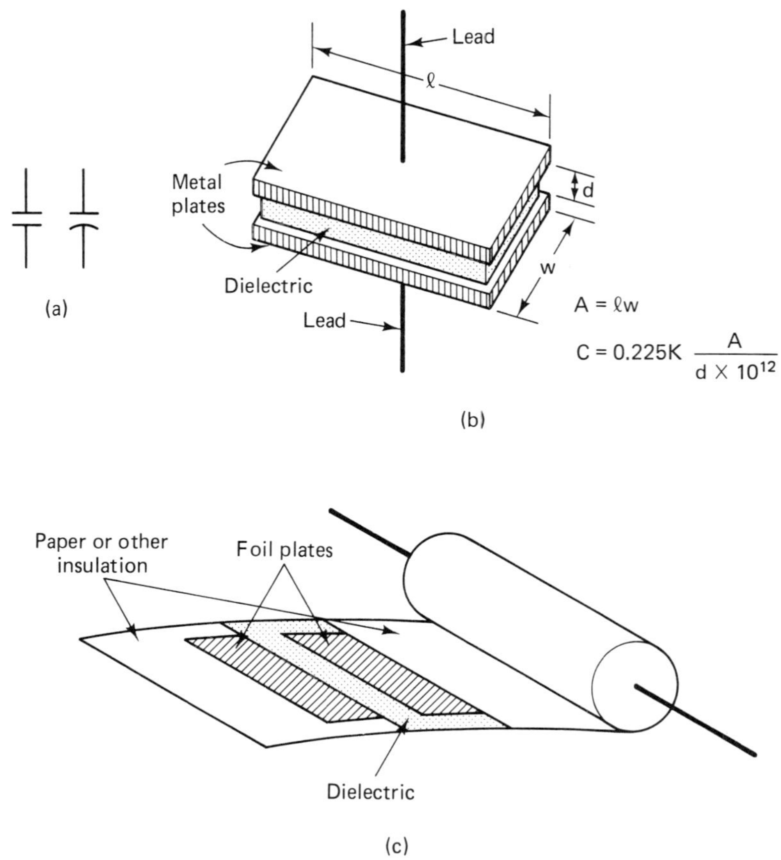

Figure 11–1 Capacitor: (a) schematic symbols; (b) physical construction; (c) tubular type.

The schematic symbols for a capacitor (Figure 11-1a) resemble a one-cell battery symbol. The curved line on one symbol indicates the outer conductor, which is usually connected to the ground potential. A stylized example of a capacitor is shown in Figure 11-1b, where two metallic plates are separated by an insulating material called the *dielectric.*

11-1a.1 Dielectric

There are several insulating materials used for the dielectric of a capacitor such as air, oil, paper, glass, various plastics, and certain chemical compounds. Each of these materials has a different composition and is assigned a numerical constant which in some part determines the capacitance of a capacitor. Table 11-1 shows some materials and their assigned constants:

TABLE 11-1
Dielectric Constants

Material	*Constant*
Air	1.0
Polystyrene	2.5
Mylar	3.0
Paper	3.5
Mica	6
Glass	6–10

11-1a.2 The Farad

The unit of measure for capacitance is called the *farad,* named after Michael Faraday, who was one of the early experimenters working with capacitance and inductance.

The farad is defined as the ability to store 1 coulomb in 1 second with an applied voltage of 1 volt and is expressed with the formula

$$C = \frac{Q}{E}$$

where C = capacitance
Q = charge of 1 coulomb (6.25×10^{18} electrons)
E = electromotive force or voltage

The farad is much too large to use in most electrical/electronic circuits; therefore, smaller units, the microfarad and picofarad, are used with capacitors. A *microfarad* is equal to one millionth of a farad and is indicated as μF. A *picofarad* is equal to one millionth of a millionth of a farad and is indicated as pF or sometimes as $\mu\mu$F. The following example will show this more clearly:

$$1 \text{ farad} = 1{,}000{,}000 \text{ or } 1 \times 10^{6} \text{ microfarads}$$
$$1\ \mu\text{F} = 0.000001 \text{ or } 1 \times 10^{-6} \text{ farad}$$
$$1 \text{ farad} = 1{,}000{,}000{,}000{,}000 \text{ or } 1 \times 10^{12} \text{ picofarads}$$
$$1 \text{ pF} = 0.000000000001 \text{ or } 1 \times 10^{-12} \text{ farad}$$

Typical values of capacitance are 2200 μF, 100 μF, 10 μF, 1 μF, 0.1 μF, 0.01 μF, 0.001 μF, 1000 pF, 330 pF, and 10 pF.

11-1a.3 Factors Determining the Amount of Capacitance

The amount of capacitance of a capacitor can be found using the formula

$$C = \frac{0.225KA}{d \times 10^{12}}$$

where K = numerical value of the dielectric constant used
A = area of the plates
d = distance between the plates

Referring to Figure 11-1b, the area of the plates is the product of their length and width (i.e., $A = lw$). The area is directly proportional to the capacitance. This means that a large plate area has greater capacitance than a smaller plate area, which has less capacitance. The distance between the plates is inversely proportional to the amount of capacitance. The closer the plates are (without touching), the greater the capacitance, as opposed to a larger distance with less capacitance.

To give an example of these factors, assume that a particular capacitor has plates that are 100 in. long and 2 in. wide, separated by a distance of 0.0001 in. with a dielectric of paper ($K = 3.5$). The capacitance can be calculated with the formulas

$$\begin{aligned} A &= lw \\ &= 100 \text{ in.} \times 2 \text{ in.} \\ &= 200 \text{ in.}^2 \end{aligned}$$

$$\begin{aligned} C &= \frac{0.225KA}{d \times 10^{12}} \\ &= \frac{0.225 \times 3.5 \times 200 \text{ in.}^2}{0.0001 \text{ in.} \times 10^{12}} \\ &= 1{,}575{,}000 \times 10^{-12} \\ &= 1.575 \times 10^{-6} \\ &= 1.575\ \mu\text{F} \end{aligned}$$

The table below shows how the capacitance is affected by a change in the various factors. Initially, there is no change. First there is an increase in d, then a decrease in d. Next there is an increase in A, and then a decrease in A. Finally, there is an increase in K.

Change	0.225	×	K	×	A	/	d	=	C
None	0.225		3.5		200		0.0001		1.575 μF
Increase	0.225		3.5		200		0.001	=	0.1575 μF
Decrease	0.225		3.5		200		0.00001	=	15.75 μF
Increase	0.225		3.5		400		0.0001	=	3.15 μF
Decrease	0.225		3.5		100		0.0001	=	0.7875 μF
Increase	0.225		6		200		0.0001	=	2.7 μF

11-1a.4 Factors Affecting the Physical Size of Capacitors

From the previous discussion the size of the plates of the capacitor used was 2 in. wide and 100 in. long. A capacitor of this size would not be very practical in today's modern circuit. Various methods are used to keep the area of the plates the same dimensions, but reduce the overall physical size of the capacitor so that it is easy to handle and be usable in circuits. One such method is shown in Figure 11-1c. The plates consist of two thin metal foil strips that can be rolled into tubular form. The dielectric material is placed between the foil and with additional insulation is rolled into the tube. External conductors or leads are connected to the foil, and the entire unit is sealed with wax, plastic, or a metal can.

You will find two capacitors with the same value of capacitance, but one will be larger than the other. The reason for this is that the larger capacitor will have a higher voltage rating. Voltage can increase to a point where it will "arc" or breakover between two conductors. The larger the distance between the conductors, the more voltage is required to break over. The dielectric can be made thicker to increase the voltage breakover point, but the capacitance decreases. Therefore, the area of the plates must be increased to maintain the same value of capacitance, which results in a larger capacitor.

Generally speaking, the *voltage rating* of a capacitor determines its physical size. A 10-μF capacitor rated at 10 V dc is much smaller than a 10-μF capacitor rated at 100 V dc; however, capacitors in the picofarad range can be relatively small and have a voltage rating of several thousand volts.

11-1b CAPACITORS IN SERIES AND PARALLEL CIRCUITS

The total capacitance of a circuit can be found with a method similar to that for solving for total resistance and inductance except that the functions are reversed. When capacitors are in series you could use the reciprocal formula as shown in Figure 11-2a. For example, if $C_1 = 1$ μF, $C_2 = 2$ μF, and $C_3 = 3$ μF, then

$$\begin{aligned}
C_T &= \frac{1}{\dfrac{1}{C_1} + \dfrac{1}{C_2} + \dfrac{1}{C_3}} \\
&= \frac{1}{\dfrac{1}{1\ \mu\text{F}} + \dfrac{1}{2\ \mu\text{F}} + \dfrac{1}{3\ \mu\text{F}}} \\
&= \frac{1}{\dfrac{6+3+2}{6}} \\
&= \frac{1}{\dfrac{11}{6}} \\
&= \frac{1}{1.833} \\
&= 0.546\ \mu\text{F}
\end{aligned}$$

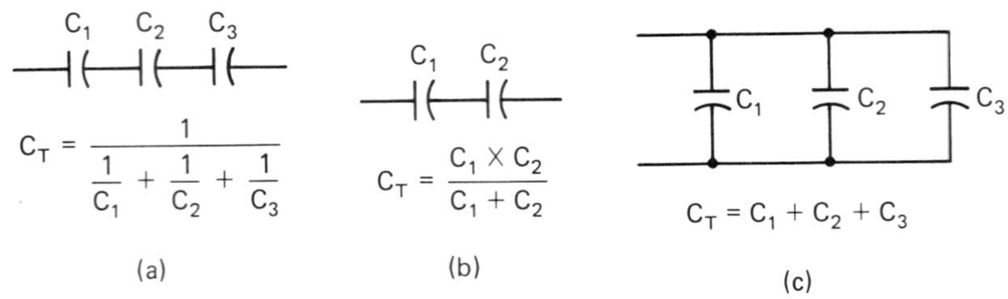

Figure 11–2 Capacitor combinations: (a) three in series; (b) two in series; (c) in parallel.

When only two capacitors are in series you can use the product-over-sum formula, as shown in Figure 11-2b. For example, if $C_1 = 2\ \mu F$ and $C_2 = 3\ \mu F$, then

$$\begin{aligned} C_T &= \frac{C_1 \times C_2}{C_1 + C_2} \\ &= \frac{2\ \mu F \times 3\ \mu F}{2\ \mu F + 3\ \mu F} \\ &= \frac{6\ \mu F^2}{5\ \mu F} \\ &= 1.2\ \mu F \end{aligned}$$

If all the capacitors in a series circuit are of the same value, the total capacitance can be found by dividing the total number of capacitors into the value of one capacitor. The total capacitance for capacitors in a series circuit will always be less than the capacitor with the smallest value. In effect, it is as if the distance between the plates has increased and the capacitance has decreased.

When the capacitors are in a parallel circuit, they are simply added together as shown in Figure 11-2c. For example, if $C_1 = 1\ \mu F$, $C_2 = 2\ \mu F$, and $C_3 = 3\ \mu F$, then

$$\begin{aligned} C_T &= C_1 + C_2 + C_3 \\ &= 1\ \mu F + 2\ \mu F + 3\ \mu F \end{aligned}$$

The effect here is the same result as if the plate area of a single capacitor had been increased and the capacitance increases also.

11-1c CAPACITANCE IN A DC CIRCUIT

The capacitor is able to store voltage. Figure 11-3 shows a step-by-step analysis of the electron flow for charging and discharging a capacitor. Initially, there is no charge on the capacitor and switch S_1 is in position B (Figure 11-3a). The neutral charge on the capacitor is shown as 0 V. Switch S_1 is moved to position A (Figure 11-3b). The capacitor is now connected across the battery. Electrons on the upper plate of the capacitor are attracted to the positive terminal of the battery. This creates a difference of potential between the plates and an electrostatic field begins to develop. Electrons are attracted from the negative terminal of the battery toward the bottom plate of the capacitor in an attempt to complete the circuit's current path. These electrons cannot pass through the dielectric; therefore, they begin to accumulate on the bottom plate. The voltage charge on the plates of the capacitor increases

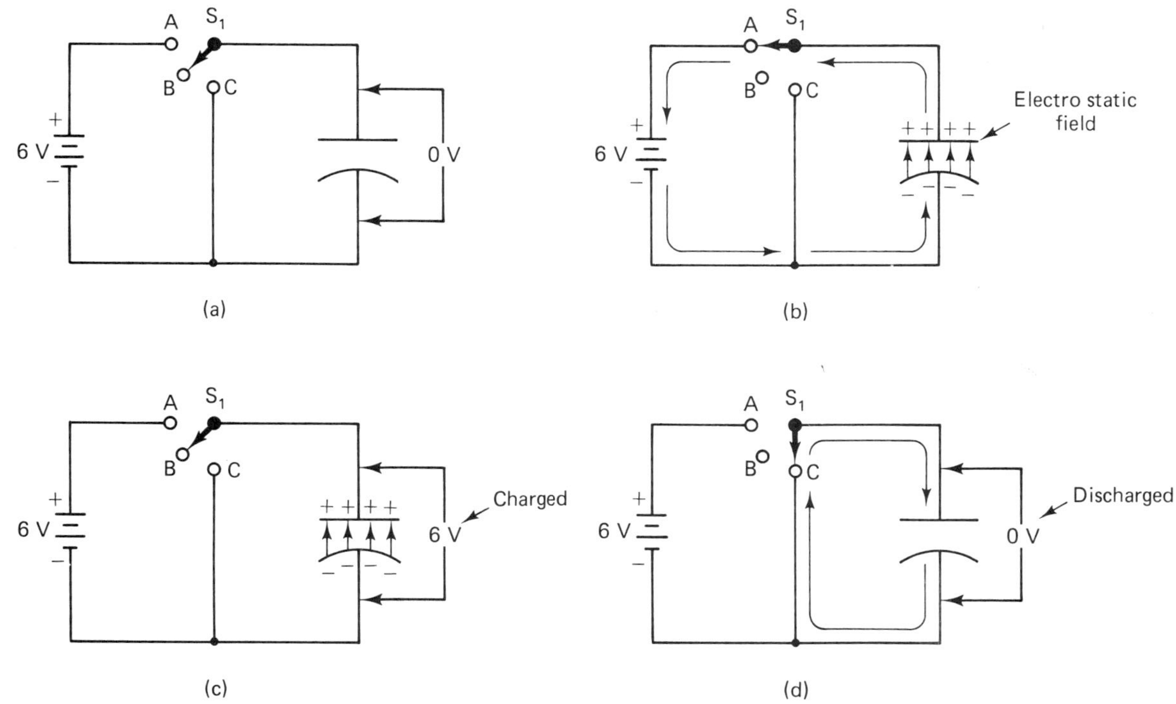

Figure 11–3 Capacitor operation in a dc circuit: (a) neutral charge; (b) charging capacitor; (c) charged capacitor; (d) discharging capacitor.

and begins to oppose the voltage of the battery. The instant the switch is placed in position A the maximum amount of current flows in the circuit. As the capacitor begins to charge up toward the total voltage of the battery (+6 V), more electrons accumulate on the negative plate, which creates an increasing repelling force and causes the charging current to decrease. Finally, a point is reached where the capacitor is fully charged (+6 V) and the voltage across its plates is equal to the voltage of the battery. There is no current flow in the circuit at this time, and in effect the capacitor is blocking the current.

When the switch is placed in position B (Figure 11-3c), the capacitor remains charged at +6 V. The capacitor is said to be storing a charge and the electrostatic field across its plates is still present. If this voltage is great enough, it would present a safety hazard to humans. Anyone coming in contact with the leads of the capacitor could be severely shocked.

If the switch is moved to position C (Figure 11-3d), an electrical path is provided from one plate of the capacitor to the other plate. Electrons from the bottom plate will flow toward the positive charges on the top plate. These charges become neutralized and the capacitor becomes discharged. The voltage across the capacitor is again 0 V.

The movement of electrons during the charging and discharging of the capacitor does not constitute a current flow because there is no current flow when the capacitor is fully charged and fully discharged. Theoretically, in a purely capacitive circuit, the current leads the applied voltage by 90°. As a reminder of this action for capacitors the name *ICE* is used, where *I* the current leads or comes before *E* (for electromotive force or voltage) in a capacitive circuit *C*.

11-1c.1 *RC* Time Constant

A circuit with a resistor and a capacitor (*RC*) has a voltage time delay similar to the current time delay of a circuit, with a resistor and inductor (*RL*) that was presented in Section 9-1c.1. An analysis of an *RC* time constant is shown in Figure 11-4. When switch S_1 is placed in position A (Figure 11-4a), electrons begin to flow through the resistor toward the positive terminal of the battery. A voltage drop is developed across the resistor. Initially, the current in the circuit is maximum and the voltage across the resistor V_R is maximum, as indicated in the graphical representation of Figure 11-4b. As the voltage across the capacitor (V_C)

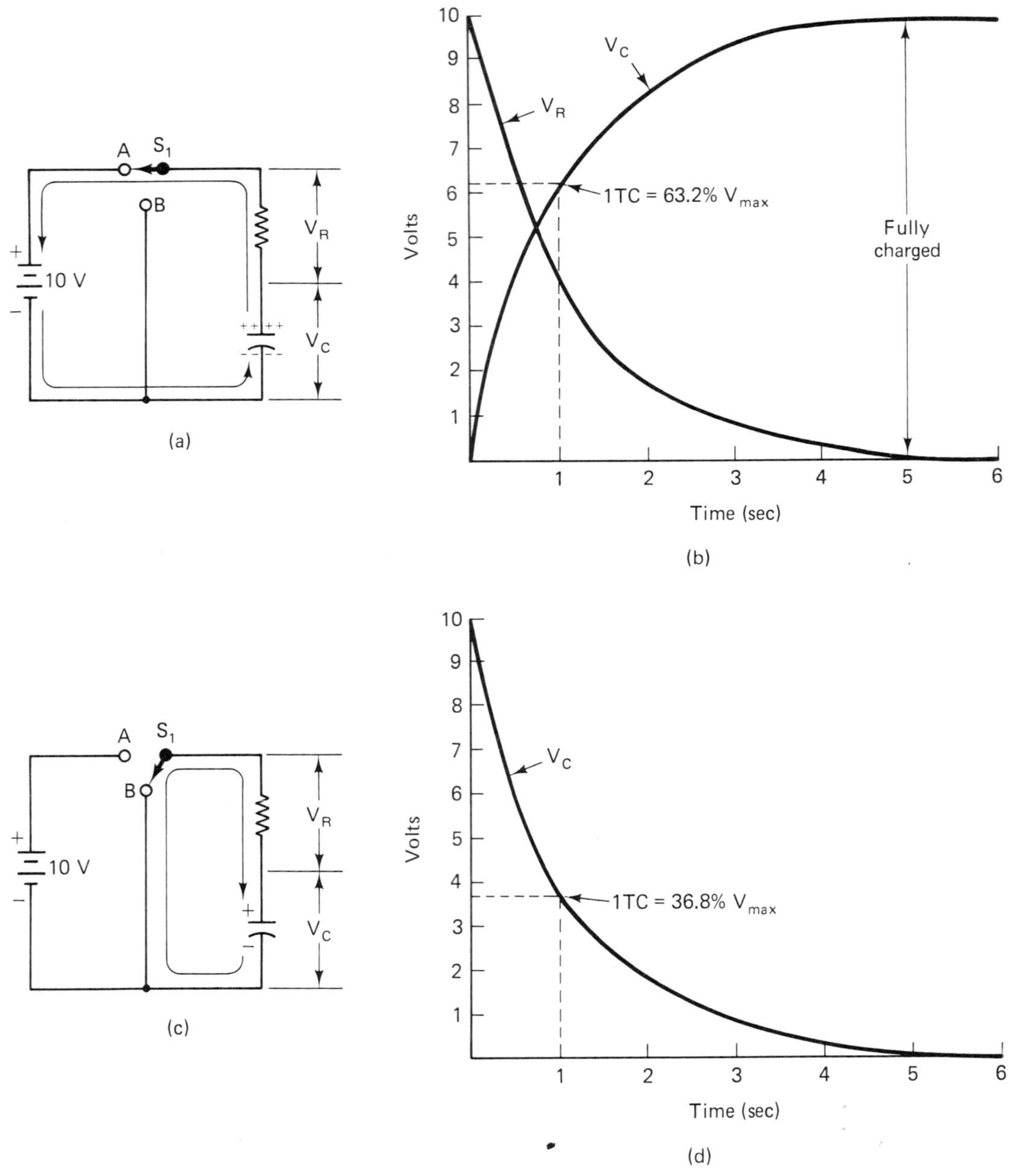

Figure 11–4 *RC* time constant: (a) charging voltages; (b) graphical representation; (c) discharging voltage; (d) graphical representation.

increases, it opposes the voltage of the battery and less current flows in the circuit. As the current decreases, so does V_R. In other words, as V_C increases, V_R decreases.

The time it takes the voltage on the capacitor to reach 63.2% of the total applied voltage is called the time constant for an *RC* circuit. Five time constants are required to charge the capacitor fully to the maximum applied voltage. The voltage buildup on the capacitor for each succeeding time constant is 63.2% of the remaining applied voltage. Assume that 10 V is applied to the *RC* circuit. The total amount of voltage charge on the capacitor and the voltage drop on the resistor for each time constant is shown below.

Time Constant	V_C *(v)*	V_R *(v)*
1	6.32	3.68
2	8.63	1.37
3	9.49	0.51
4	9.81	0.19
5	9.93	0.07

When the switch is placed in position B (Figure 11-4c), the electrons stored at the bottom plate of the capacitor will flow through the resistor to neutralize the positive charge on the top plate. Since the same resistor is in the circuit when the capacitor is charging up, the capacitor will discharge at the same time constant rate. After the first time constant, the voltage across the capacitor has decreased to 36.8% of the total voltage ($V_{\max} - 63.2\% V_{\max} = 36.8 V_{\max}$). Since the source voltage of the circuit is V_C and is decreasing, the voltage across the resistor V_R is also decreasing at the same rate. The capacitor will discharge at the 63.2%/time constant rate of the remaining charge, and after five time constants the capacitor will be completely discharged, as shown in Figure 11-4d.

The time constant of an *RC* circuit is simply the product of the value of the resistor and the capacitor. For example, if $R = 1\ \text{k}\Omega$ and $C = 1\ \mu\text{F}$, the time constant (TC) can be found thus:

$$
\begin{aligned}
\text{TC} &= RC \\
&= 1\ \text{k}\Omega \times 1\ \mu\text{F} \\
&= 1 \times 10^{3} \times 1 \times 10^{-6} \\
&= 1 \times 10^{-3}\ \text{s} \\
&= 1\ \text{ms}
\end{aligned}
$$

11-1c.2 *RC* Time Constant and Square-Wave Pulses

Square-wave pulses applied to an *RC* circuit have an effect similar to those for the *LR* circuit, which was covered in Section 9-1c.2. The effects of an *RC* time constant in a capacitive circuit can be seen with the use of square-wave pulses, as shown in Figure 11-5. The *RC* time constant of the circuit is 10 μs. An input frequency of 50 kHz has a pulse width of 10 μs, the same time as the time constant. The voltage on the capacitor can only charge up to 63.2% before the input voltage reverses polarity; therefore, the voltage on the capacitor begins to discharge in the opposite direction, and vice versa. The voltage across the capacitor ap-

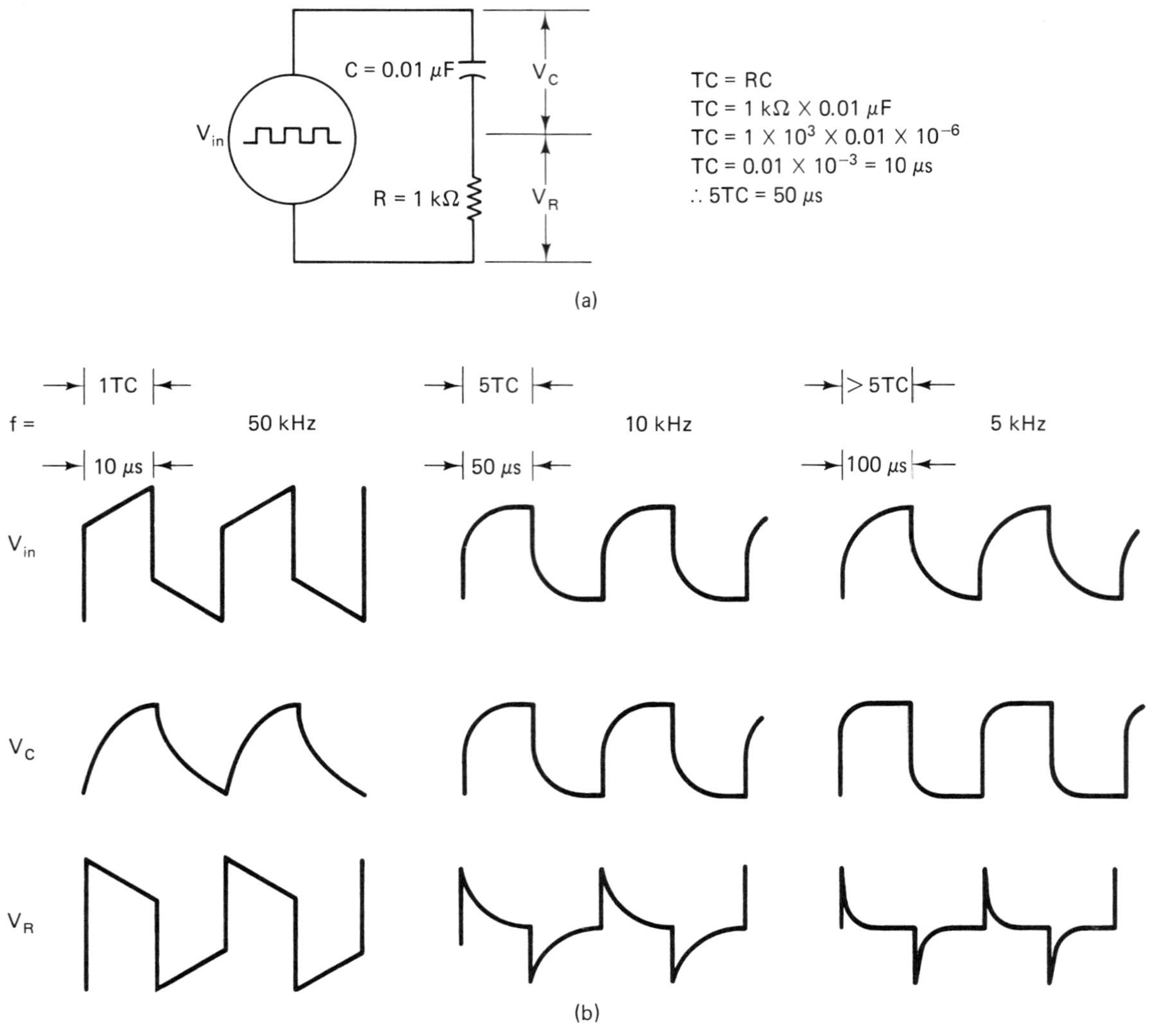

Figure 11–5 *RC* time constant and pulse shaping: (a) schematic diagram; (b) voltage waveforms.

pears as a sawtooth waveform. Current is constantly flowing through the circuit and there will always be a voltage drop across the resistor.

If the PRR of the input voltage is decreased to 10 kHz, the pulse width becomes 50 μs, which is equal to five time constants of the circuit. Now the capacitor is allowed to become fully charged before the polarity of the input voltage changes. The waveform V_C resembles more of a square wave, whereas the waveform of V_R rises sharply and immediately begins to decay. About the time V_R decays to 0 V, the input pulse change polarity and the process starts over.

If the input voltage is decreased again to 5 kHz, the pulse width becomes 100 μs. The input pulse is now 10 times longer than the circuit time constant. Voltage V_C is more of a square wave now, because it has more than sufficient time to charge fully, but V_R resembles a spike waveform. Now the current flows for very little time in the circuit and the resulting voltage drop across the resistor has a spike appearance.

The output across the capacitor is referred to as an integrator since its waveform is proportional to the time interval of the input voltage. The output across the resistor is called the differentiator because its waveform is proportional to the rate of change of the input voltage.

11-1d TYPES OF CAPACITORS

There are many types of capacitors that are used for a variety of applications. Figure 11-6 shows a summary of the time constant formula, the characteristic charge curve, schematic symbols, and some examples of the physical construction of capacitors.

The most common type of capacitor is the tubular type (Figure 11-6d), which may range in size from $\frac{1}{2}$ in. to over 6 in. in length. Their diameter can vary from an $\frac{1}{8}$ in. to over 2 in. Wire leads for circuit connections protrude from each end. The dielectric for tubular capacitors can be paper, oil-impregnated paper, glass ribbon, ceramic, or plastic film such as Mylar and polycarbonate. Usually, the capacitance value and voltage rating is printed on its side.

Some capacitors may use a color code to indicate their value (Figure 11-6e). The colors usually have the same numerical value as that of the resistor color code. Reading from left to right, the first color indicates the maximum operating temperatures under which the capacitor will perform satisfactorily. The next two colors are the first and second significant digits. The third color is the multiplier, and the fourth color is the tolerance.

Small capacitors in rectangular and disk form are in the picofarad range (Figure 11-6f and g). The capacitance value is usually printed on the side, and if enough space is available, the voltage rating is also. Their dielectric is usually mica, ceramic, or some type of plastic.

When large values of capacitance are needed in a circuit, electrolytic capacitors (Figure 11-6h) are used. To increase the capacitance and still keep the size of the capacitor as small as possible, a dielectric with a higher contant is used. This dielectric is called aluminum oxide and is formed during the fabrication of the capacitor. Tantalum is also used as the dielectric for electrolytic capacitors. The dielectric may be in dry form or a paste. It is important that most electrolytic capacitors be connected into a circuit in a specific manner. The electrostatic field must be in the proper direction across the plates or else a large current will flow which produces a gas within the capacitor and can result in a harmful explosion. The leads or terminals of an electrolytic capacitor are marked with a + and a − symbol to indicate the proper polarity with which to connect it into a circuit. The + terminal is connected to the more positive potential in the circuit and the − terminal is connected to the more negative potential. The schematic symbol of an electrolytic capacitor will also indicate polarity (Figure 11-6c). *Remember to observe polarity when you connect or replace an electrolytic capacitor in a circuit.*

Capacitors with air as the dielectric (Figure 11-6i) are usually used to tune radio and television frequencies. The plates are circular in shape and are mounted on a shaft. There are fixed plates called *stator plates,* and every other plate is movable, which is called *rotor plates.* This capacitor is called a *variable capacitor,* where the rotor plates are rotated and mesh with the stator plates and the capacitance can be varied. Maximum capacitance occurs with full mesh and minimum capacitance occurs with no mesh. A variable capacitor is indicated with an arrow through its schematic symbol (Figure 11-6c).

The *trimmer capacitor* (Figure 11-6j) is also a very small variable capacitor. Its dielectric may be air or mica. It is used to "fine adjust" frequency circuits and may be used with the tuning capacitor men-

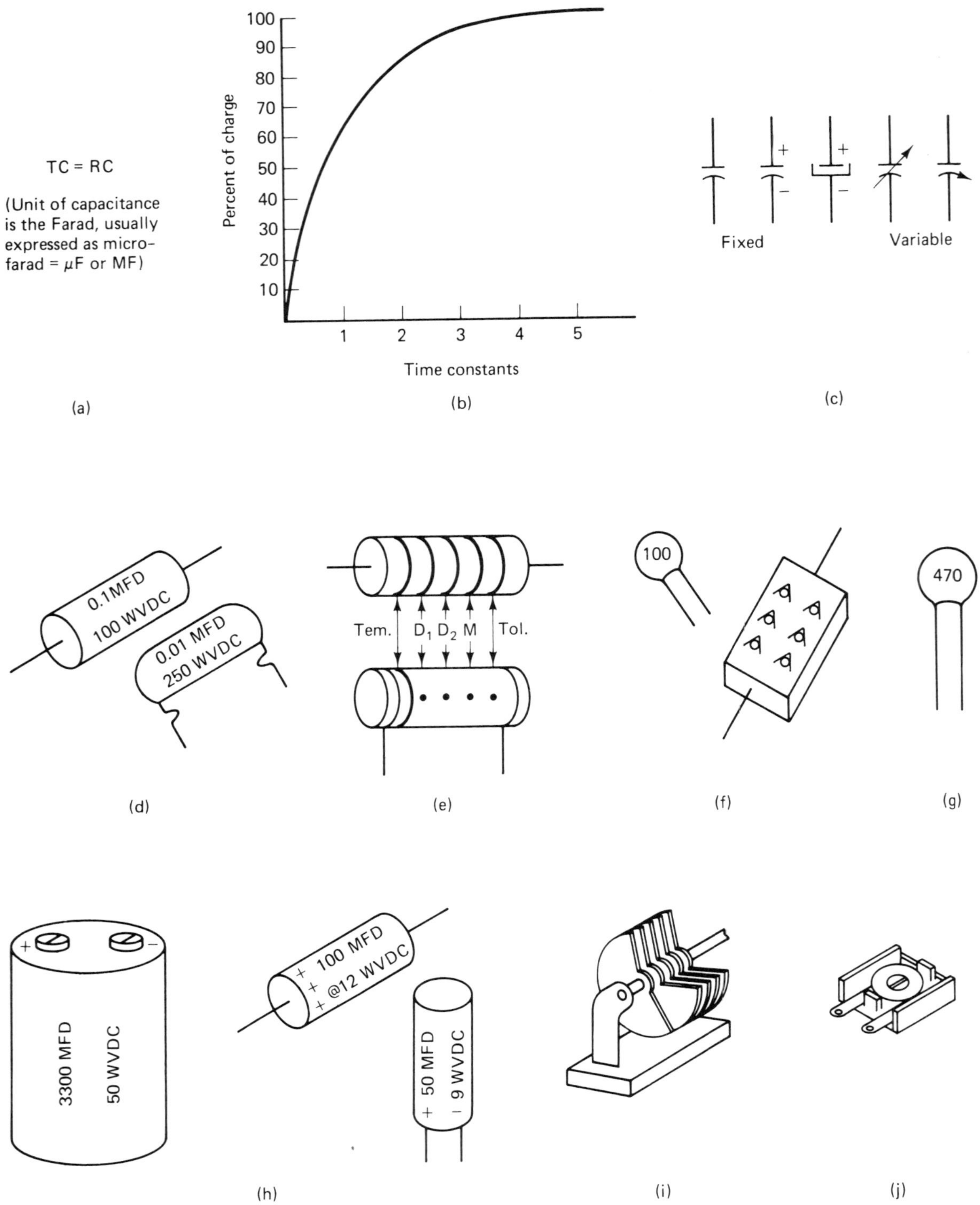

Figure 11–6 Capacitors: (a) time-constant formula; (b) voltage-charge characteristic curve; (c) symbols; (d) tubular (paper); (e) color coded; (f) mica; (g) disk; (h) electrolytic; (i) tuning (air); (j) trimmer (air).

tioned above. The trimmer capacitor is indicated by the schematic symbol with an arrow on one of its plates (Figure 11-6c).

These are the general types of capacitors, but you may encounter many other types of physical arrangements. Usually, electronic supply houses have manuals or catalogs that will identify the various types of capacitors.

11-1e CAPACITIVE REACTANCE

When a capacitor is placed in an ac circuit the current flows in one direction and charges up the capacitor toward one polarity. Then the current changes direction and the capacitor discharges and charges up toward the opposite polarity. The time it takes the capacitor to charge and discharge presents an opposition to the flow of current in the circuit. This opposition to ac voltage caused by the capacitor is called *capacitive reactance* and uses the symbol X_c. The amount of capacitive reactance offered by a capacitor is inversely proportional to frequency and its capacitance. Its unit of measurement is given in ohms.

Capacitive reactance is represented by the formula

$$X_c = \frac{1}{2\pi fc}$$

where 2π is equal to 6.28. If we find the reciprocal of 2π, ($1/2\pi = 0.159$), the formula can be simplified:

$$X_c = \frac{0.159}{fC}$$

For example, a capacitor with 0.1 μF is placed into an ac circuit with a frequency of 1 kHz. The capacitive reactance at this time can be calculated thus:

$$\begin{aligned} X_c &= \frac{0.159}{fC} \\ &= \frac{0.159}{1\text{ kHz} \times 0.1\ \mu\text{F}} \\ &= \frac{0.159}{1 \times 10^3 \times 0.1 \times 10^{-6}} \\ &= \frac{0.159}{0.1 \times 10^{-3}} \\ &= 1.59 \times 10^3 \\ &= 1590\ \Omega \\ &= 1.59\ \text{k}\Omega \end{aligned}$$

If the frequency increases, the capacitive reactance will decrease. For example, if $f = 5$ kHz, then

$$\begin{aligned} X_c &= \frac{0.159}{fC} \\ &= \frac{0.159}{5\text{ kHz}\ 0.1\ \mu\text{F}} \\ &= \frac{0.159}{5 \times 10^3 \times 0.1 \times 10^{-6}} \\ &= \frac{0.159}{0.5 \times 10^{-3}} \\ &= 0.318 \times 10^3\ \Omega \\ &= 318\ \Omega \end{aligned}$$

By simple algebra, the basic capacitive reactance formula can be rearranged to find f and C when they are unknown:

$$C = \frac{1}{2\pi f X_c} \qquad f = \frac{1}{2\pi C X_c}$$

Since the capacitor charges and discharges in an ac circuit, it appears as if the current is being passed by the capacitor. In other words, it is said that a capacitor blocks dc but passes ac. The current that appears to pass through a capacitor flows in one direction for one alternation and then in the other direction for the second alternation. The current that apparently flows through a capacitor can be found using Ohm's law, thus:

$$I_c = \frac{V_c}{X_c}$$

11-1f CAPACITANCE IN AN AC CIRCUIT

Using Ohm's law to find current, voltage drops, and power in an ac circuit with capacitance is similar to the procedure you used with the circuits involving inductance. In the inductive circuit the current lags the applied voltage, but in the capacitive circuit the current leads the applied voltage.

11-1f.1 Voltage and Current Phase Angles

In a simple series RC circuit as shown in Figure 11-7a, the voltage drop across the resistor V_R, and the current i, of the circuit are in phase. For this reason, they are used as the reference point for the circuit as shown

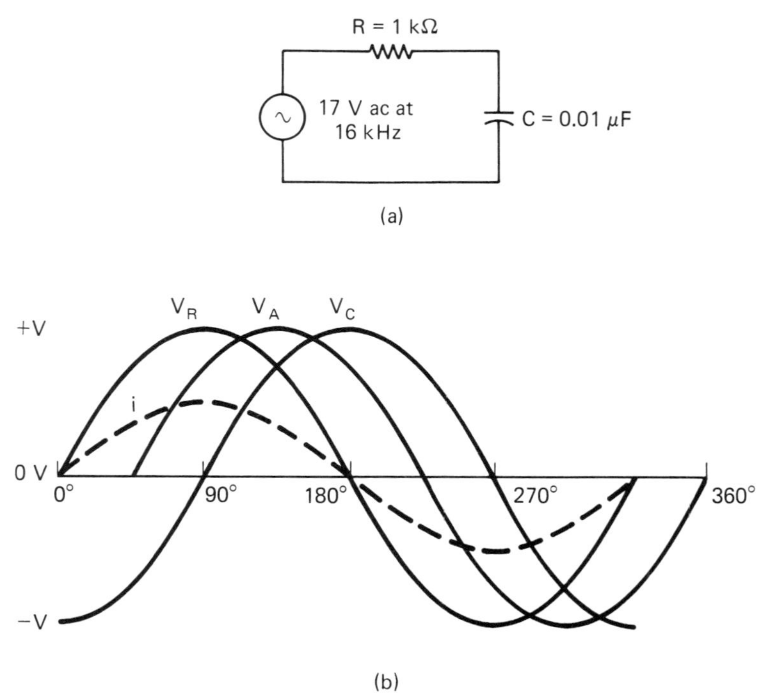

Figure 11–7 Voltage and current relationships in a series *RC* circuit: (a) schematic diagram; (b) waveform phase relationships.

in Figure 11-7b. The phase angle difference between i and V_c will be $90°$. The applied voltage V_A is the algebraic sum of V_R and V_c and is out of phase with these two voltages. Again, trying to analyze the waveforms of a circuit with a reactive component such as an inductor or capacitor is very difficult.

11-1f.2 Voltage Phasor Analysis

The voltage phasor triangle shown in Figure 11-8 makes it easier to understand the voltage phase differences in a capacitive circuit. Notice that the phasor for V_c is in a downward position. This is because the current I in the circuit leads the voltage across the capacitor by $90°$. Since I_R and V_R are in phase, the V_c phasor lags the V_R phasor by $90°$.

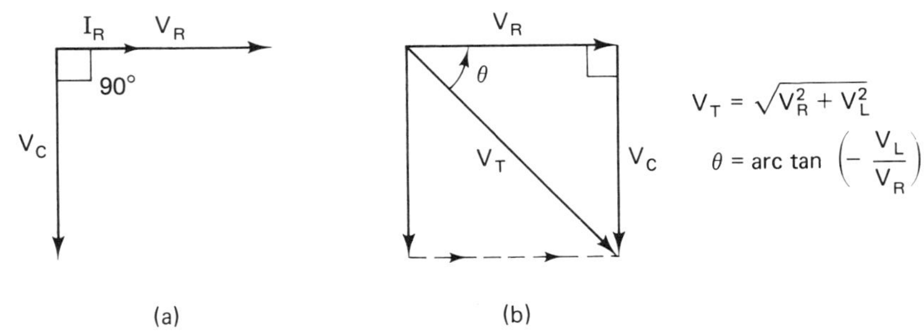

Figure 11–8 Voltage phase triangle: (a) typical phasor relationship; (b) transposed for easier calculation.

If phasor V_C is projected to the right as shown in Figure 11-8b, the hypotenuse of the formed triangle becomes the V_T phasor. Now the value of V_T can be calculated using the formula $V_T = \sqrt{V_R^2 + V_c^2}$. For example, if $V_R = 12$ V and $V_c = 12$ V, then

$$\begin{aligned} V_T &= \sqrt{V_R^2 + V_c^2} \\ &= \sqrt{12^2\ V + 12^2\ V} \\ &= \sqrt{144 + 144} \\ &= \sqrt{288} \\ &= 16.97\ V \end{aligned}$$

The applied voltage is approximately 17 V. The values can be expressed as peak to peak, peak, or rms, but in all cases the values must be consistent.

The angle θ between V_R and V_c is found by solving for the arc tangent of the triangle using the formula, $\theta = \arctan(-V_c/V_R)$. For example:

$$\begin{aligned} \theta &= \arctan\left(\frac{-V_c}{V_R}\right) \\ &= \arctan\left(\frac{-12\ V}{12\ V}\right) \\ &= \arctan -1 \\ &= -45° \end{aligned}$$

The negative sign indicates that the phase angle is for a capacitive circuit.

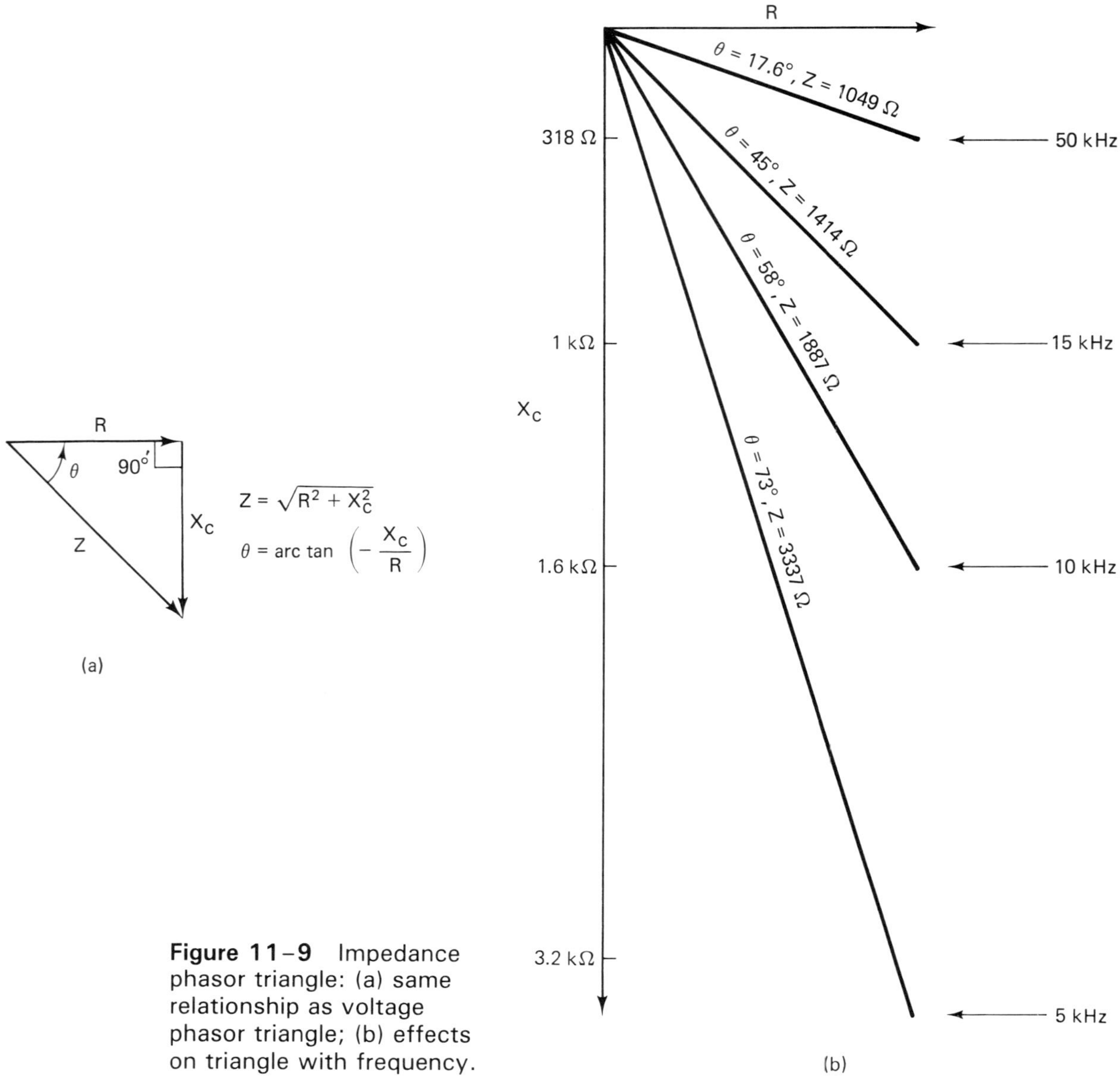

Figure 11–9 Impedance phasor triangle: (a) same relationship as voltage phasor triangle; (b) effects on triangle with frequency.

11-1f.3 Impedance Triangle

The impedance triangle for a capacitive circuit is similar to the impedance for an inductive circuit, except that it is in a downward position as shown in Figure 11-9a. If the value of the capacitor and frequency of the circuit is known, X_c can be calculated. Once X_c is known, the total impedance of the circuit can be found using the formula $Z = \sqrt{R^2 + X_c^2}$. For example, if R = 1 kΩ and X_c = 1 kΩ, then

$$\begin{aligned} Z &= \sqrt{R^2 + X_c^2} \\ &= \sqrt{1\ \text{k}\Omega^2 + 1\ \text{k}\Omega^2} \\ &= \sqrt{1 \times (10^3)^2 + 1 \times (10^3)^2} \\ &= \sqrt{1 \times 10^6 + 1 \times 10^6} \\ &= \sqrt{2 \times 10^6} \\ &= 1.414 \times 10^3 \\ &= 1.414\ \text{k}\Omega \end{aligned}$$

The current of the circuit is found by using Ohm's law:

$$\begin{aligned} I_T &= \frac{V_T}{Z} \\ &= \frac{17\text{ V}}{1.414\text{ k}\Omega} \\ &= \frac{17}{1.414 \times 10^3} \\ &= 0.012 \\ &= 0.012 \times 10^{-3} \\ &= 12\text{ mA} \end{aligned}$$

The voltage drops across each component can be checked thus:

$$\begin{aligned} V_R &= I_T R \\ &= 12\text{ mA} \times 1\text{ k}\Omega \\ &= 12 \times 10^{-3} \times 1 \times 10^3 \\ &= 12\text{ V} \end{aligned} \qquad \begin{aligned} V_c &= I_T X_c \\ &= 12\text{ mA} \times 1\text{ k}\Omega \\ &= 12 \times 10^{-3} \times 1 \times 10^3 \\ &= 12\text{ V} \end{aligned}$$

The impedance phase angle is found using the formula $\theta = \arctan(-X_c/R)$. Notice that if the frequency decreases, the phase angle increases because X_c increases, as shown in Figure 11-9b.

11-1f.4 *RC* Parallel AC Circuit

When a resistor and a capacitor are placed in parallel with an ac voltage, the voltage across both devices are in phase, as shown in Figure 11-10. The current through the resistor, i_R, is in phase with the applied

Figure 11–10 *RC* parallel circuit: (a) schematic diagram; (b) voltage–current relationships; (c) current phasor triangle.

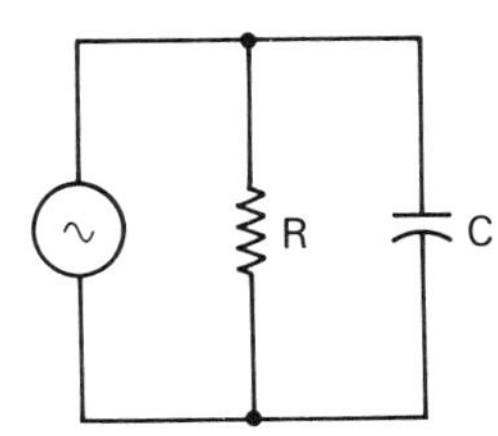

(a)

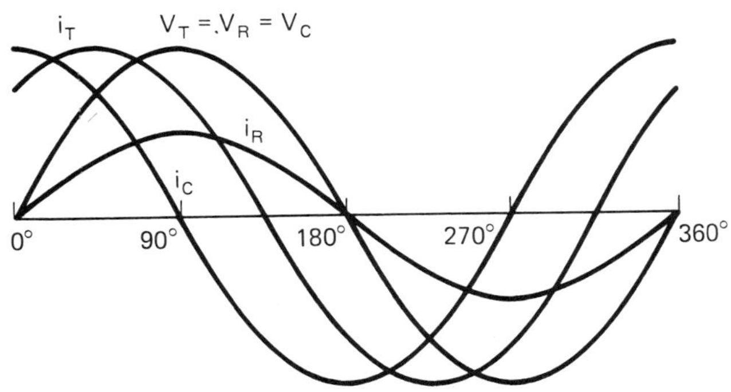

(b)

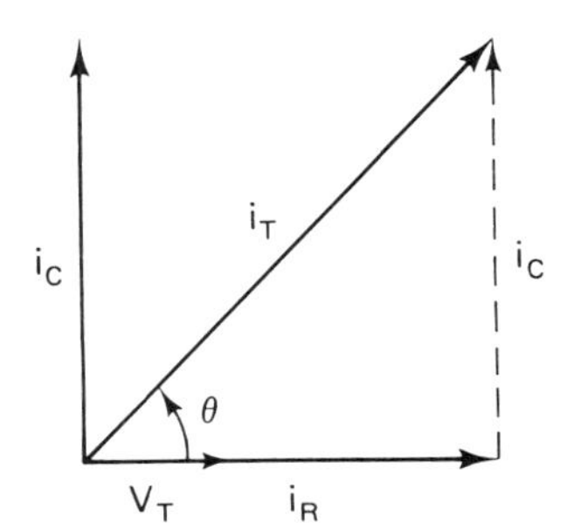

(c)

voltage, but the current in the capacitor branch is out of phase with the applied voltage. Both currents combine to form the total current, which is out of phase with the applied voltage. Trying to analyze these current relationships as shown in Figure 11-10b is quite difficult; therefore, the current phasor triangle shown in Figure 11-10c is used. Notice that this phasor triangle is in the upward position, because the current in a capacitive circuit leads the applied voltage.

To find the total current in the circuit, you must first calculate the current through each branch using Ohm's law. For example, let $R = 1\ \text{k}\Omega$, $X_L = 1\ \text{k}\Omega$, and $V_T = 12$ V; then

$$\begin{aligned} I_R &= \frac{V_T}{R} \\ &= \frac{12\ \text{V}}{1\ \text{k}\Omega} \\ &= \frac{12}{1 \times 10^3} \\ &= 12 \times 10^{-3}\ \text{A} \\ &= 12\ \text{mA} \end{aligned} \qquad \begin{aligned} i_L &= \frac{V_T}{X_L} \\ &= \frac{12\ \text{V}}{1\ \text{k}\Omega} \\ &= \frac{12}{1 \times 10^3} \\ &= 12 \times 10^{-3}\ \text{A} \\ &= 12\ \text{mA} \end{aligned}$$

Now the total current can be found by using the formula

$$\begin{aligned} i_T &= \sqrt{i_R^2 + i_L^2} \\ &= \sqrt{12\ \text{mA}^2 + 12\ \text{mA}^2} \\ &= \sqrt{12 \times (10^{-3})^2 + 12 \times (10^{-3})^2} \\ &= \sqrt{144 \times 10^{-6} + 144 \times 10^{-6}} \\ &= \sqrt{288 \times 10^{-6}} \\ &= 16.97 \times 10^{-3}\ \text{A} \\ &= 16.97\ \text{mA} \end{aligned}$$

The phase angle between i_c and V_T is found by using the formula

$$\begin{aligned} \theta &= \arctan \frac{i_c}{I_R} \\ &= \arctan \frac{12\ \text{mA}}{12\ \text{mA}} \\ &= \arctan 1 \\ &= 45^\circ \end{aligned}$$

As you can see, the formulas used for *RL* and *RC* circuits are the same. The only difference between them is the direction of the phasor triangle. Another important point to remember is that when the frequency of an *RC* circuit increases, X_c decreases and the phase angle approaches 0°. If the frequency decreases, X_c increases and the phase angle approaches 90°.

11-1f.5 Power and Capacitive Reactance

A capacitor in an ac circuit will charge up when the current is flowing for one alternation. This stored energy is then returned to the circuit when the current reverses on the next alternation. The capacitor will now charge up in the opposite direction and then return this stored energy back to the circuit on the next alternation. Theoretically, no energy is lost, but because of some resistance in the circuit, a small amount of power is lost in the form of heat.

The power formulas associated with inductive circuits are also used with capacitive circuits. The apparent power and the true power can be found using the following formulas. For example, let $V_T = 12$ V, $i_T = 12$ mA, and $\cos\theta = 45°$; then

$$\begin{aligned} P_A &= V_T i_T \\ &= 12\text{ V} \times 12\text{ mA} \\ &= 12 \times 12 \times 10^{-3} \\ &= 144 \times 10^{-3}\text{ W} \\ &= 144\text{ mW} \end{aligned} \qquad \begin{aligned} P_T &= V_T i_T \cos\theta \\ &= 12\text{ V} \times 12\text{ mA} \times \cos 45° \\ &= 12 \times 12 \times 10^{-3} \times 0.707 \\ &= 101.8 \times 10^{-3}\text{ W} \\ &= 101.8\text{ mW} \end{aligned}$$

The power in a parallel RC circuit can be found in the same manner as a parallel RL circuit using these formulas:

$$P_R = V_T i_R \qquad P_c = V_T i_c \qquad P_T = \sqrt{P_R^2 + P_c^2}$$

11-1g CAPACITIVE FREQUENCY FILTERS

The capacitor is very similar to an inductor as a reactive device with respect to frequency changes and therefore is also useful in frequency filter circuits, as shown in Figure 11-11. Basically, a passive frequency filter circuit is a voltage divider with the input voltage applied across two components and the output voltage taken across one component.

With a basic RC low-pass filter (Figure 11-11a), the output is taken across the capacitor. Remember, the voltage drops in a series circuit are directly proportional to the resistance or reactance of the devices. If the resistance of one device increases, its voltage also increases, but the voltage on the other device will have to decrease. At low frequencies the X_c of the capacitor is very high, the current in the circuit is low and most of the input voltage is dropped across it. As the frequency increases, X_c decreases, the current increases, but the voltage drop across the capacitor decreases. The cutoff frequency occurs at 0.707 V_{max} of the output voltage (Figure 11-11b). At higher frequencies the X_c becomes so small that the capacitor appears as a short circuit and nearly all of the voltage is dropped by the resistor.

With an RC high-pass filter the output voltage is taken across the resistor as shown in Figure 11-11c. At low frequencies, X_c is high, the current is low, and most of the voltage is dropped across the capacitor, which leaves less voltage across the resistor. As the frequency increases, X_c decreases, more current flows in the circuit, but less voltage is dropped across the capacitor and more voltage is dropped across the resistor. A point is reached where X_c is so small that the capacitor appears to be shorted and nearly all of the input voltage is dropped across the resistor.

The cutoff frequency of this circuit also occurs at 0.707 V_{max} of the output voltage (Figure 11-11d). At the half-power point, $X_c = R$ and the

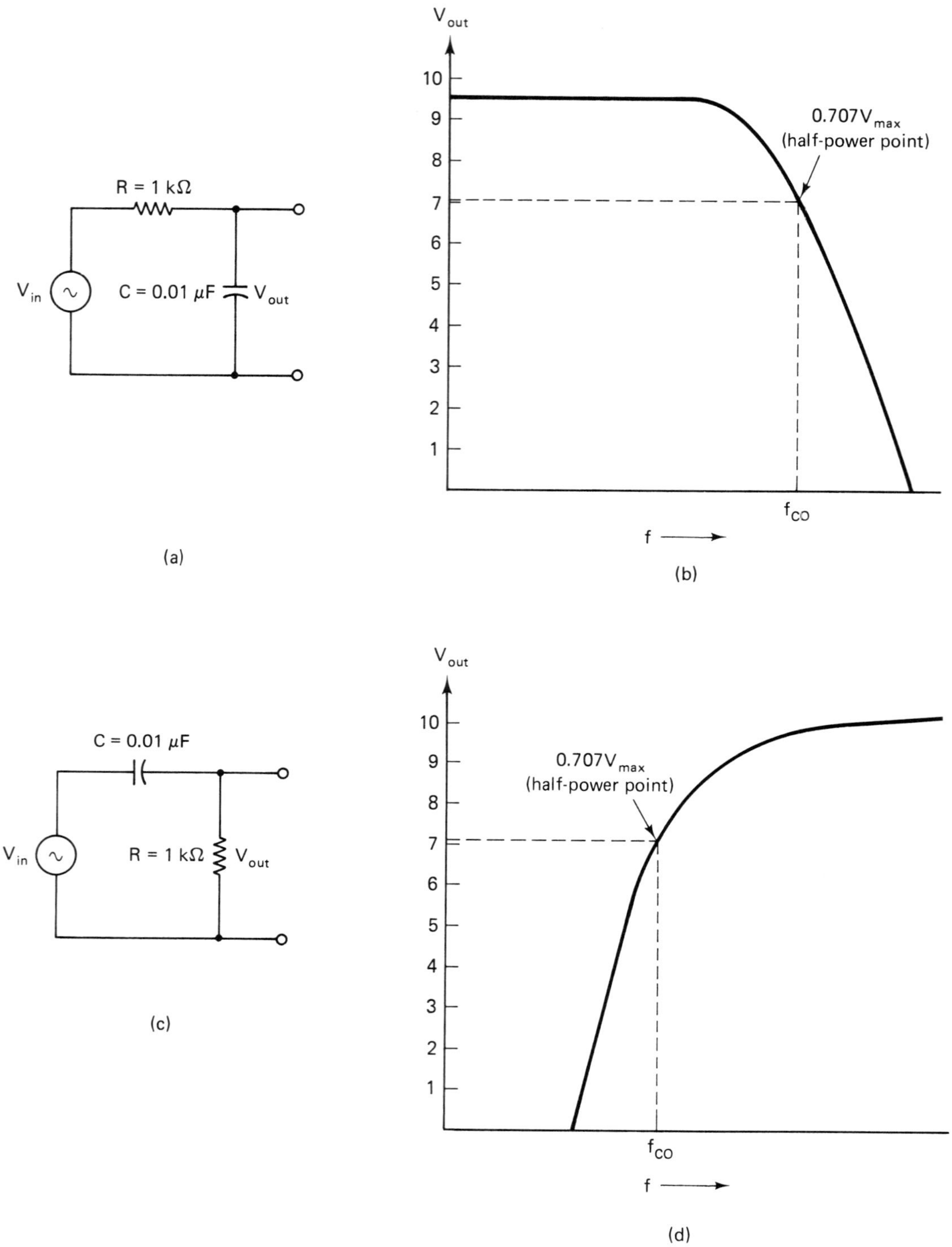

Figure 11–11 Capacitive filters: (a) low-pass circuit; (b) low-pass graph; (c) high-pass circuit; (d) high-pass graph.

phase angle is 45°. The cutoff frequency can be found by the formula $f = 1/2\pi RC$. For example, if R = kΩ and C = μF, then

$$
\begin{aligned}
f_{co} &= \frac{1}{2\pi RC} \\
&= \frac{1}{6.28 \times 1 \times 10^{3} \times 1 \times 10^{-6}} \\
&= \frac{1}{6.28 \times 10^{-3}} \\
&= 0.159 \times 10^{3}\ \text{Hz} \\
&= 159\ \text{Hz}
\end{aligned}
$$

11-1h SUMMARY OF *RC* Circuit Formulas

Series Circuit

$$C_T = \frac{1}{\frac{1}{C_1} + \frac{1}{C_2} + \frac{1}{C_3}}$$

$$C_T = \frac{C_1 \times C_2}{C_1 + C_2}$$

$$X_c = \frac{1}{2\pi fC} = \frac{0.159}{fC}$$

$$f = \frac{1}{2\pi CX_c} = \frac{0.159}{CX_c}$$

$$\text{C} = \frac{1}{2\pi fX_c} = \frac{0.159}{fX_c}$$

$$Z = \sqrt{R^2 + X_c^2}$$

$$i_T = \frac{V_T}{Z}$$

$$V_R = I_T R$$

$$V_c = I_T X_c$$

$$V_T = \sqrt{V_R^2 + V_c^2}$$

$$P_A = V_T i_T$$

$$P_T = P_A \cos\theta$$

$$\text{PF} = \frac{P_T}{P_A} \cos\theta$$

$$\text{TC} = RC$$

$$\theta = \arctan\left(\frac{-X_c}{R}\right) = -\frac{V_c}{V_R}$$

$$f_{co} = \frac{1}{2\pi RC} = \frac{0.159}{RC}$$

Parallel Circuit

$$C_T = C_1 + C_2 + C_3$$

$$V_T = V_R = V_c$$

$$X_c = \frac{1}{2\pi fC} = \frac{0.159}{fC}$$

$$i_R = \frac{V_T}{R}$$

$$i_c = \frac{V_T}{X_c}$$

$$i_T = \sqrt{i_R^2 + i_c^2}$$

$$Z = \frac{V_T}{i_T}$$

$$P_R = V_R i_R$$

$$P_c = V_c i_c$$

$$P_T = \sqrt{P_R^2 + P_c^2}$$

$$f_{co} = \frac{1}{2\pi RC} = \frac{0.159}{RC}$$

SECTION 11-2 DEFINITION EXERCISES

Write a brief description for each of the following terms.

1. Passive device ______________________________

____

2. Capacitance ______________________________

____

3. Dielectric ______________________________

____

4. Capacitor ______________________________

____

5. Farad ______________________________

____

6. Microfarad ______________________________

____

7. Picofarad ______________________________

____

8. Capacitor voltage rating ______________________________

____

9. *RC* time constant ______________________________

____

10. *RC* integrator ______________________________

____

11. *RC* differentiator ______________________________

____

12. Tubular capacitor ______________________________

____

13. Disk capacitor ______________________________

____

14. Electrolytic capacitor ______________________________

____

15. Variable capacitor ______________________________

____

16. Trimmer capacitor ______________________________

____

17. Capacitive reactance ______________________________

____

18. ICE ______________________________

____

19. Voltage divider ______________________________

____

20. Passive filter ______________________________

____

SECTION 11-3 EXERCISES AND PROBLEMS

Complete this section before beginning the next section.

1. Draw the schematic diagram for the following capacitors.

a. tubular or disk **b.** electrolytic **c.** variable

2. List the factors that determine the capacitance of a capacitor.

a.

b.

c.

3. Solve for the total capacitance of three capacitors in series from the data given.

a. $C_1 = 2\ \mu F$, $C_2 = 3\ \mu F$, $C_3 = 4\ \mu F$

$C_T =$ __________

b. $C_1 = 0.1\ \mu F$, $C_2 = 0.01\ \mu F$, $C_3 = 0.02\ \mu FC_3$

$C_T =$ __________

c. $C_1 = 10\ \mu F$, $C_2 = 10\ \mu F$, $C_3 = 10\ \mu F$

$C_T =$ __________

4. Solve for the total capacitance of two capacitors in series from the data given.

a. $C_1 = 3.3\ \mu F$, $C_2 = 2.2\ \mu F$

$C_T =$ __________

b. $C_1 = 0.02\ \mu F$, $C_2 = 0.033\ \mu F$

$C_T =$ __________

c. $C_1 = 5\ \mu F$, $C_2 = 5\ \mu F$

$C_T =$ __________

5. Solve for the total capacitance of three capacitors in parallel from the data given.

a. $C_1 = 2\ \mu F$, $C_2 = 4\ \mu F$, $C_3 = 8\ \mu F$

$C_T =$ __________

b. $C_1 = 0.01\ \mu F$, $C_2 = 0.02\ \mu F$, $C_3 = 0.05\ \mu F$

$C_T =$ __________

c. $C_1 = 0.1\ \mu F$, $C_2 = 0.002\ \mu F$, $C_3 = 0.015\ \mu F$

$C_T =$ __________

6. Fill in the following blanks.

a. 1 farad = __________ microfarads

b. 1 farad = __________ picofarads

c. 1 μF = __________ farad

d. 1 pF = __________ farad

7. Fill in the following blanks.

a. If the plates of a capacitor are moved farther apart, the capacitance will __________.

b. If the area of the plates of a capacitor is reduced, the capacitance will __________.

c. If the plates of a capacitor are moved closer together, the capacitance will __________.

d. If the area of the plates of a capacitor is increased, the capacitance will __________.

e. If the dielectric constant of a capacitor is increased, the capacitance will __________.

8. Calculate the following time constants for the values given.

a. $R = 1\ k\Omega$, $C = 0.2\ \mu F$

TC = __________

b. $R = 2.2\ k\Omega$, $C = 0.01\ \mu F$

TC = __________

c. $R = 100\ k\Omega$, $C = 50\ \mu F$

TC = __________

9. A particular RC circuit has a maximum voltage across it of 15 V. When it is first turned on the capacitor begins to charge-up toward the maximum voltage. Calculate the amount of charge the capacitor has for each of the time constants.

Time Constant	*Amount of Charge*
1	______
2	______
3	______
4	______
5	______

10. Calculate the capacitive reactance for the data given.

a. $f = 1$ kHz
$C = 5\ \mu F$

$X_C =$ ____________

b. $f = 10$ kHz
$C = 0.01\ \mu F$

$X_C =$ ____________

c. $f = 40$ kHz
$C = 0.001\ \mu F$

$X_C =$ ____________

11. Find the unknown values from the data given.

a. $X_C = 318\ \Omega$
$C = 0.05\ \mu F$

$f =$ ____________

b. $X_C = 1325\ \Omega$
$C = 2\ \mu F$

$f =$ ____________

c. $X_C = 16\ \Omega$
$f = 2$ kHz

$C =$ ____________

d. $X_C = 10.6\ k\Omega$
$f = 5$ kHz

$C =$ ____________

12. Indicate the direction of change in X_C (increase or decrease) for the specific change in the factor.

Factor Change	*X_C Change*
f decreases	______
C increases	______
f increases	______
C decreases	______

13. Draw the *RC* circuit shown in Figure 11-7a in the space provided. Solve for the unknowns using the data given.

a. $V_{\text{in}} = 50$ V at 10 kHz, $R = 1$ kΩ, $C = 0.015$ μF

X_C = ________ Z = ________ i_T = ________

V_R = ________ V_C = ________ V_T = ________

θ = ________ P_A = ________ P_T = ________

PF = ________

b. $V_{\text{in}} = 100$ V at 100 Hz, $R = 2.2$ kΩ, $C = 0.2$ μF

X_C = ________ Z = ________ i_T = ________

V_R = ________ V_C = ________ V_T = ________

θ = ________ P_A = ________ P_T = ________

14. In a *RC* series circuit, $R = 1$ kΩ and $C = 0.02$ μF. Solve for X_c, Z, and θ with the various frequencies given.

f = 10 kHz X_C = ________ Z = ________ θ = ________

f = 20 kHz X_C = ________ Z = ________ θ = ________

f = 5 kHz X_C = ________ Z = ________ θ = ________

15. With the data calculated in Question 14, draw three impedance triangles with all the data shown on the same graph similar to the one shown in Figure 11-9b (let $\frac{1}{4}$ in. equal 100).

16. Draw a *RC* parallel circuit as shown in Figure 11-10a in the space provided. Solve for the unknowns with the data given.

a. V_{in} = 100 V at 12 kHz
R = 2.2 kΩ
C = 0.005 μF

i_R = ____________

i_C = ____________

i_T = ____________

θ = ____________

P_A = ____________

P_T = ____________

PF = ____________

b. V_{in} = 60 V at 25 kHz
R = 5 kΩ
C = 0.002 μF

i_R = ____________

i_C = ____________

i_T = ____________

θ = ____________

P_A = ____________

P_T = ____________

PF = ____________

17. Draw a *RC* low-pass filter with its corresponding graph as shown in Figure 11-11a and b.

18. Draw a *RC* high-pass filter with its corresponding graph as shown in Figure 11-11c and d.

19. Solve for the cutoff frequency and the voltage at the half-power point for the data given.

a. $V_{in} = 100\ V_{p\text{-}p}$
$R = 100\ \Omega$
$C = 0.002\ \mu F$

f_{co} = ____________

V_{out} = ____________

b. $V_{in} = 60\ V_{p\text{-}p}$
$R = 500\ \Omega$
$C = 50\ pF$

f_{co} = ____________

V_{out} = ____________

20. Match the description in column A with the correct term in column B.

Column A

____ **a.** The property of temporarily storing a charge between two insulated conductors when they are at a difference of potential

____ **b.** A device that introduces capacitance into an electric circuit

____ **c.** The unit of capacitance

____ **d.** The insulating material between the plates of a capacitor

____ **e.** The force that exists between the plates of a charged capacitor

____ **f.** The opposition that a capacitor offers to ac

____ **g.** Opposition to a change in voltage, including both resistance and reactance

____ **h.** Guide for remembering that *E* lags *I* in a capacitor

____ **i.** The time for a capacitor to charge to 63.2% of a maximum voltage

____ **j.** The actual power in a circuit because of capacitance reactance

____ **k.** The frequency at which the output of a filter is at the half-power point

____ **l.** The angle that *E* lags *I* in an ac circuit with capacitance

Column B

1. farad
2. capacitor
3. dielectric
4. electrostatic field
5. capacitance
6. ICE
7. impedance
8. true power
9. capacitive reactance
10. time constant
11. phase angle
12. cutoff frequency

SECTION 11-4 EXPERIMENTS

EXPERIMENT 1. UNDERSTANDING THE FUNCTION OF A CAPACITOR

Objective: To demonstrate how a capacitor affects the operation of a dc circuit and an ac circuit.

Introduction: A dc voltage source is applied to an *RC* series circuit. Initially, there is a voltage drop across the resistor while the capacitor is charging up. When the capacitor is fully charged, no current flows and there is no voltage drop across the resistor. An audio signal generator is applied to the circuit and it is shown how different values of capacitance affect X_C, which in turn is directly proportional to the voltage drop across the capacitor.

Materials Needed:

1 1-kΩ resistor at 0.5 W
1 10-μF capacitor at 25 WV dc
1 1-μF capacitor at 25 WV dc
1 0.1-μF capacitor at 25 WV dc
1 0.01-μF capacitor at 25 WV dc

Equipment Required:

1 0 to 15-V dc power supply
1 Audio signal generator
1 Multimeter or DVM
1 Breadboard with leads or wires for constructing circuit.

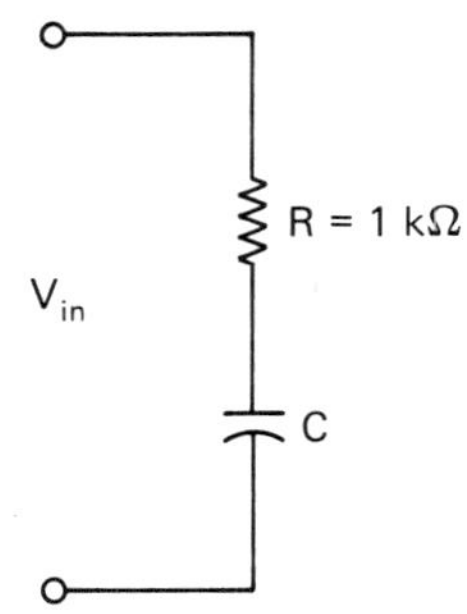

Figure 11-12

Procedure:

1. Connect the circuit shown in Figure 11-12. Use the 10-μF capacitor and observe polarity.
2. Connect the voltmeter across the resistor.
3. Set the dc power supply to +12 V.
4. Connect the power supply across the V_{in} terminals. The voltmeter should give a reading and then go to zero. The voltage across *R* = ____________ .
5. Measure the voltage across the capacitor. The voltage across *C* = ____________ .
6. Disconenct the power supply. How much voltage is being stored by *C* ? ____________ .

7. Momentarily place a wire across the circuit terminals to discharge the capacitor. Be sure not to touch the terminals with your hands. Hold the shorting wire by the insulation.
8. Connect the audio signal generator across the circuit and set it for 1 kHz at 12 $V_{rms.}$
9. Calculate X_C for the 10-μF capacitor and place the value in the proper place in the following table.

C (μF)	V_C	V_R	$X_C = \dfrac{1}{2\pi fC}$
10			
1			
0.1			
0.01			

10. Measure V_C and V_R and place the values in the proper place in the table.
11. Change C for the remaining values given in the table and repeat steps 8 through 10 for each change.

Fill-in Questions:

1. A capacitor ____________ dc current.

2. A capacitor appears to ____________ ac current.

3. When the capacitance of a capacitor decreases, X_C ____________ .

4. The ac voltage drop across a capacitor is ____________ proportional to ____________ .

EXPERIMENT 2. *RC* TIME CONSTANT AND SQUARE-WAVE PULSES

Objective; To show the effects of different *RC* time constants and examine the change of waveforms on square-wave pulses.

Introduction: In this experiment you will observe the voltage rise time and fall time for an *RC* circuit. You will also observe the pulse shape change for square-wave pulses.

Materials Needed:

1 SPDT switch
1 1-MΩ resistor at 0.5 W
1 470-kΩ resistor at 0.5 W
1 100-kΩ resistor at 0.5 W
1 1-kΩ resistor at 0.5 W
1 10-μF capacitor at 25 WV dc
1 1-μF capacitor at 25 WV dc
1 0.1-μF capacitor at 25 WV dc
1 0.01-μF capacitor at 25 WV dc

Equipment Required:

1 0- to 15-V dc power supply
1 Square-wave generator
1 Multimeter or DVM
1 Oscilloscope
1 Breadboard with leads or wires for constructing circuit

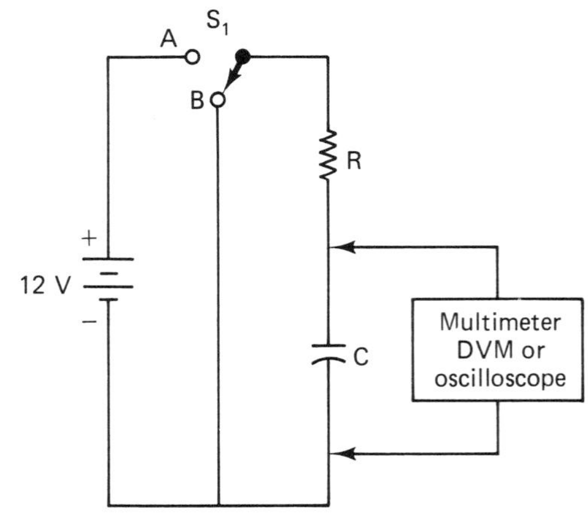

Figure 11–13

Procedure:

1. Construct the circuit shown in Figure 11-13. Use the first values given in the following table for C and R.

C (μF)	R (Ω)	TC (s)	5 TC
10	1 M	______	______
10	470 k	______	______
1	1 M	______	______
0.1	100 k	______	______

2. Calculate the value of one RC time constant and place value in the proper place in the table.
3. Calculate the value of five RC time constants and place the value in the proper place in the table.
4. If possible, place an oscilloscope across C. A voltmeter will also work well.
5. Place S_1 to position A and observe the rising voltage across C. About how long does it take for the capacitor to charge up fully?

6. Place S_1 to position B and observe the falling voltage across *C*. About how long does it take to completely discharge the capacitor?
7. Use the other values for *C* and *R* in Table 11-2 and repeat steps 2 through 6.
8. Disconnect the power supply and S_1 from the circuit.
9. Connect the circuit shown in Figure 11-5a.
10. Set the square-wave generator for 50 kHz and observe the waveforms across *C*, *R*, and V_{in}.
11. Set the square-wave generator for 10 kHz and observe the waveforms across *C*, *R*, and V_{in}.
12. Set the square-wave generator for 5 kHz and observe the waveforms across *C*, *R*, and V_{in}.

Fill-In Questions:

1. An *RC* time constant is the product of *R* ____________ *C*.
2. For one time constant the voltage across a capacitor will rise to ____________ of V_{max}.
3. It takes ____________ time constants for a capacitor to fully charge.
4. It takes ____________ time constants for a capacitor to discharge fully.
5. If the input voltage to an *RC* series circuit has a pulse width the same time as the circuit time constant, the waveform across the capacitor will appear as a ____________ .
6. If the input voltage to an *RC* series circuit has a pulse width greater than five times the circuit time constant, the waveform across the resistor will appear as a ____________ .

EXPERIMENT 3. *RC* LOW- AND HIGH-PASS FILTERS

Objective: To demonstrate the characteristics of *RC* low- and high-pass filters and provide the reader with the skills in measuring ac reactive circuits.

Introduction: This experiment is similar to the experiment on *RL* low- and high-pass filters. You will construct the filter circuits, perform voltage measurements, and construct frequency response graphs.

Materials Needed:

1 1-kΩ resistor at 0.5 W
1 0.01-μF capacitor at 25 wvdc

Equipment Required:

1 Sine-wave generator
1 Dual-trace oscilloscope
1 Breadboard and leads or wire for constructing circuit

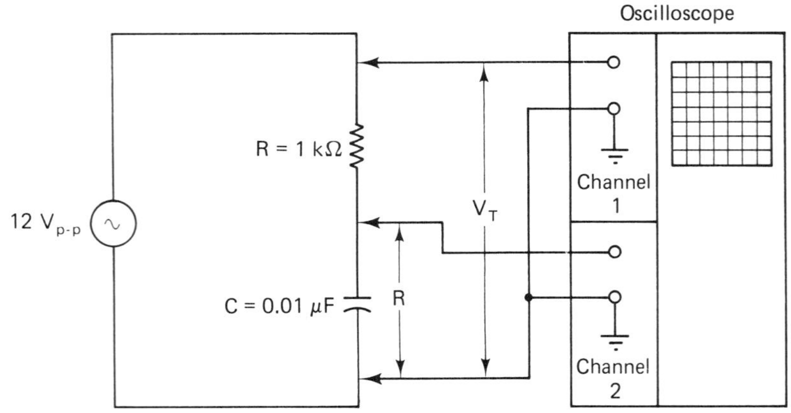

(a)

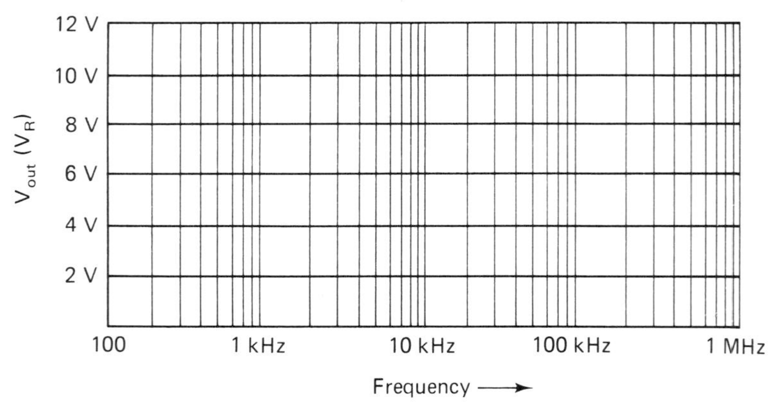

Figure 11–14 (b)

Procedure:

1. Construct the circuit shown in Figure 11-14a. This is a low-pass filter circuit. Make sure that both ground leads of the oscilloscope are connected together. If you connect both leads of each channel across each component, you may short out one of the components.
2. Adjust the generator for an output of 0.5 kHz. Observe channel 1 of the oscilloscope for an amplitude of 12 $V_{p\text{-}p}$ of reading.
   Important! Every time you change the frequency of the generator, be sure that the amplitude is maintained the same to give accurate results.
3. Using only one channel of the oscilloscope, measure the voltage across *C* and *R* and enter these readings into their proper place in

the following table. Also calculate X_c and record the value in the table.

f (kHz)	Calculate X_C	Measure: V_C	V_R
0.5	______	______	______
1	______	______	______
10	______	______	______
16	______	______	______
25	______	______	______
50	______	______	______
100	______	______	______
200	______	______	______
500	______	______	______

4. Repeat steps 2 and 3 for the other frequency values given in the table.
5. Using the data entries for V_C in the table, construct an output voltage versus frequency curve on the blank graph shown in Figure 11-14b.
6. Turn off the power to the circuit. Exchange the R and C components so that the resistor is now on the bottom. This is a high-pass filter circuit. Apply power to the circuit.
7. Using one channel of the oscilloscope, measure V_R for the frequencies given in the following table and record the readings in the proper place.

f (kHz)	V_R	f (kHz)	V_R
0.5	______	50	______
1	______	100	______
10	______	200	______
16	______	500	______
25	______		

8. In the space provided below, construct an output voltage versus frequency graph for the data of V_R in the table.

9. Calculate the cutoff frequency for these circuits using the formula

$$f_{co} = \frac{0.159}{CR} = \underline{\qquad\qquad} \text{Hz}$$

10. Calculate the amplitude of the output voltage at the cutoff frequency using the formula

$$V_{fco} = 0.707 V_{max} = \underline{\qquad\qquad} V_{p\text{-}p}$$

Fill-In Questions:

1. When the frequency applied to an *RC* series circuit increases, X_c ____________ ; therefore, V_c ____________ .
2. When the frequency applied to an *RC* series circuit decreases, X_c ____________ ; therefore, V_c ____________ .
3. The cutoff frequency for these circuits is approximately ____________ kHz.
4. The output voltage amplitude of these circuits at the cutoff frequency is about ____________ $V_{p\text{-}p}$.
5. For an *RC* low-pass filter circuit, the output is taken across the ____________ .
6. For an *RC* high-pass filter circuit, the output is taken across the ____________ .

SECTION 11-5 TROUBLESHOOTING APPLICATION: TESTING CAPACITORS

There are four possible causes for a malfunctioning capacitor. It can become open, shorted, leaky, or have a change of value in the capacitance. The best way to check capacitors is with the use of a capacitor checker. It is an instrument that is relatively easy to use. You select the capacitance range of the capacitor, connect the capacitor to the leads, and the display will indicate the exact capacitance. This instrument may also test the capacitor for current leakage between the plates. In many cases a capacitor checker is not available; therefore, a multimeter or DVM can be used to perform a go/no go test on a capacitor. Figure 11-15 shows how to test a capacitor with a multimeter.

Using an ohmmeter, you can test a capacitor for an open, short, and leaky condition. As the ohmmeter is initially connected to a capacitor, the pointer should deflect or the digital display will indicate a momentary current flow. When the capacitor charges up to the voltage across the ohmmeter leads, current stops flowing and the meter should indicate infinity (Figure 11-15a). If the meter leads are reversed and placed on the capacitor (Figure 11-15b), the pointer deflection will be nearly twice as much as before, because the voltage charge on the capacitor is now in series with the voltage of the meter leads. An alternate test would be to place the ohmmeter across the capacitor leads (Figure 11-15a). After the capacitor is charged, remove the ohmmeter. Set the meter to read low dc volts and place it back on the capacitor leads with the same polarity (Figure 11-15c). The meter should indicate the voltage of the charged capacitor, and then it will decrease as the capacitor discharges through the meter. The malfunction of a capacitor can be determined thus:

Open: no meter deflection or reading (the capacitor will not store a charge).

Short: the meter will indicate zero resistance.

Leaky: the meter will indicate some resistance less than infinity.

Large-value electrolytic capacitors will take a long time to charge up to the total voltage on the ohmmeter leads, so you will have to be patient in testing these capacitors. In some cases new electrolytic capacitors have excessive leakage current. This leaky condition may continue for some time after the capacitor is placed into a working circuit. Eventually, this leakage current becomes minimum and the circuit will operate properly. In other words, an electrolytic capacitor has to be "conditioned or ripened" in the circuit.

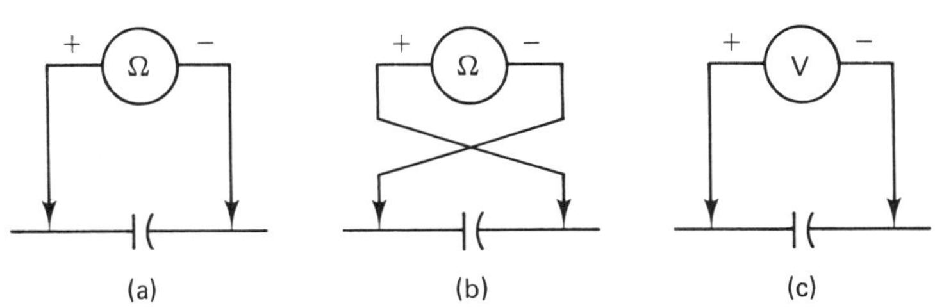

Figure 11–15 Testing a capacitor.

SECTION 11-6 INSTANT REVIEW

- A passive device is an electrical component that may control but does not create or amplify energy.
- Capacitance is the ability to store an electrical charge in an insulating material between two conductors with a difference of potential.
- The dielectric is the insulating material that separates the plates of a capacitor.
- A capacitor is a device for introducing capacitance into an electrical circuit.
- The farad is the unit of capacitance.
- A microfarad is one millionth of a farad (1×10^{-6}F).
- A picofarad is one millionth of a millionth of a farad (1×10^{-12} F).
- The amount of capacitance of a capacitor is directly proportional to the area of its plates and dielectric constant, and inversely proportional to the distance between the plates.
- The size of a capacitor is determined mainly by its voltage rating.
- Total capacitance for capacitors in series is found with the reciprocal method or product-over-sum method.
- Total capacitance for the same value capacitors in series is found by dividing the value of one capacitor by the total number of capacitors.
- Total capacitance for capacitors in parallel is found by adding up the individual capacitors.
- One time constant for an *RC* circuit is the time it takes the capacitor to charge up to 63.2% of the maximum applied voltage.
- It takes five time constants to fully charge or fully discharge a capacitor.
- The formula for finding a time constant is TC = *RC*.
- In an integrator circuit, the output voltage is taken across the capacitor for an *RC* circuit.
- In a differentiator circuit, the output is taken across the resistor for an *RC* circuit.
- Capacitors come in many sizes and shapes: tubular, disk, can, variable (air), trimmer, and electrolytic, which may be the polarized type.
- "ICE" means that the current leads the voltage (electromotive force) in a capacitive circuit.
- Capacitive reactance (X_C) is the opposition a capacitor offers to ac voltage and is measured in ohms.
- If frequency decreases, X_C increases, and vice versa.
- If capacitance decreases, X_C increases, and vice versa.
- Impedance (Z) is the total opposition to current flow in an ac circuit, including resistance and capacitive reactance.
- In a series *RC* circuit, the vectorial sum of V_R and V_C equals the applied voltage.
- In a series *RC* circuit, the vectorial sum of R and X_C equals the impedance.

- In a series *RC* circuit, $i_T = i_R = i_C$ and is found by V_T/Z.
- In a parallel *RC* circuit, $V_T = V_R = V_C$.
- In a parallel *RC* circuit, the vectorial sum of i_R and i_C equals i_T.
- Apparent power is V times I.
- A reactive component such as a capacitor gives some energy back to the circuit.
- True power is V times I times cos θ, which is the phase angle of the voltage and current in a reactive circuit.
- The power factor (PF) is the true power divided by the apparent power.
- The total power of a parallel *RC* circuit is the vectorial sum of P_R and P_C.
- An *RC* circuit can also be used as a low-pass or high-pass filter.
- The cutoff frequency (f_{co}) for a filter circuit occurs at the frequency where the output voltage falls to $0.707V_{max}$.
- A capacitor blocks dc current but appears to pass ac current.

SECTION 11-7 SELF-CHECKING QUIZ

Circle the most correct answer for each question.

1. The property that stores a charge between two conductors with a difference of potential is called:

a. dielectric constant
b. capacitive reactance
c. capacitance
d. battery potential

2. The relationship of voltage and current in a pure capacitive circuit is that current:

a. leads voltage by 90°
b. lags voltage by 90°
c. lags voltage by 180
d. leads voltage by 45°

3. The unit of capacitance is the:

a. henry **b.** farad
c. ohm **d.** watt

4. If the plates of a capacitor are moved farther apart, its capacitance will:

a. increase **b.** decrease
c. remain the same **d.** none of the above

5. Two capacitors are in series. One capacitor has 20 μF and the other has 30 μF. The total capacitance is about:

a. 50 μD **b.** 600 μF
c. 12 μF **d.** 10 μF

6. The area of the plates of a capacitor are decreased. Its capacitance will:

a. increase **b.** decrease
c. remain the same **d.** none of the above

7. Three capacitors are in parallel. $C_1 = 5$ μF, $C_2 = 15$ μF, and $C_3 = 20$ μF. The total capacitance is about:

a. 3.2 μF **b.** 13.33 μF
c. 40 μF **d.** 60 μF

8. In a series *RC* circuit, $R = 2.2$ kΩ and $C = 0.03$ μF. The time constant for this circuit is approximately:

a. 2.23 ms **b.** 0.66 ms
c. 0.014 ms **d.** 66 μs

9. A capacitor with 470 pF is equal to:

a. 0.470 μF
b. 0.000470 F
c. 0.000470 μF
d. 47×10^{-9} F

10. If the square-wave voltage pulse width applied to an *RC* series circuit is longer than five times the circuit's time constant, the voltage waveform appearing across the resistor would resemble a:

a. triangle waveform
b. square waveform
c. sawtooth waveform
d. spike waveform

11. Which one of the following is *not* true of capacitive reactance:

a. directly proportional to capacitance
b. inversely proportional to capacitance
c. is measured in ohms
d. has no value for a dc circuit

12. A particular capacitor with 0.02 μF is placed into a circuit with a frequency of 50 kHz. Its X_c is approximately:

a. 159 Ω
b. 318 Ω
c. 2500 Ω
d. 40 Ω

13. In a series *RC* circuit, $R = 3.3$ k Ω and $X_c = 6.8$ k Ω. The impedance of this circuit is approximately:

a. 10.1 kΩ
b. 7.6 kΩ
c. 57.1 kΩ
d. 3.5 kΩ

14. When the frequency of a circuit increases, the X_c in the circuit:

a. increases
b. decreases
c. remains the same
d. becomes inductive

15. In a parallel *RC* circuit, $R = 5$ kΩ, $X_c = 2.5$ kΩ, and $V_T = 100$ V. The total current of the circuit is:

a. 10.8 mA
b. 60 mA
c. 44.72 mA
d. 20 mA

16. If $P_A = 50$ W and $P_T = 42$ W, the PF is:

a. 1.19
b. 0.84
c. 0.707
d. 1.05

17. With an *RC* low-pass series filter the output is taken across the:

a. capacitor
b. resistor
c. both capacitor and resistor
d. none of the above

18. In a series *RC* circuit, $R = 4.7$ kΩ and $C = 0.01$ μF. The f_{co} for this circuit is approximately:

a. 4.7 kHz
b. 338 Hz
c. 3.4 kHz
d. 212 Hz

19. Which of the following is *not* true for a capacitor?

a. blocks dc current
b. apparently passes ac current
c. stores a charge
d. can be used to increase or amplify current

20. If an ohmmeter test shows that a capacitor has 50 kΩ between its leads, it can be assumed that the capacitor is:

a. open
b. shorted
c. leaky
d. okay

ANSWERS TO EXPERIMENTS AND SELF-CHECKING QUIZ

Experiment 1: **(1)** blocks **(2)** pass **(3)** increases **(4)** directly, X_c

Experiment 2: **(1)** times **(2)** 63.2% **(3)** five **(4)** five **(5)** sawtooth **(6)** spike

Experiment 3: **(1)** decreases, decreases **(2)** increases, increases **(3)** 15.9 **(4)** 8.5 **(5)** capacitor **(6)** resistor

Self-Checking Quiz: **(1)** c **(2)** a **(3)** b **(4)** b **(5)** c **(6)** b **(7)** c **(8)** d **(9)** c **(10)** d **(11)** a **(12)** a **(13)** b **(14)** b **(15)** c **(16)** b **(17)** a **(18)** c **(19)** d **(20)** c

Unit 12

RLC Circuits

INTRODUCTION

Combinations of resistors, inductors, and capacitors in electric/electronic circuits, known as *RLC* circuits, are very important to other phases of electronics. When you select a specific radio or television program, *LC* tuning circuits are used. Frequency filters are used in audio and video production and in nearly every type of electronic communication.

UNIT OBJECTIVES

Upon completion of this unit, you will be able to:

1. Describe the conditions of an *RLC* circuit.
2. Define resonance and show how $X_L = X_C$.
3. Show the effect of X_L and X_C with a frequency change.
4. Use the formula for calculating the resonant frequency of a circuit.
5. Use the formulas for calculating total impedance, current, and power of an *RLC* circuit.
6. Use phasors for describing current and voltage phase angle relationships.
7. Explain the difference between an *RLC* series circuit and an *RLC* parallel circuit.
8. Show the action of a tank circuit.
9. List some applications of a tank circuit.
10. Define the terms *bandwidth, bandpass, bandstop, bandreject,* and the *Q* of a circuit.
11. Identify and explain the operation of the basic types of filters: low-pass, high-pass, bandpass, and bandstop or notch filter.
12. Troubleshoot an *RLC* series circuit.

SECTION 12-1 FUNDAMENTAL CONCEPTS

12-1a IMPEDANCE AND FREQUENCY OF AN *RLC* CIRCUIT

When a resistor, inductor, and capacitor are placed into an ac circuit there are three basic conditions that may exist. The reactance of the circuit can be balanced, appear more inductive, or appear more capacitive. Figure 12-1 shows a comparison of the reactance of an *RLC* series circuit. The inductive reactance and capacitive reactance are plotted on the same graph. They are not to scale, but are shown only to indicate their relationship to frequency.

We may use a simple statement to remember the current and voltage relationship of an inductor and capacitor: "ELI the ICE man," which means that voltage leads current in an inductive circuit and current leads voltage in a capacitive circuit. This is the reason X_L and X_C oppose each other on the graph. The reactance curve shows the effect of X_L increasing as frequency increases, while X_C decreases.

If the frequency applied to the circuit is such that X_L and X_C are the same value, the circuit is said to be *resonant* (Figure 12-1b). Notice that the distance between 0 Ω and the X_L line and the X_C line are the same. Since they are equal and opposite, they cancel each other and the total impedance of the circuit is only the remaining resistance. In other words, *resonance* is a condition of an *RLC* circuit where $X_L = X_C$.

If the frequency decreases and goes below resonance, the X_L decreases and X_C increases (Figure 12-1c). The circuit is now more capaci-

Figure 12–1 Reactance versus frequency: (a) *RLC* series circuit; (b) at resonant frequency; (c) below resonant frequency; (d) above resonant frequency.

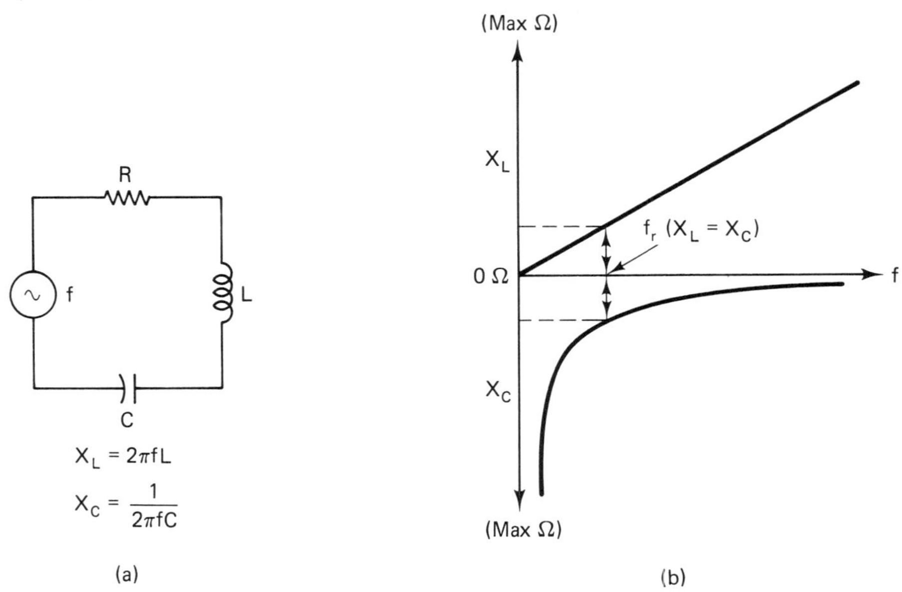

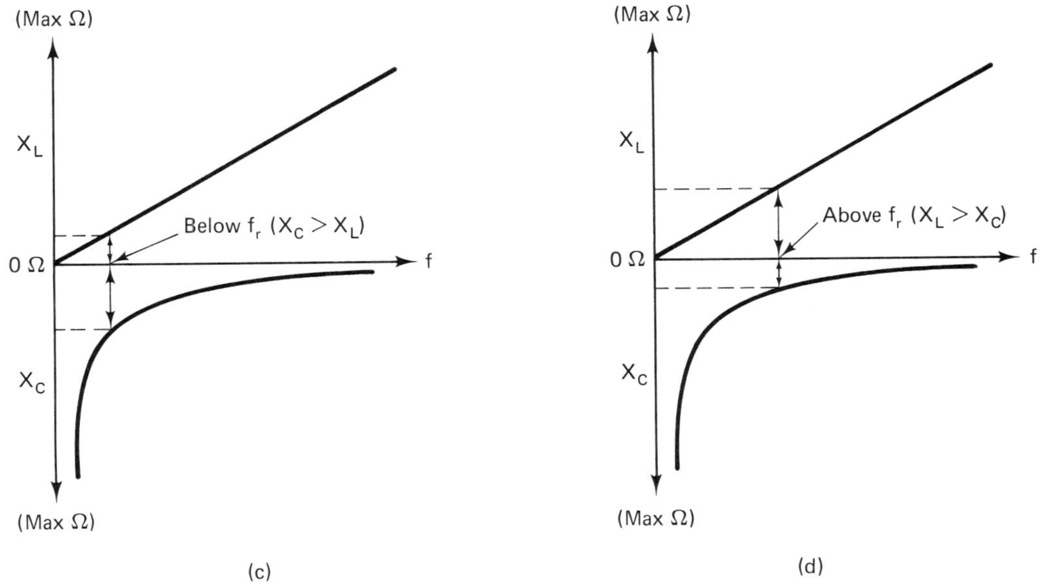

Figure 12–1 (cont.)

tive and offers more opposition to current flow. If the frequency increases and goes above resonance, the X_L increases and the X_C decreases (Figure 12-1d). The circuit is now more inductive and offers more opposition to current flow.

The frequency at which $X_L = X_C$ is called the *resonant frequency* can be found by using the formula

$$f_R = \frac{1}{2\pi\sqrt{LC}} \quad \text{or} \quad f_R = \frac{0.159}{\sqrt{LC}}$$

For example, if $L = 1$ H and $C = 2\ \mu$F, then

$$f_R = \frac{0.159}{\sqrt{LC}} = \frac{0.159}{\sqrt{1 \times 2 \times 10^{-6}}} = \frac{0.159}{\sqrt{2 \times 10^{-6}}} = \frac{0.159}{1.414 \times 10^{-3}}$$
$$= 0.112 \times 10^3 = 112\ \text{Hz}$$

12-1b *RLC* SERIES CIRCUIT

A simple *RLC* series circuit consists of a resistor, inductor, and capacitor (Figure 12-2). The reactances of this type of circuit make the calcula-

Figure 12–2 *RLC* series circuit.

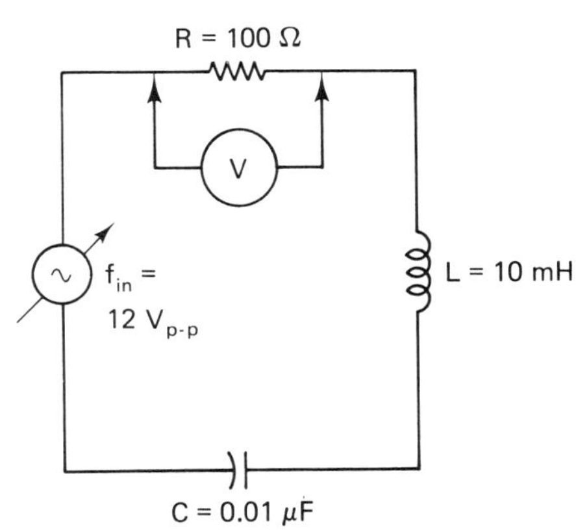

tions for finding the parameters a little more complex. The use of phasors help to show the action that is taking place within the circuit.

12-1b.1 Reactance and Impedance Phasors

The total impedance of an *RLC* series circuit can be found by combining the reactances and resistance as shown in Figure 12-3. As mentioned before, when $X_L = X_C$, the reactances cancel each other and only the resistance offers opposition to current flow in the circuit. The phase angle is 0° (Figure 12-3a). For example, if $X_C = 1.5\ \text{k}\Omega$, $X_L = 1.5\ \text{k}\Omega$, and $R = 1\text{k}\ \Omega$, then

$$\begin{aligned} Z &= \sqrt{R^2 + (X_L - X_C)^2} \\ &= \sqrt{(1\ \text{k}\Omega)^2 + (1.5\ \text{k}\Omega - 1.5\ \text{k}\Omega)^2} \\ &= \sqrt{1\ \text{k}^2\Omega} \\ &= 1\ \text{k}\Omega \end{aligned}$$

If the voltage applied to the circuit is 12 V, the current in the circuit can be found with the formula

$$\begin{aligned} i_T &= \frac{V_T}{Z} \\ &= \frac{12\ \text{V}}{1\ \text{k}\Omega} \\ &= \frac{12}{1 \times 10^3} \\ &= 12 \times 10^{-3} \\ &= 12\ \text{mA} \end{aligned}$$

The power of the circuit can now be found using the formula

$$\begin{aligned} P_T &= V_T i_T \cos\theta \\ &= 12\ \text{V} \times 12\ \text{mA} \times 1 \\ &= 144 \times 10^{-3} \\ &= 144\ \text{mW} \end{aligned}$$

If one reactance is larger than the other, the lesser reactance is subtracted from the greater reactance and then the total impedance is calculated. When X_C is greater than X_L, X_L is subtracted from X_C, as shown in Figure 12-3b. The formula shows that X_C is subtracted from X_L, resulting in a negative value for the remaining X_C. This is correct and indicates that the phase angle is counterclockwise from the reference R phasor. For example, if $X_C = 2.5\ \text{k}\Omega$, $X_L = 1.5\ \text{k}\Omega$, and $R = 1$ kΩ, then

$$\begin{aligned} Z &= \sqrt{R^2 + (X_L - X_C)^2} \\ &= \sqrt{1\ \text{k}\Omega^2 + (1.5\ \text{k}\Omega - 2.5\ \text{k}\Omega)^2} \\ &= \sqrt{(1 \times 10^3)^2 + (-1\ \text{k}\Omega \times 10^3)^2} \\ &= \sqrt{1 \times 10^6 + (-)1 \times 10^6} \\ &= \sqrt{-2 \times 10^6} \\ &= -1.414 \times 10^3 \\ &= -1.414\ \text{k}\Omega \end{aligned}$$

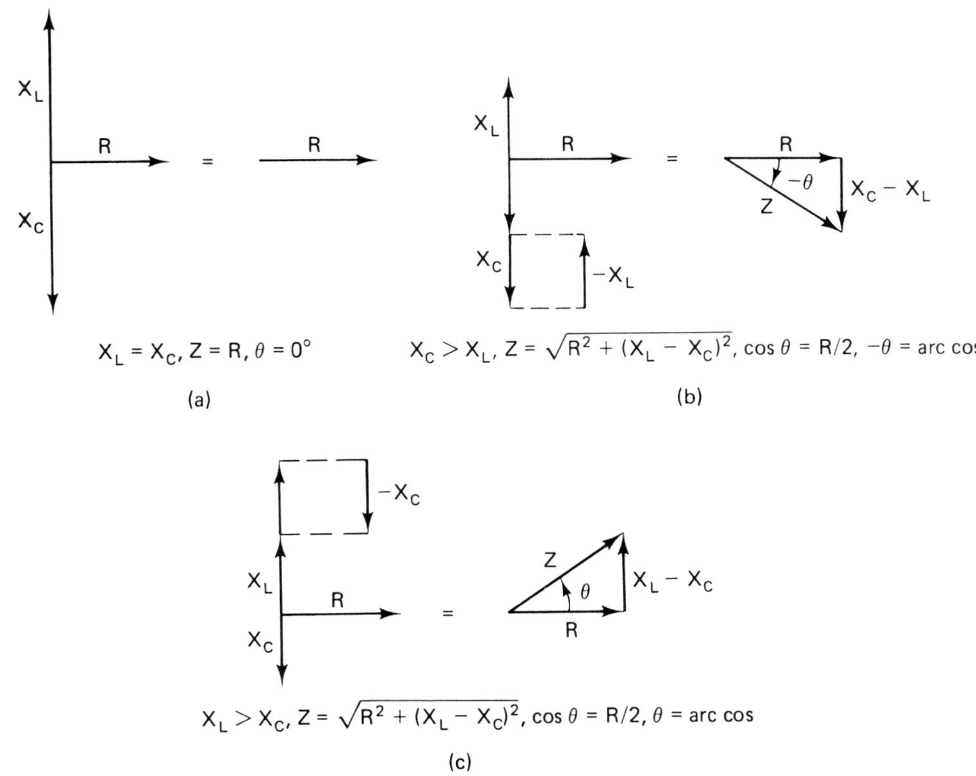

Figure 12–3 Impedance phasors for an *RLC* series circuit: (a) at resonance; (b) below resonance; (c) above resonance.

The phase angle is found thus:

$$\cos\theta = \frac{R}{Z} = \frac{1\ \text{k}\Omega}{1.414\ \text{k}\Omega} = 0.707$$

$$-\theta = \arccos 0.707 = -45°$$

If the applied voltage is 12 V, the current in the circuit can be calculated with the formula

$$i_T = \frac{V_T}{Z} = \frac{12\ \text{V}}{1.414\ \text{k}\Omega} = \frac{12}{1.414 \times 10^3} = 8.49 \times 10^{-3} = 8.49\ \text{mA}$$

The power consumed in the circuit can be found by the formula

$$P_T = V_T i_T \cos\theta = 12\ \text{V} \times 8.49\ \text{mA} \times 0.707 = 72.03 \times 10^{-3} = 72.03\ \text{mW}$$

When X_L is greater than X_C, X_C is subtracted from X_L, as shown in Figure 12-3c. For example, if $X_L = 3$ kΩ, $X_C = 1.5$ kΩ, and $R = 1$ kΩ, then

$$
\begin{aligned}
Z &= \sqrt{R^2 + (X_L - X_C)} \\
&= \sqrt{1\text{ k}\Omega + (3\text{ k}\Omega - 1.5\text{ k}\Omega)^2} \\
&= \sqrt{(1 \times 10^3)^2 + (1.5 \times 10^3)^2} \\
&= \sqrt{1 \times 10^6 + 2.25 \times 10^6} \\
&= \sqrt{3.25 \times 10^6} \\
&= 1.8 \times 10^3 \\
&= 1.8\text{ k}\Omega
\end{aligned}
$$

The phase angle is found thus:

$$
\begin{aligned}
\cos\theta &= \frac{R}{Z} \\
&= \frac{1\text{ k}\Omega}{1.8\text{ k}\Omega} \\
&= 0.555
\end{aligned}
\qquad
\begin{aligned}
\theta &= \text{arccos} \\
&= 0.555 \\
&= 56.3\,^\circ
\end{aligned}
$$

If the voltage applied to the circuit is 12 V, the total current of the circuit can now be found using the formula

$$
\begin{aligned}
i_T &= \frac{V_T}{Z} \\
&= \frac{12\text{ V}}{1.8\text{ k}\Omega} \\
&= \frac{12}{1.8 \times 10^3} \\
&= 6.67 \times 10^{-3} \\
&= 6.67\text{ mA}
\end{aligned}
$$

The power is found using the formula

$$
\begin{aligned}
P_T &= V_T i_T \cos\theta \\
&= 12\text{ V} \times 6.67\text{ mA} \times 0.555 \\
&= 12 \times 6.67 \times 10^{-3} \times 0.555 \\
&= 44.42 \times 10^{-3} \\
&= 44.42\text{ mW}
\end{aligned}
$$

12-1b.2 Effects of Series Resonance

There are a few important points to remember about an *RLC* series circuit when the applied frequency is the resonant frequency. When the circuit is at resonance:

1. $X_L = X_C$.
2. The total impedance will be minimum and is equal to the resistance of the circuit.
3. The current in the circuit will be maximum.
4. The phase angle is equal to 0°.

Also remember:

1. If the applied frequency decreases below the resonant frequency, the circuit becomes capacitive.
2. If the applied frequency increases above the resonant frequency, the circuit becomes inductive.

12-1c *RLC* PARALLEL CIRCUIT

When the resistor, inductor, and capacitor are placed in parallel the reactances have a different effect on the circuit's impedance and current and must be calculated differently.

12-1c.1 Current Phasors in a Parallel Circuit

In an *RLC* parallel circuit as shown in Figure 12-4, the voltage is in phase across all the components. The circuit is analyzed from the standpoint of the various currents. These currents are calculated from formulas similar to the ones used in other reactive circuits. For example, if $V_T = 12$ V, $X_L = 2$ kΩ, and $X_C = 1$ kΩ, then

$$i_R = \frac{V_T}{R} = \frac{12\text{ V}}{1\text{ k}\Omega} = \frac{12}{1 \times 10^3} = 12 \times 10^{-3} = 12\text{ mA}$$

$$i_L = \frac{V_T}{X_L} = \frac{12\text{ V}}{2\text{ k}\Omega} = \frac{12}{2 \times 10^3} = 6 \times 10^{-3} = 6\text{ mA}$$

$$i_C = \frac{V_T}{X_C} = \frac{12\text{ V}}{1\text{ k}\Omega} = \frac{12}{1 \times 10^3} = 12 \times 10^{-3} = 12\text{ mA}$$

The phasor representation of these currents is shown in Figure 12-4b and c. The reactive currents i_L and i_C oppose each other; therefore, there is a canceling effect. In this case i_C is greater than i_L and the circuit is considered to be capacitive. The total current leads the applied voltage.

Figure 12–4 *RLC* parallel circuit: (a) schematic diagram; (b) current phasors; (c) resultant total current.

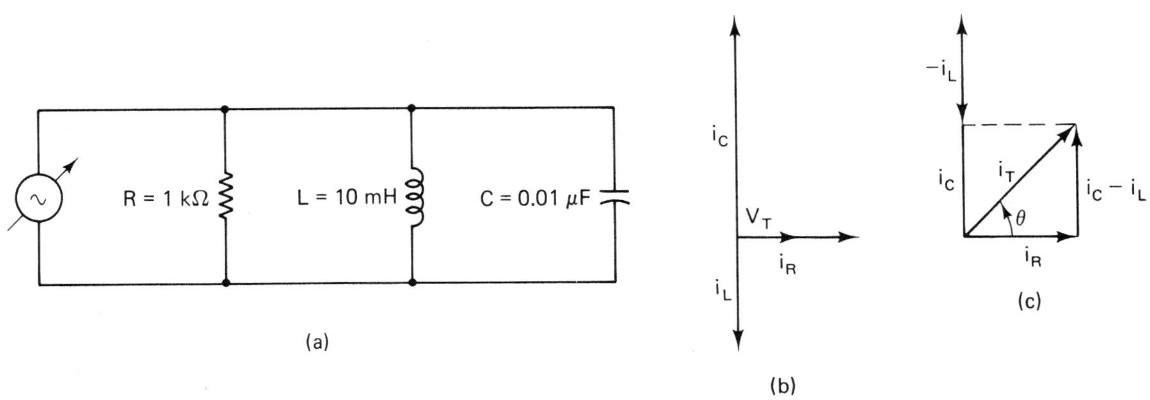

The total current is found by the formula

$$\begin{aligned} i_T &= \sqrt{i_R^2 + (i_L - i_C)^2} \\ &= \sqrt{(12\text{ mA})^2 + (6\text{ mA} - 12\text{ mA})^2} \\ &= \sqrt{(12 \times 10^{-3})^2 + (6 \times 10^{-3})^2} \\ &= \sqrt{144 \times 10^{-6} + 36 \times 10^{-6}} \\ &= \sqrt{180 \times 10^{-6}} \\ &= 13.42 \times 10^{-3} \\ &= 13.42\text{ mA} \end{aligned}$$

Now that the total current is known, the total impedance can be found, thus:

$$\begin{aligned} Z &= \frac{V_T}{i_T} \\ &= \frac{12\text{ V}}{13.42\text{ mA}} \\ &= \frac{12}{13.42 \times 10^{-3}} \\ &= 0.894 \times 10^{-3} \\ &= 894\ \Omega \end{aligned}$$

The phase angle of the circuit is found by dividing the current through the resistor by the remaining reactive current of $i_L - i_C$, which is the tangent of θ; therefore,

$$\begin{aligned} \tan\theta &= \frac{i_R}{i_L - i_C} \\ &= \frac{12\text{ mA}}{12\text{ mA} - 6\text{ mA}} \\ &= \frac{12\text{ mA}}{6\text{ mA}} \\ &= 2 \end{aligned}$$

and θ is found thus:

$$\begin{aligned} \theta &= \arctan 2 \\ &= 63.43° \end{aligned}$$

The power of the circuit can be found by the formula

$$\begin{aligned} P_T &= V_T i_T \cos\theta \\ &= 12\text{ V} \times 13.42\text{ mA} \times \cos 63.43° \\ &= 12 \times 13.42 \times 10^{-3} \times 0.44729 \\ &= 72.03 \times 10^{-3} \\ &= 72.03\text{ mW} \end{aligned}$$

If the current through the inductor is greater, the total current will lag the applied voltage and the circuit will be considered inductive. In this case the phase angle will be negative-going or in a counterclockwise direction.

When the circuit is at resonance, $X_L = X_C$. With this condition $i_L = i_C$ and they cancel each other. The total current will equal i_R. In other words, with respect to the signal generator, only the resistor is connected in the circuit. The current flowing between the inductor and the capacitor is at maximum because they give energy back to the circuit. The current being supplied by the generator is minimum because only the resistor appears in the circuit. This effect will become more apparent in Section 12-1c.2. The phase angle will be 0° and the total impedance is equal to the value of the resistor.

12-1c.2 The Tank Circuit

In numerous applications of electronics an inductor and a capacitor are placed in parallel. The effects of this type of circuit is used for filtering specific frequencies and for selecting desired frequencies used in radio and television. This circuit is referred to as a tank circuit. A *tank circuit* is a circuit with an inductor and capacitor in parallel that is capable of storing electrical energy over a band of frequencies where the maximum concentration of energy is at the resonant frequency. It is important for the technician to have a basic understanding of the action of a tank circuit. Figure 12-5 shows the action of a tank circuit.

Initially, the tank circuit is connected to a dc voltage source and the capacitor charges up to maximum voltage with a positive charge on the top plate and an excess of electrons on the bottom plate (Figure 12-5a). The switch S_1 is opened and the electrons stored on the bottom plate begin to flow through the inductor toward the top plate of the capacitor (Figure 12-5b). When the electrons flow through the inductor, a flux field is built up around it. As the capacitor reaches its normal discharge

Figure 12–5 Tank circuit: (a) voltage applied; (b) voltage removed and *C* discharges; (c) *C* charged in opposite direction; (d) discharge in other direction; (e) charged in original direction; (f) voltage decay.

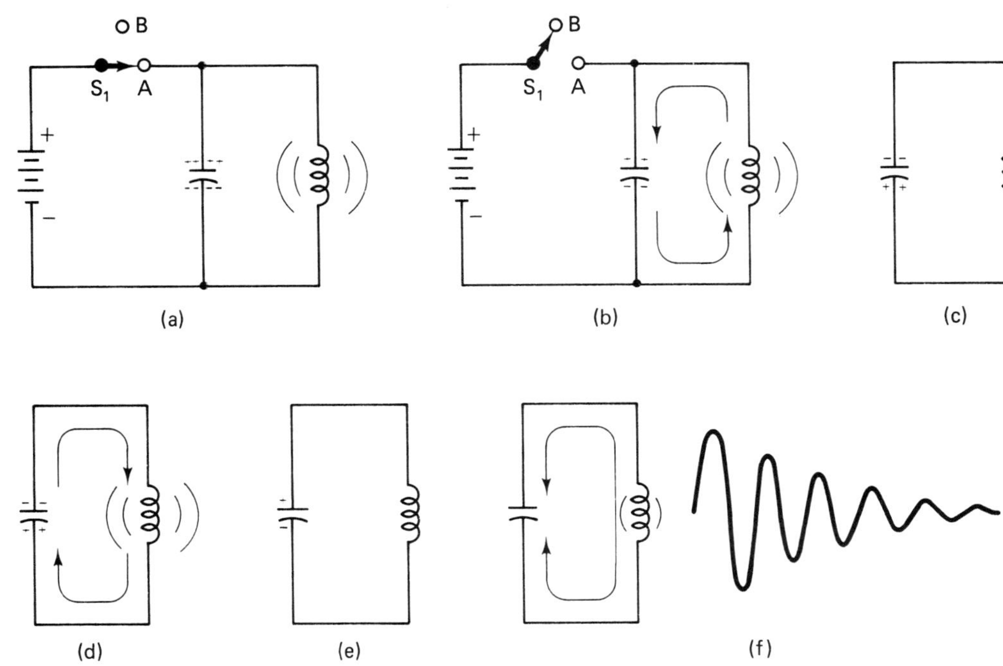

level, the flux field about the inductor collapses and forces more electrons into the circuit in the same direction, which causes an excess of electrons to be stored on the top plate of the capacitor (Figure 12-5c). Now the capacitor discharges in the opposite direction in an attempt to neutralize the charge on it (Figure 12-5d). Again, a flux field is created about the inductor which collapses and forces more electrons to be stored on the bottom plate of the capacitor (Figure 12-5e). In an ideal situation this charging and discharging action would continue forever and could be called a "perpetual action machine." A sine-wave voltage waveform would be developed with the changing of the direction of the current in the circuit. However, there is resistance in the inductor windings and the wire leads of the capacitor, which consumes energy in the form of lost power. This causes the capacitor to be charged less and less with each change of direction of the current. The voltage waveform decreases to zero and is referred to as the *damping effect* (Figure 12-5f).

12-1c.3 Tuning Circuits

Figure 12-6 shows two examples of tank circuits used as tuning circuits. Radio and television stations transmit information into the air or atmosphere in the form of electromagnetic waves. Each station is assigned a particular operating frequency. The radio or television receiver set has tuning circuits which allow the reception of the transmitted information.

In Figure 12-6a, the radio-frequency electromagnetic waves transmitted by a radio or television station travel through the air and cut across the antenna of the receiver, which induces minute voltages in the primary winding of the transformer. These small voltages induce voltages in the secondary winding of the transformer. A variable capacitor is connected to the secondary winding to form a tank circuit. It is

Figure 12–6 Tuning circuits: (a) tuning a frequency from an antenna; (b) tuning for a specific frequency between two circuits.

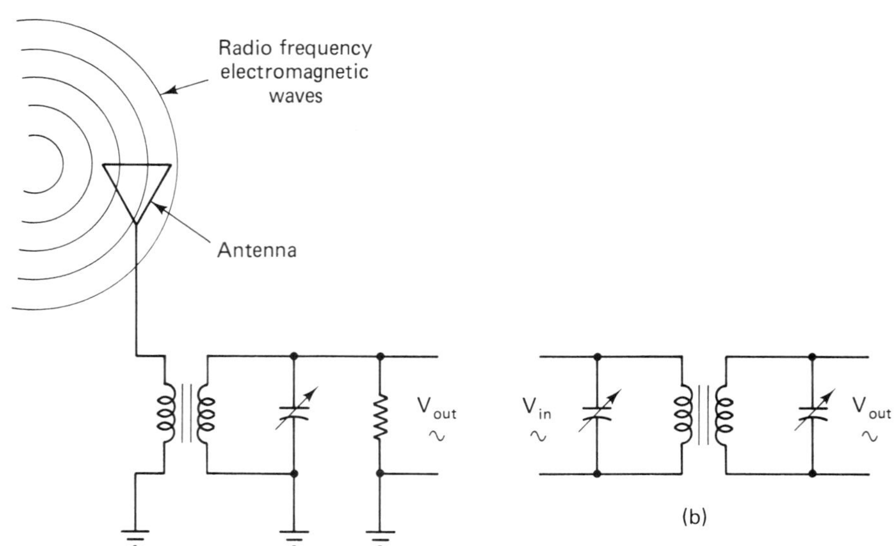

adjusted to change the resonant frequency of the circuit, thereby selecting or tuning in the desired station.

The signal of the tuned station is processed and amplified to provide video and audio information. It is necessary to transfer the maximum signal voltage from one processing circuit to another, and tuning circuits are used for this purpose, as shown in Figure 12-6b. When the variable capacitor of the primary winding is adjusted to the desired resonant frequency, maximum current flows in the primary tank circuit. This, in turn, induces a maximum voltage into the secondary winding, which is also tuned to the same resonant frequency.

12-1c.4 Effects of Parallel Resonance

There are a few important points to remember about an *RLC* parallel circuit when the applied frequency is the resonant frequency. When the circuit is at resonance:

1. $X_L = X_C$.
2. Maximum current flows between the inductor and the capacitor. Because these components give energy back to the circuit, no current is consumed and it appears as if the reactance is infinite. Impedance is maximum.
3. The only opposition to current flow from the standpoint of the generator is the resistor; therefore, current being supplied (also called line current) by the generator is minimum.
4. The phase angle is equal to 0°.

Also remember:

1. If the applied frequency decreases below the resonant frequency, X_L decreases, more current flows in this branch, and the circuit becomes inductive.
2. If the applied frequency increases above the resonant frequency, X_C decreases, more current flows in this branch, and the circuit becomes capacitive.

12-1d BANDWIDTH AND *Q* OF AN *RLC* CIRCUIT

Tuned circuits are used primarily as frequency-selection circuits or filters. Normally, a circuit is tuned to the desired resonant frequency for a maximum output. However, several frequencies above and below the resonant frequency will have sufficient amplitude to be passed by the circuit. This range of frequencies above and below the resonant frequency is called the *bandwidth.* Figure 12-7a shows an example of bandwidth.

The graph shows that V_{out} is plotted against frequency. The highest output occurs at the resonant frequency, which is designated as f_r. This may also be called the *center frequency.* The half-power points determine the upper- and lower-frequency limits. The lower frequency that occurs at the lower half-power point is designated f_L. The higher frequency that occurs at the higher half-power point is designated f_H.

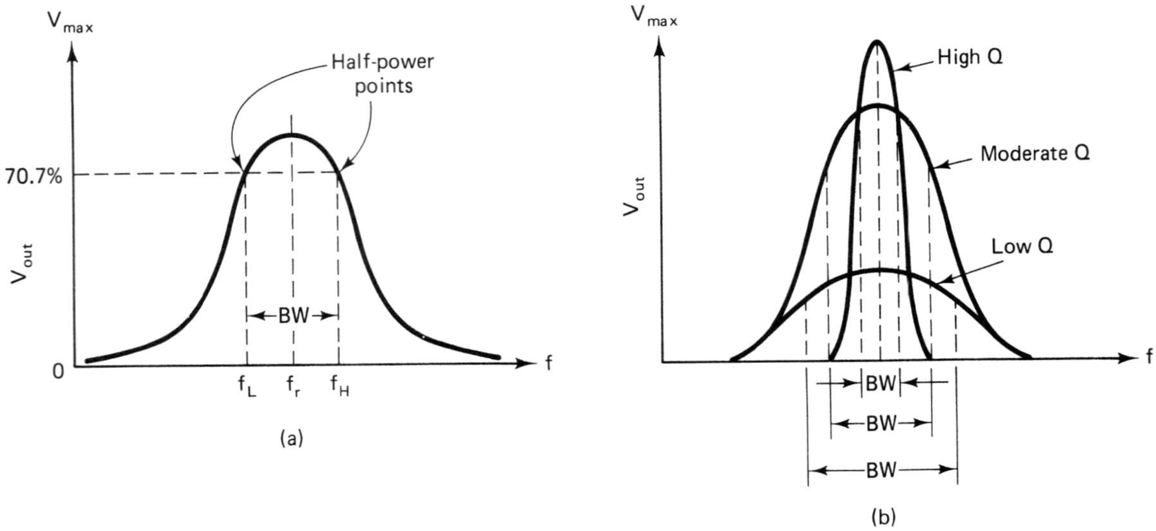

Figure 12–7 Bandpass curve: (a) typical bandpass curve; (b) bandwidth versus *Q*.

The bandwidth, designated BW, is the band of frequencies that extend from f_L to f_H. For example, if $f_L = 2$ kHz and $f_H = 6$ kHz, then

$$\begin{aligned} \text{BW} &= f_H - f_L \\ &= 6\text{ kHz} - 2\text{ kHz} \\ &= 4\text{ kHz} \end{aligned}$$

The *bandwidth* of a circuit can be more specifically defined as the range of frequencies that have an amplitude greater than $0.707V_{\text{max}}$ of the output voltage. For a tuned circuit these occur at the half-power points above and below the resonant frequency.

The bandwidth of a tuned circuit is affected by the amount of resistance* of the inductor. The relationship of the amount of X_L to resistance determines the quality of an inductor to be used in a tuned circuit. This quality affects the selectivity of the circuit and is referred to as the *Q factor,* figure of merit, or magnification factor. The *Q* factor is expressed as a formula, thus:

$$Q = \frac{X_L}{R}$$

For example, if $X_L = 10\text{ k}\Omega$ and $R = 100\ \Omega$, then

$$\begin{aligned} Q &= \frac{X_L}{R} \\ &= \frac{10\text{ k}\Omega}{100\ \Omega} \\ &= \frac{10 \times 10^3}{100} \\ &= 0.1 \times 10^3 \\ &= 100 \end{aligned}$$

**Note:* At very high frequencies the electrons in the wire coil are forced to travel near the surface. This is called the "skin effect" and presents a higher ac resistance.

The value of Q is only a figure of merit and has no units associated with it. The less ac resistance an inductor has, the greater will be its Q. An inductor with a high Q has more selectivity and the output from a circuit will be greater, as shown in Figure 12-7b. If an inductor has more ac resistance, its Q is less, the output of the circuit is less (which spreads the half-power points), and the bandwidth increases. This relationship of Q to bandwidth can be expressed as a formula:

$$BW = \frac{f_R}{Q}$$

For example, if f_R 100 kHz and $Q = 100$, then

$$\begin{aligned} BW &= \frac{f_R}{Q} \\ &= \frac{100\text{ kHz}}{100} \\ &= \frac{100 \times 10^3}{100} \\ &= 1 \times 10^3 \\ &= 1000\text{ Hz} \end{aligned}$$

The actual bandwidth would be 500 Hz above and below the resonant frequency, thus:

$$\begin{aligned} f_L &= f_R - \frac{BW}{2} \\ &= 100\text{ k}\Omega - \frac{1000}{2} \\ &= 100\text{ k}\Omega - 500 \\ &= 100\text{ k}\Omega - 0.5\text{ k}\Omega \\ &= 99.5\text{ kHz} \end{aligned} \qquad \begin{aligned} f_H &= f_R + \frac{BW}{2} \\ &= 100\text{ k}\Omega + \frac{1000}{2} \\ &= 100\text{ k}\Omega + 500 \\ &= 100\text{ k}\Omega + 0.5\text{ k}\Omega \\ &= 100.5\text{ kHz} \end{aligned}$$

The following examples show what occurs when an inductor with a Q of 10 is used in the same circuit:

$$\begin{aligned} BW &= \frac{f_R}{Q} \\ &= \frac{100\text{ kHz}}{10} \\ &= \frac{100 \times 10^3}{10} \\ &= 10 \times 10^3 \\ &= 10\text{ kHz} \end{aligned}$$

Therefore,

$$\begin{aligned} f_L &= f_R - \frac{BW}{2} \\ &= 100\text{ k}\Omega - \frac{10\text{ k}\Omega}{2} \\ &= 100\text{ k}\Omega - 5\text{ k}\Omega \\ &= 95\text{ kHz} \end{aligned} \qquad \begin{aligned} f_H &= f_R + \frac{BW}{2} \\ &= 100\text{ k}\Omega + \frac{10\text{ k}\Omega}{2} \\ &= 100\text{ k}\Omega + 5\text{ k}\Omega \\ &= 105\text{ kHz} \end{aligned}$$

By rearranging the bandwidth formula, f_R and Q can be found. For example,

$$f_R = Q \times \text{BW} = 10 \times 10\,\text{kHz} = 100\,\text{kHz} \qquad Q = \frac{f_R}{\text{BW}} = \frac{100\,\text{kHz}}{10\,\text{kHz}} = 10$$

So far we have been discussing the Q of a coil, but if resistance is added to a tuned circuit, the total Q of the circuit is reduced. Adding resistance to a tuned circuit may be done to increase the bandwidth of the circuit. The Q of a tuned circuit can never be less than the Q of the inductor.

The Q of a series resonant circuit can be found by measuring the applied voltage or input voltage and the voltage across the inductor or capacitor and solving for the formula

$$Q = \frac{V_L}{V_{\text{in}}} \quad \text{or} \quad \frac{V_C}{V_{\text{in}}}$$

For example, if $V_{\text{in}} = 5\ \text{V}_{\text{p-p}}$ and $V_L = 32\ \text{V}_{\text{p-p}}$, then

$$Q = \frac{V_L}{V_{\text{in}}} = \frac{32}{5} = 6.4$$

This is a typical voltage-gain formula and indicates the relationship of the Q of the circuit as a magnification factor. Similarly, if the Q of the circuit and the input voltage are known, you can solve for the voltage across the inductor or capacitor. For example, if $Q = 6.4$ and $V_{\text{in}} = 10\ \text{V}_{\text{p-p}}$, then

$$V_L \text{ or } V_C = QV_{\text{in}} = 6.4 \times 10\ \text{V}_{\text{p-p}} = 64\ \text{V}_{\text{p-p}}$$

The Q of a parallel resonant circuit is little more difficult to find on a practical basis, but can be found with the formula

$$Q = \frac{i_{\text{Tank}}}{i_{\text{in}}}$$

One important point must be made with a parallel resonant circuit. At resonance, the current in the tank circuit is maximum and circulates between the inductor and the capacitor; therefore, very little current is required from the line or source supplying the tank circuit. When the circuit is not at resonance, maximum line current flows. The current graph representing these conditions will have a dip or minimum value at resonance, the opposite of a series resonance circuit. You will be able to understand a more practical application of this action with the bandstop filter in the next section.

12-1e *RLC* FREQUENCY FILTERS

An ideal frequency filter output will go to zero voltage at the desired cutoff frequency. So far we have studied *RL* and *RC* low-pass and high-pass filters. These filters have a slope at their cutoff point, and the output voltage either falls off or rises gradually. When both reactive components, the inductor and the capacitor, are used in a filter, the cutoff frequency is more clearly defined and the slope of the rise or fall of the output is more steep. In this section we show improved low-pass and high-pass filters and present bandpass and bandstop filters.

12-1e.1 Improved Low-Pass Filter

An improved low-pass filter is shown in Figure 12-8a. The frequency response curve is shown in Figure 12-8b. An inductor is in series with a capacitor. The resistor across the capacitor develops the output voltage V_{out}. At low frequency X_L is small and most of the current can flow through R, which drops most of the input voltage. With this condition X_C is large and takes very little current away from R. At the half-power

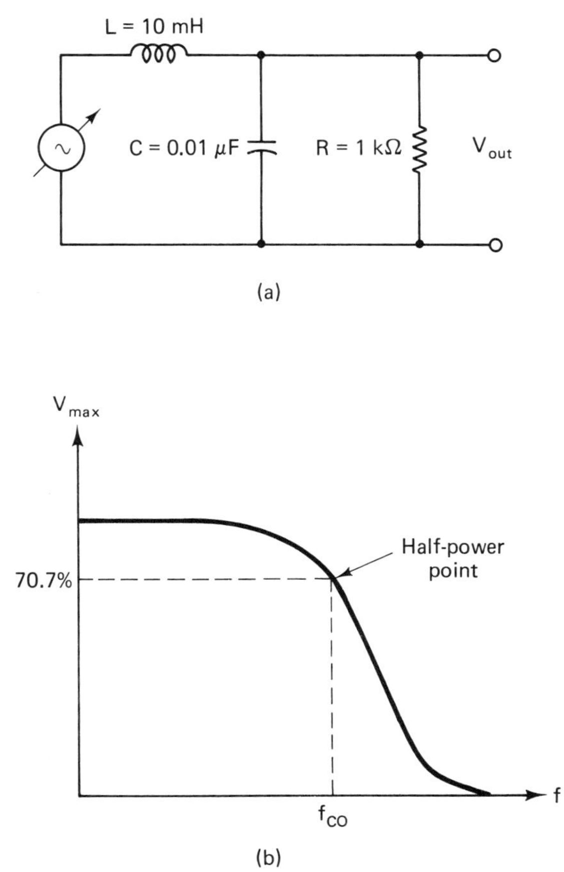

Figure 12–8 Improved low-pass filter: (a) schematic diagram; (b) frequency response curve.

point X_L begins to increase and drops some of the input voltage; therefore, less current flows through R and V_{out} begins to decrease. At the same time, X_C has decreased and some of the current flows through C. At high frequencies X_L is much larger and drops a large percentage of the input voltage. Any current that does flow through L will flow through C because X_C is less. The capacitor "shunts" the current away from the resistor and V_{out} will be practically zero.

12-1e.2 Improved High-Pass Filter

An improved high-pass filter is shown in Figure 12-9a. The frequency response curve is shown in Figure 12-9b. A capacitor is in series with an inductor. The output voltage V_{out} is developed by the resistor across the inductor. At low frequencies, X_C is very large and most of the input voltage is dropped across the capacitor. With this condition, X_L is very low and any current that passes C will flow through L, thereby bypassing R. The output voltage will be nearly zero. At the half-power point X_C has decreased to the point where more current flows toward L and R. Since the frequency has increased, X_L has increased, less current flows through L, and more current flows through R, which begins to develop an output voltage. At high frequencies X_C becomes very low and passes a larger current to L and R. X_L has increased to a point where most of the current will flow through R and develop a maximum V_{out}.

Figure 12–9 Improved high-pass filter: (a) schematic diagram: (b) frequency response curve.

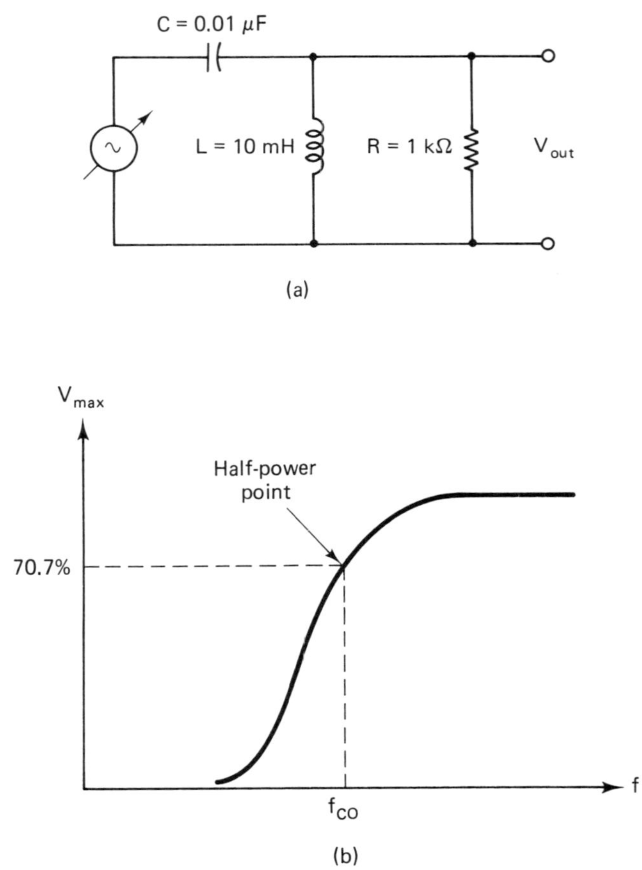

12-1e.3 Bandpass Filters

Bandpass filters pass only a narrow band of frequencies located on each side of the resonant frequency. Figure 12-10 shows two types of bandpass filters. A series bandpass filter circuit (Figure 12-10a) consists of L and C in series. The resistor is the load that develops the output voltage. At the resonant frequency $X_L = X_C$ and give energy back to the circuit. The current is maximum at this time and develops a maximum IR drop across R. When the input frequency decreases below f_R, X_C increases and reduces the current in the circuit. Less voltage is developed across R. At low frequencies X_C is so great that most of the input voltage is dropped across C and little or no current flows through R, which results in little or no V_{out}. When the input frequency increases above f_R, X_L increases and reduces the current in the circuit. Less voltage is developed across R. At high frequencies X_L is so large that most of the input voltage is dropped across L and little or no current flows through R, again resulting in little or no V_{out}.

A parallel bandpass filter circuit (Figure 12-10b) consists of C and L in parallel with the load resistor. At resonance the circulating current of the tank circuit of C and L is maximum and does not require any line current. The current from the generator flows through R and develops a maximum V_{out}. When the input frequency goes below f_R, X_L decreases and more current flows through L. Less current flows

Figure 12–10 Bandpass filter: (a) series resonant circuit; (b) parallel resonant circuit; (c) frequency response curve.

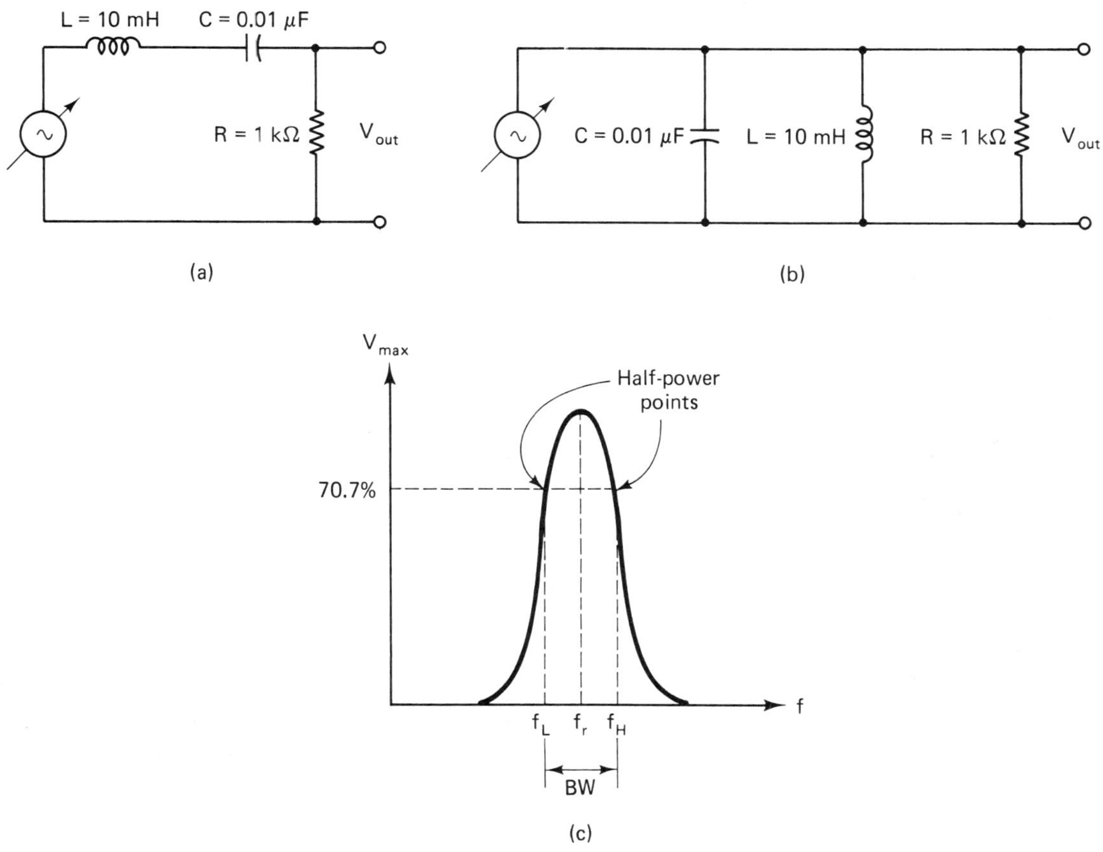

through R and V_{out} decreases. When the input frequency goes above f_R, X_C decreases and more current flows through C. Again, less current flows through R and V_{out} decreases. When the input frequency is low enough and high enough, X_L and X_C become so low that most of the current from the generator flows through them; therefore, V_{out} becomes zero. The frequency response curve for a bandpass filter circuit is shown in Figure 12-10c.

12-1e.4 Bandstop Filters

Bandstop filters pass all frequencies except a narrow band of frequencies located on each side of the resonant frequency. These types of filters may also be called *bandreject filters* or *notch filters.* Figure 12-11 shows two types of bandstop filters. A series bandstop filter (Figure 12-11a) consists of a tank circuit in series with the load resistor. At the resonant frequency $X_L = X_C$, which presents a high impedance in series with R. The line current is minimum and the voltage drop across R is minimum. When the input frequency decreases below f_R, X_L decreases and more line current is allowed to flow through R, which develops a higher

Figure 12–11 Bandstop filter: (a) series resonant circuit; (b) parallel resonant circuit; (c) frequency response curve.

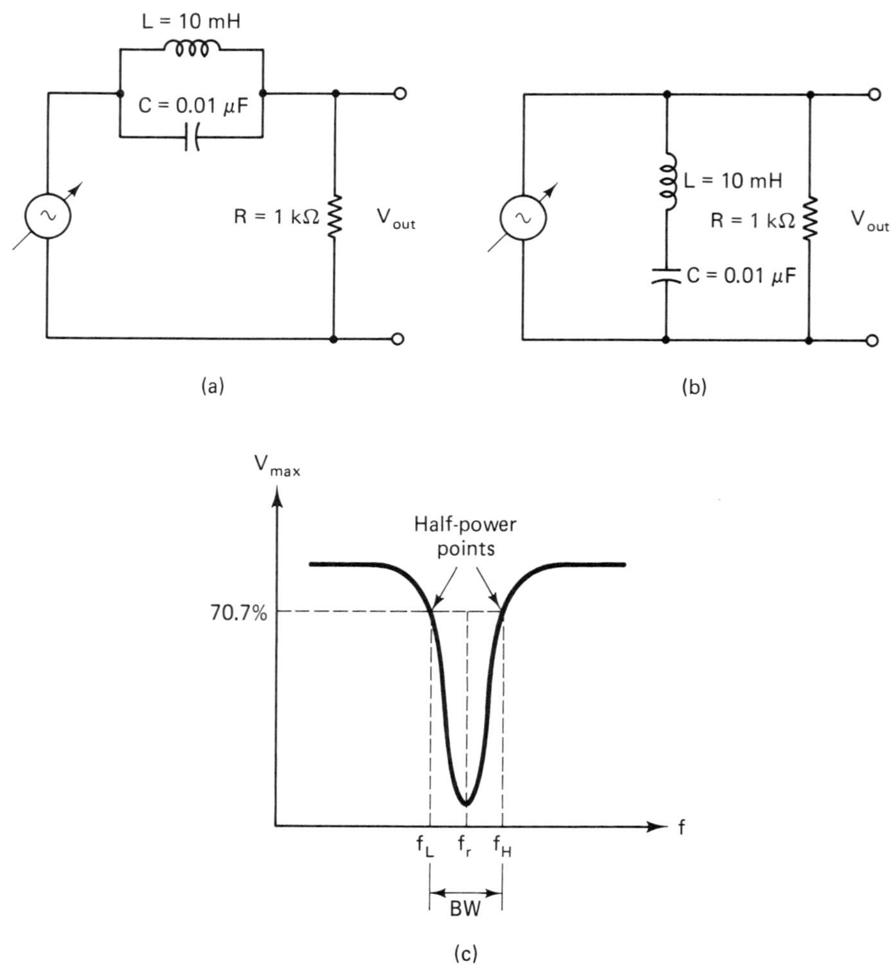

V_{out}. At low frequencies X_L becomes so low that most of the generator current can flow through R, developing a maximum V_{out}. When the input frequency increases above f_R, X_C decreases and allows more line current to flow through R. Again a higher V_{out} is developed. At high frequencies X_C becomes low, so that most of the generator current flows through R, which causes a maximum V_{out}.

A parallel bandstop filter (Figure 12-11b) has a series LC circuit in parallel with the load resistor. At resonance the LC series branch presents a low impedance to the generator and most of the line current flows through L and C, in effect, "shunting or shorting" the current away from R, resulting in a very low V_{out}. When the input frequency increases or decreases from f_R, the impedance of this branch increases and more current flows through R, creating a larger V_{out}. At extreme low and high frequencies the impedance of the LC branch is so high that nearly all of the line current flows through R, which develops a maximum V_{out}. The frequency response curve for a bandstop filter is shown in Figure 12-11c.

12-1e.5 Finding Half-Power Points and Bandwidth

In a practical filter circuit the actual values of the components may be within tolerance but be off sufficiently to create a noticeable error. The following procedures can be used to find the actual cutoff frequency or resonant frequency of a filter circuit.

Low-Pass Filter

1. Measure the maximum V_{out} well below the calculated f_{co}.
2. With the V_{out} being monitored, increase the input signal until the V_{out} drops to 70.7% of the maximum voltage. Now read the frequency on the input signal generator.

Example: Maximum V_{out} = 5 V at 1 kHz. At 3.5 V (5 V × 0.707), the generator indicates 20 kHz, which is the f_{co}.

High-Pass Filter

1. Measure the maximum V_{out} well above the calculated f_{co}.
2. With the V_{out} being monitored, decrease the input signal until the V_{out} drops to 70.7% of the maximum voltage. Now read the frequency on the generator.

Example: Maximum V_{out} = 8 V at 50 kHz. At 5.7 V (8 V × 0.707), the generator indicates 30 kHz, which is the f_{co}.

Bandpass Filter

1. Adjust the input signal generator for a maximum V_{out}. This is referred to as "peaking" the circuit. The frequency indicated by the signal generator is the f_r.
2. With the V_{out} being monitored, decrease the frequency of the generator until V_{out} drops to 70.7% of the maximum voltage. The generator now indicates f_L.

3. With the V_{out} still being monitored, increase the frequency of the generator until V_{out} again drops to 70.7% of the maximum voltage. The generator now indicates f_H. The bandwidth of frequencies passes are those between f_L and f_H.

Example: Maximum $V_{out} = 10$ V at 1 kHz. At 7 V (10×0.707), the generator indicates 600 Hz, which is the f_L. Also, at 7 V, the generator indicates 1.4 kHz, which is the f_H. The bandwidth is from 600 Hz to 1.4 kHz, which is 800 Hz wide.

Bandstop Filter

1. Adjust the signal generator for a minimum V_{out}. The frequency indicated by the generator is the f_r.
2. Decrease the frequency of the generator until V_{out} stabilizes at a maximum value.
3. Now increase the frequency of the generator until V_{out} drops to 70.7%. The frequency indicated by the generator is f_L.
4. Increase the frequency of the generator until V_{out} at a maximum value again.
5. Now decrease the frequency of the generator until V_{out} drops to 70.0%. The frequency indicated by the generator is f_H. The bandwidth of frequencies stopped or rejected are those between f_L and f_H.

Example: $V_{out} = 1$ V at 60 Hz, which is the f_r. At 10 V the generator indicates 30 Hz. At 7 V the generator indicates 50 Hz, which is the f_L. Again at 10 V the generator indicates 80 Hz. At 7 V the generator indicates 70 Hz, which is the f_H. The bandwidth is from 50 to 70 Hz, which is 20 Hz wide.

The point where V_{out} drops to 70.7% of maximum is also called a half-power point. It is also referred to as the -3-dB point. For example, if $V_{out} = 20$ V across a 1-kΩ load, the maximum P_{out} can be calculated thus:

$$P_{out} = \frac{V^2}{R} = \frac{20 \times 20}{1000} = 0.4 \text{ W}$$
$$= 400 \text{ mW}$$

The half-power point would be 400 mW/2 = 200 mW. Substituting these values in the decibel formula gives

$$\text{db} = 10 \log \frac{P_1}{P_2} = 10 \log \frac{200 \text{ mW}}{400 \text{ mW}} = 10 \log 0.5$$
$$= (\text{the log of 0.5 given by a calculator} = -0.30103); \text{therefore,}$$
$$= 10 \times -0.30103 = -3.0103$$
$$= -3$$

In other words, the half-power point occurs 3 dB down from maximum V_{out}.

12-1f SUMMARY OF *RLC* CIRCUIT FORMULAS

Series Circuit

$$f_R = \frac{1}{2\pi\sqrt{LC}} = \frac{0.159}{\sqrt{LC}}$$

$$Z = \sqrt{R^2 + (X_L - X_C)^2}$$

$$i_T = \frac{V_T}{Z}$$

$$i_R = i_L = i_C$$

$$V_R = i_T R$$

$$V_L = i_T X_L$$

$$V_C = i_T X_C$$

$$V_T = \sqrt{V_R^2 + (V_L - V_C)^2}$$

$$P_T = V_T i_T \cos\theta$$

$$\cos\theta = \frac{R}{Z}$$

$$\theta = \arccos$$

$$Q = \frac{X_L}{R} = \frac{V_L}{V_{\text{in}}} = \frac{V_C}{V_{\text{in}}} = \frac{i_{\text{Tan}} K}{i_{\text{in}}} = \frac{f_R}{\text{BW}}$$

$$\text{BW} = f_H - f_L$$

$$\text{BW} = \frac{f_r}{Q}$$

$$f_R = Q\,\text{BW}$$

$$f_L = f_R - \frac{\text{BW}}{2}$$

$$f_H = f_R + \frac{\text{BW}}{2}$$

Parallel Circuit

$$f_R = \frac{1}{2\pi\sqrt{LC}} = \frac{0.159}{\sqrt{LC}}$$

$$V_T = V_R = V_L = V_C$$

$$i_R = \frac{V_T}{R}$$

$$i_L = \frac{V_L}{X_L}$$

$$i_C = \frac{V_C}{X_C}$$

$$i_T = \sqrt{i_R^2 + (i_L - i_C)^2}$$

$$Z = \frac{V_T}{i_T}$$

$$\tan\theta = \frac{i_R}{i_L - i_C}$$

$$\theta = \arctan$$

$$P_T = V_T i_T \cos\theta$$

$$Q = \frac{X_L}{R} = \frac{V_L}{V_{\text{in}}} = \frac{V_C}{V_{\text{in}}} = \frac{i_{\text{Tan}} K}{i_{\text{in}}} = \frac{f_R}{\text{BW}}$$

$$\text{BW} = f_H - f_L$$

$$\text{BW} = \frac{f_r}{Q}$$

$$f_R = Q\,\text{BW}$$

$$f_L = f_R - \frac{\text{BW}}{2}$$

$$f_H = f_R + \frac{\text{BW}}{2}$$

SECTION 12-2 DEFINITION EXERCISES

Write a brief description for each of the following terms.

1. "ELI the ICE man" ______________________________

____

2. Resonance ______________________________

____

3. Resonant frequency (f_R) ______________________________

____

4. Tank circuit ______________________________

____

5. Tuned circuit ______________________________

____

6. Perpetual action machine ______________________________

____

7. Damping effect ______________________________

____

8. Bandwidth ______________________________

____

9. Center frequency ______________________________

____

10. Q of a coil ______________________________

____

11. Q of a circuit ______________________________

____

12. Magnification factor ______________________________

____

13. Bandpass filter ______________________________

____

14. Bandstop filter ______________________________

____

15. Bandreject filter ____________________

____

16. Notch filter ____________________

____

SECTION 12-3 EXERCISES AND PROBLEMS

Complete this section before beginning the next section.

1. Solve the following problems for the resonant frequency.

a. $L = 2$ H
$C = 1\ \mu$F

$f_R =$ __________

b. $L = 10$ mH
$C = 0.1\ \mu$F

$f_R =$ __________

c. $L = 30$ mH
$C = 0.002\ \mu$F

$f_R =$ __________

2. Draw the circuit shown in Figure 12-2 in the space provided.

Let $f_{in} = 15.9$ kHz. Solve for the following:

$X_L =$ __________ $X_C =$ __________ $Z =$ __________

$i_T =$ __________ $V_R =$ __________ $V_L =$ __________

$V_C =$ __________ $\theta =$ __________ $P_T =$ __________

3. Referring to Question 2, solve for the following when f_{in} changes to 10 kHz.

$X_L =$ __________ $X_C =$ __________ $Z =$ __________

$i_T =$ __________ $V_R =$ __________ $V_L =$ __________

$V_C =$ __________ $\theta =$ __________ $P_T =$ __________

Is this circuit more capacitive or more inductive? __________

4. Referring to Question 2, solve for the following when F_{in} changes to 25 kHz.

X_L = ________ X_C = ________ Z = ________

i_T = ________ V_R = ________ V_L = ________

V_C = ________ θ = ________ P_T = ________

Is this circuit more capacitive or more inductive? ________

5. List the effects of series resonance.

a. **b.**

c. **d.**

6. Draw the circuit shown in Figure 12-4 in the space provided.

Let f_{in} = 15.9 kHz at 12 $V_{p\text{-}p}$. Solve for the following:

X_L = ________ X_C = ________ i_R = ________

i_L = ________ i_C = ________ i_T = ________

Z = ________ θ = ________ P_T = ________

7. Referring to Question 6, solve for the following when f_{in} changes to 10 kHz.

X_L = ________ X_C = ________ i_R = ________

i_L = ________ i_C = ________ i_T = ________

Z = ________ θ = ________ P_T = ________

Is this circuit more capacitive or more inductive? ________

8. Referring to Question 6, solve for the following when f_{in} changes to 25 kHz.

X_L = ________ X_C = ________ i_R = ________

i_L = ________ i_C = ________ i_T = ________

Z = ________ θ = ________ P_T = ________

Is this circuit more capacitive or more inductive? ________

9. List the effects of parallel resonance.

a. **b.**

c. **d.**

10. Solve for the bandwidth for the data given.

a. $f_H = 20$ kHz, $f_L = 10$ kHz — BW = ____________

b. $f_H = 33$ kHz, $f_L = 27$ kHz — BW = ____________

c. $f_H = 12.5$ kHz, $f_L = 10.25$ kHz — BW = ____________

11. Solve for the Q for the data given.

a. $X_L = 5$ kΩ, $R = 100$ Ω — Q = ____________

b. $X_L = 10$ kΩ, $R = 133$ Ω — Q = ____________

c. $X_L = 2.5$ kΩ, $R = 75$ Ω — Q = ____________

12. Find the unknown values for the data given.

a. $f_R = 2$ kHz, $Q = 5$ — BW = ____________

b. $f_R = 30$ kHz, $Q = 120$ — BW = ____________

c. $BW = 5$ kHz, $Q = 50$ — f_R = ____________

d. $f_R = 50$ kHz, BW = 5 kHz — Q = ____________

e. $f_R = 1$ kHz, BW = 60 Hz — f_L = ____________

f. $f_R = 16$ kHz, BW = 400 Hz — f_H = ____________

13. Give a brief description of the following filters.

a. Low-pass:

b. High-pass:

c. Bandpass:

d. Bandstop:

14. Indicate which factor determines the bandwidth of a tuned circuit.

____ **a.** inductance ____ **b.** capacitance ____ **c.** resistance

15. Match the description in column A with the correct term in column B.

Column A

____ **a.** A condition in a circuit where $X_L = X_C$

____ **b.** A circuit where an inductor is connected in parallel with a capacitor

____ **c.** An *RLC* circuit condition where Z = minimum

____ **d.** The quality of an inductor or circuit

____ **e.** A circuit that passes a narrow band of frequencies

____ **f.** A circuit condition where i_T = minimum

____ **g.** A circuit that blocks a narrow band of frequencies

____ **h.** Another name for f_R

Column B

1. tank circuit
2. series resonance
3. Q
4. parallel resonance
5. resonance
6. notch filter
7. center frequency
8. bandpass filter

SECTION 12-4 EXPERIMENTS

EXPERIMENT 1. SERIES RESONANCE

Objective: To demonstrate how to adjust an *RLC* series circuit for the resonant condition.

Introduction: A voltmeter is placed across the resistor in an *RLC* series circuit to monitor the *IR* drop. The signal generator is adjusted until a maximum *IR* drop is indicated. The resonant frequency is then read from the generator.

Materials Needed:
1 100-Ω resistor at 0.5 W
1 10-mH inductor
1 0.01-μF capacitor at 25 wvdc

Equipment Required:
1 Audio signal generator
1 Multimeter or DVM
1 Breadboard with leads or wires for constructing circuit

Procedure:

1. In the space below, draw the schematic diagram in Figure 12-2.

2. Construct the circuit that you have drawn.
3. Adjust the generator until there is a maximum voltage reading across R.
4. List the voltage drop across each component here: V_R = ____________, V_L = ____________, V_C = ____________, V_T = ____________. The circuit is at resonance, $X_L = X_C$, Z = minimum, and i_T is maximum. (Notice that V_L and V_C are larger than V_T.)
5. Adjust the generator to 10 kHz. Measure the voltage drop across each component and list them here: V_R = ____________, V_L = ____________, V_C = ____________. The voltage drops are directly proportional to the resistance and reactance. Is this circuit more capacitive or more inductive? ____________
6. Adjust the generator to 30 kHz. Measure the voltage drop across each component and list them here: V_R = ____________, V_L = ____________, V_C = ____________. Is the circuit now more capacitive or more inductive? ____________
7. Disconnect the circuit.

Fill-In Questions:

1. When an *RLC* series circuit is at resonance: the impedance is ____________, the current in the circuit is ____________, and the voltage across L and C is greater than ____________.
2. When the applied frequency is below resonance, X_C increases and the circuit is more ____________.
3. When the applied frequency is above resonance, X_L increases and the circuit is more ____________.

EXPERIMENT 2. PARALLEL RESONANCE

Objective: To show how to adjust an *RLC* parallel circuit for the resonant condition.

Introduction: A voltmeter is placed across the series resistor in a parallel *RLC* circuit to monitor the *IR* drop. The signal generator is adjusted until a minimum *IR* drop is indicated. The resonant frequency is then read from the generator.

Materials Needed:
1 100-Ω resistor at 0.5 W
1 10-mH inductor
1 0.01-μF capacitor at 25 WV dc

Equipment Required:
1 Audio signal generator
1 Multimeter or DVM
1 Breadboard with wires or leads for constructing circuit

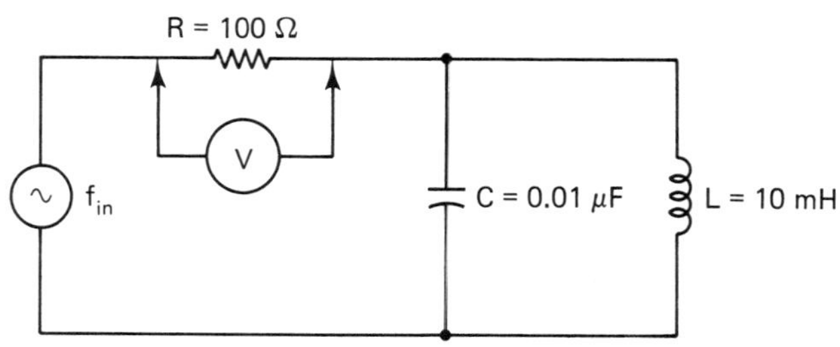

Figure 12–12

Procedure:

1. Construct the circuit shown in Figure 12-12.
2. Adjust the signal generator for a minimum voltage drop across *R*. List V_R here: ____________ .
3. Measure and list the following voltages here: V_L = ____________, V_C = ____________, V_T = ____________ . The circuit is at resonance, $X_L = X_C$, Z = maximum, and i_T = minimum. The voltage drops across *L* and $C = V_T$.
4. Adjust the generator for 10 kHz. Measure and record the following voltages here: V_R = ____________, V_L = ____________, V_C = ____________ . The circuit is now more ____________ .
5. Adjust the generator for 30 kHz. Measure and record the following voltages here: V_R = ____________, V_L = ____________, V_C = ____________ . The circuit is now more ____________ .
6. Disconnect the circuit.

Fill-In Questions:

1. When an *RLC* parallel circuit is at resonance: the impedance is ____________, the current is ____________, and V_L and V_C equals ____________.

2. When the applied frequency is below resonance, X_C increases and the circuit is more ____________.

3. When the applied frequency is above resonance, X_L increases and the circuit is more ____________.

EXPERIMENT 3. FILTER CIRCUITS

Objective: To demonstrate how tuned *RLC* circuits can be used as frequency filters.

Introduction: You will construct each circuit and apply several frequencies to them to show their characteristics. The circuits used in this experiment will be a low-pass filter, a high-pass filter, a bandpass filter, and a bandstop filter.

Materials Needed:

2 1-kΩ resistors at 0.5 W
1 10-mH inductor
1 0.01-μF capacitor at 25 WV dc
1 Breadboard with leads or wire for constructing circuits

Equipment Required:

1 Audio signal generator
1 Multimeter, DVM, or oscilloscope

Procedure:

Low-pass Filter

1. In the space below, draw the filter circuit shown in Figure 12-8a.

2. Set the generator for an output of 10 $V_{p\text{-}p}$.

3. Adjust the generator for each of the frequencies given and measure and record V_{out}.

f_{in}	V_{out}	f_{in}	V_{out}
100 Hz	______	20 kHz	______
500 Hz	______	30 kHz	______
1 kHz	______	50 kHz	______
10 kHz	______	100 kHz	______
15 kHz	______		______

4. Make a rough sketch of the frequency response curve for this circuit, indicating f_{co}, in the space below (refer to Figure 12-8b).

High-pass Filter

Procedure:

1. Draw the filter circuit shown in Figure 12-9a in the space below.

2. Set the generator for an output of 10 $V_{p\text{-}p}$.

3. Adjust the generator for each of the frequencies listed and measure and record V_{out}.

f_{in}	V_{out}	f_{in}	V_{out}
100 Hz	______	20 kHz	______
500 Hz	______	30 kHz	______
1 kHz	______	50 kHz	______
10 kHz	______	100 kHz	______
15 kHz	______		______

4. Make a rough sketch of the frequency response curve for this circuit, indicating f_{co}, in the space below (refer to Figure 12-9b).

Bandpass Filter

Procedure:

1. Draw the filter circuit shown in Figure 12-10a in the space below.

2. Set the generator for an output of 10 $V_{p\text{-}p}$.

3. Adjust the generator for each of the frequencies given and measure and record V_{out}.

f_{in} (kHz)	V_{out}	f_{in} (kHz)	V_{out}
1	______	25	______
5	______	30	______
10	______	35	______
15	______	50	______
20	______	100	______

4. Make a rough sketch of the frequency response curve for this circuit, indicting f_R, in the space below (refer to Figure 12-10b).

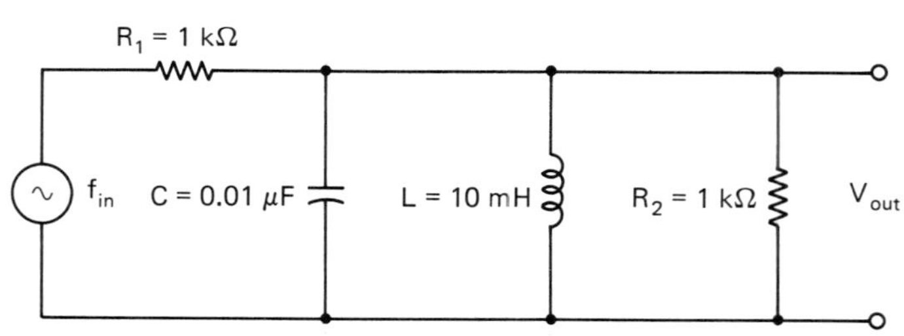

Figure 12–13

5. Construct the circuit shown in Figure 12-13.
6. Set the generator for an output of 10 $V_{p\text{-}p}$.
7. Adjust the generator for each of the frequencies given and measure and record V_{out}. Maximum V_{out} for this circuit will be +5 V.

f_{in} (kHz)	V_{out}	f_{in} (kHz)	V_{out}
1	______	25	______
5	______	30	______
10	______	35	______
15	______	50	______
20	______	100	______

8. Make a rough sketch of the frequency response curve for this circuit, indicating f_r, in the space provided (refer to Figure 12-10b).

Bandstop Filter

Procedure:

1. Draw the filter circuit shown in Figure 12-11a in the space provided.
2. Set the generator for an output of 10 $V_{p\text{-}p}$.
3. Adjust the generator for each of the frequencies given and measure and record V_{out}.

f_{in} (kHz)	V_{out}	f_{in} (kHz)	V_{out}
1	______	25	______
5	______	30	______
10	______	35	______
15	______	50	______
20	______	100	______

4. Make a rough sketch of the frequency response curve for this circuit, indicating f_R, in the space provided (see Figure 12-11b).

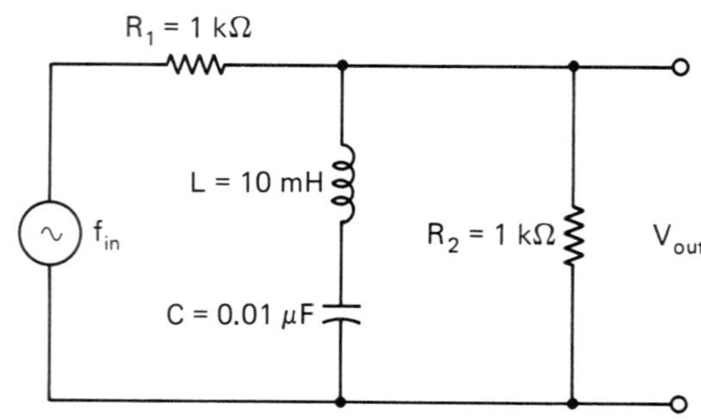

Figure 12–14

5. Construct the circuit shown in Figure 12-14.
6. Set the generator for an output of 10 $V_{p\text{-}p}$.
7. Adjust the generator for each of the frequencies given and measure and record V_{out}.

f_{in} (kHz)	V_{out}	f_{in} (kHz)	V_{out}
1	______	25	______
5	______	30	______
10	______	35	______
15	______	50	______
20	______	100	______

8. Make a rough sketch of the frequency response curve for this circuit, indicating f_R, in the space provided (refer to Figure 12-11b).

Fill-In Questions:

1. A circuit that passes frequencies below the f_{co} is called a ______________ filter.
2. A circuit that passes frequencies above the f_{co} is called a ______________ filter.
3. A ______________ filter passes only a narrow group of frequencies.

4. A ____________ filter passes all frequencies but rejects only a narrow group of frequencies.

SECTION 12-5 TROUBLESHOOTING APPLICATION: TESTING AN *RLC* SERIES CIRCUIT

Testing known voltages in a circuit is probably the easiest method for determining a problem and its cause. With tuned circuits, when a component changes value, the tuned or balanced condition is upset and the voltages and current will change. Any change in L and C will affect the resonant frequency. From the given formula,

$$f_R = \frac{1}{2\pi\sqrt{LC}}$$

if L or C increases, the f_R will be lower, and similarly, if L or C decreases, the f_R will be higher. When a circuit with reactive components is not tuned to its resonant frequency, the impedance will be more than minimum.

In an RLC series circuit, $X_L = X_C$, the Z is minimum, and the current is maximum. Any change in a component's value will increase Z and cause i_T to decrease. The voltage drops across each component will usually decrease.

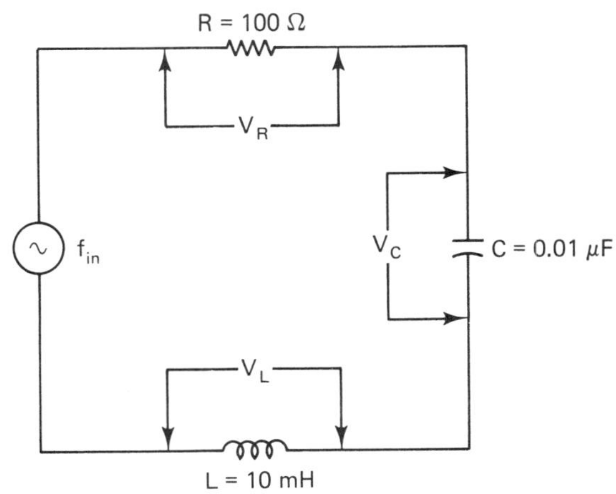

Figure 12–15 Troubleshooting an *RLC* series circuit.

Construct the circuit shown in Figure 12-15. Place the voltmeter across R and adjust the signal generator for maximum voltage. The circuit is now tuned. Place the voltmeter across the generator terminals and set the amplitude for 10 $V_{p\text{-}p}$. Using the following table as a guide, introduce the circuit condition, measure the indicated voltages, and compare your results with the V_R, V_L, V_C, and comment columns. To decrease C, place another 0.01-μF capacitor in series with the existing capacitor in the circuit. To increase C, place the two capacitors in parallel. To decrease L, place another inductor in series with the inductor in the circuit. To increase L, place the two inductors in series.

Circuit Condition	V_R ($V_{p\text{-}p}$)	V_C ($V_{p\text{-}p}$)	V_L ($V_{p\text{-}p}$)	*Comment*
Normal at f_R	3	37	46	Normal voltages
Open in circuit	0	0	0	No i_T, no IR drops
Short or decrease in R	Decrease	Increase	Increase	Increase in i_T and IR drops
Short or decrease in C	Decrease	Decrease	Decrease	Decrease in i_T and IR drops
Increase in C	Decrease	Decrease	Decrease	Same as above
Decrease in L	Decrease	Decrease	Decrease	Same as above
Increase in L	Decrease	Decrease	Increase	Decrease in i_T, but increase in X_L and maybe increase in V_L

SECTION 12-6 INSTANT REVIEW

- At resonance $X_L = X_C$.
- Below resonance X_C increases and X_L decreases.
- Above resonance X_L increases and X_C decreases.
- The formula for f_R:

$$f_R = \frac{1}{2\pi\sqrt{LC}}$$

- If L or C increases, f_R decreases.
- If L or C decreases, f_R increases.
- In a series resonant circuit:

$$\begin{aligned}&Z = \text{minimum and} = R\\ &i_T = \text{maximum}\\ &\text{phase angle } \theta = 0\\ &\text{power factor} = 1\end{aligned}$$

- Below resonance the circuit appears capacitive. Above resonance the circuit appears inductive.
- In a parallel resonant circuit:

$$\begin{aligned}&Z = \text{maximum}\\ &i_T = \text{minimum}\\ &i_{\text{tank}} = \text{maximum}\\ &\text{phase angle } \theta = 0\\ &\text{power factor} = 1\end{aligned}$$

- Below resonance the circuit appears inductive. Above resonance the circuit appears capacitive.
- Phasors represent phase relationships of R, X_L, and X_C.
- Phasors represent phase relationships of current and voltage.
- A tank circuit is an inductor and capacitor in parallel.

- Resonant circuits are referred to as tuned circuits.
- Tuning circuits are used to select radio and TV stations.
- Bandwidth is the group of frequencies located between the half-power points (i.e., BW $= f_H - f_L$).
- The Q or quality of an inductor is related to its dc resistance and expressed as $Q = X_L/R$.
- A high Q of a coil or circuit means the dc resistance is very low. Similarly, a low Q means high resistance.
- A circuit with a high Q has a narrow bandwidth.
- A circuit with a low Q has a wide bandwith.
- A low-pass filter passes all frequencies below the frequency cutoff point (half-power point) and blocks all frequencies above this point.
- A high-pass filter passes all frequencies above the frequency cutoff point and blocks all frequencies below this point.
- Capacitors and inductors in the same filter circuit create a steeper falloff or rise to the slope of the response curve.
- A bandpass filter passes only a very narrow group of frequencies centered around the resonant frequency.
- A bandstop filter passes all frequencies except a narrow group of frequencies centered around the resonant frequency.
- A bandstop filter may also be called a bandreject or notch filter.
- The resonant frequency is also called the center frequency.

SECTION 12-7 SELF-CHECKING QUIZ

Circle the most correct answer for each question.

1. A circuit is said to be resonant when:

a. $X_L > X_C$ **b.** $X_L < X_C$
c. $X_L = X_C$ **d.** none of the above

2. If the voltage and current of a circuit are in phase, the circuit is:

a. more inductive
b. purely resistive
c. more capacitive
d. all of the above

3. In a tuned circuit, L = 25 mH and C = 0.003 μF. The resonant frequency of the circuit is approximately:

a. 1.93 MHz **b.** 17.5 kHz
c. 6.36 kHz **d.** 1.93 GHz

4. If the frequency increases above f_R, X_C:

a. increases **b.** decreases
c. remains the same **d.** becomes zero

5. In an RLC series circuit, R = 1500 Ω, L = 20 mH, C = 0.01 μF, and f_R = 30 kHz. The impedance of the circuit is about:

a. 3.57 kΩ **b.** 4.77 kΩ
c. 758 Ω **d.** none of the above

6. If the applied voltage to the circuit in Question 5 is 24 V_{rms}, the i_T is approximately:

a. 12.6 mA **b.** 32 mA
c. 16.4 mA **d.** 6.7 mA

7. The true power for the circuit given in Questions 5 and 6 is about:

a. 160 mW b. 305 mW
c. 68 mW d. 148 mW

8. When an *RLC* parallel circuit is at resonance:

a. impedance is maximum
b. line current is maximum
c. the circuit appears capacitive
d. all of the above

9. In a tank circuit:

a. only resistance is present
b. energy is transferred back and forth between *C* and *L*
c. the current is minimum
d. all of the above

10. In an *RLC* parallel circuit, $R = 1\ \text{k}\Omega$, $L = 5$ mH, $C = 0.005\ \mu$F, and $f_{\text{in}} = 60$ kHz at 12 rms. The i_T of the circuit is equal to approximately:

a. 10 mA b. 16 mA
c. 20 mA d. 41 mA

11. The true power for the circuit in Question 10 is:

a. 240 mW b. 144 mW
c. 195 mW d. 127 mW

12. If the frequency increases in an *RLC* parallel resonant circuit, the circuit will appear:

a. more capacitive
b. more inductive
c. less resistive
d. balanced

13. An inductor with 50 mH and 75 Ω dc resistance has a frequency of 20 kHz applied to it. The *Q* of the coil is:

a. 106 b. 56
c. 84 d. 34

14. A circuit with an f_R of 60 kHz has a *Q* of 50. This circuit's bandwidth is:

a. 3000 b. 1200
c. 600 d. 300

15. The f_H or upper-limit frequency of the bandwidth for the circuit in Question 14 is:

a. 59.4 kHz b. 61.2 kHz
c. 60.15 kHz d. 60.6 kHz

16. If you wanted to block frequencies up to 20 kHz from a circuit, you could use a:

a. bandstop filter
b. low-pass filter
c. high-pass filter
d. bandpass filter

17. A bandpass filter uses a tuned circuit.

a. True b. False

18. If resistance is added to a bandpass filter:

a. the amplitude of f_R increases
b. its bandwidth decreases
c. its bandwidth increases
d. the amplitude and bandwidth are not affected

19. To tune an *RLC* series circuit for resonance, you would adjust the frequency generator until:

a. the voltage across *C* is minimum

b. the current is maximum

c. the resistance is zero

d. none of the above

20. When testing a series resonant circuit, you find that V_C is low. A possible cause of the trouble is:

a. an open in *L*

b. a decrease in value of *C*

c. an open in *R*

d. none of the above

ANSWERS TO EXPERIMENTS AND SELF-CHECKING QUIZ

Experiment 1: **(1)** minimum, maximum, V_T **(2)** capacitive **(3)** inductive

Experiment 2: **(1)** maximum, minimum, V_T **(2)** inductive **(3)** capacitive

Experiment 3: **(1)** low-pass **(2)** high-pass **(3)** bandpass **(4)** bandstop, or bandreject, or notch

Self-Checking Quiz: **(1)** c **(2)** b **(3)** b **(4)** b **(5)** a **(6)** d **(7)** c **(8)** a **(9)** b **(10)** c **(11)** b **(12)** a **(13)** c **(14)** b **(15)** d **(16)** c **(17)** a **(18)** c **(19)** b **(20)** b

Index

D

E

F

G

H

I

J

K

L

M

N

O

P

Q

R

S

T

V

W

X

Y